Additional Pure Mathematics

L. HARWOOD CLARKE

THIRD EDITION

prepared by
F. G. J. NORTON
Head of the Mathematics Department,
Rugby School

with answers

HEINEMANN EDUCATIONAL BOOKS
LONDON

Heinemann Educational Books Ltd
22 Bedford Square, London WC1B 3HH

LONDON   EDINBURGH   MELBOURNE   AUCKLAND
SINGAPORE   KUALA LUMPUR   NEW DELHI
IBADAN   NAIROBI   JOHANNESBURG
PORTSMOUTH (NH)   KINGSTON

ISBN 0 435 51187 4

First published 1965
Reprinted three times
Second Edition 1970
Reprinted four times
Third Edition 1980
Reprinted 1981, 1983, 1985 (twice), 1987

Set in 10/11*pt Monophoto Times by*
Eta Services (Typesetters) Ltd, Beccles, Suffolk

Printed and bound in Great Britain by
Richard Clay Ltd, Bungay, Suffolk

Foreword to the Third Edition

There have been considerable changes in many Additional Mathematics syllabuses recently, and more are planned in the next few years. The task of the Examining Boards is not made any easier by the variation in O-level Mathematics syllabuses, and students in schools and colleges may, for various reasons, have very different O-level mathematical backgrounds. Most Examining Boards feel that it is helpful to list the knowledge assumed before a start is made on the Additional Mathematics course, and this book assumes a knowledge of the material contained in the seventh edition of *Ordinary Level Mathematics*. In particular, those students starting an Additional Mathematics or A-level course which includes matrices, probability, relations and functions will find the relevant chapters of that book useful preliminary reading.

The chapters on vectors assume no prior knowledge; students always find vectors fairly difficult until they acquire confidence, and there are many easy examples and questions to give that confidence. Chapters are included on vectors in three dimensions and on the scalar product (although these are not at present in many syllabuses); the power of vector algebra is best seen in three dimensions, and the scalar product does yield easily several useful results. Those students who are not studying vectors will find they can follow the chapters on coordinate geometry (which have not been altered) easily; those who are studying vectors will see different methods of obtaining the same results.

RUGBY 1980   F. G. J. NORTON

iii

Foreword to the First Edition

This book has been written to cover the Additional Mathematics syllabus of the General Certificate of Education, the syllabus for the Alternative-Ordinary papers and that for the Subsidiary Mathematics of the Overseas Higher Certificate. The syllabuses as produced by the various Boards differ considerably in content and this book contains, in most cases, more material than is necessary for any one of these examinations. It is hoped, too, that for those offering no examination, the book will be suitable for the first year of a sixth-form course as well as satisfying the needs of those Arts students doing some general mathematical work in the Sixth Form.

The book is divided into separate sections under the headings Algebra, Calculus, Coordinate Geometry and Trigonometry, but the compartments are by no means water-tight and the idea is that the teacher should construct his own integrated course. There are two reasons for the division of the book into sections. In the first place for ease of reference in following a specific examination syllabus, and the teacher or student should have no difficulty in locating a particular topic. Secondly, to give the teacher more elasticity. To lay down the order in which the subject matter should be taught is to my mind wrong and each teacher will have his own views and own pet theories about precedence.

Calculus is used freely through the Coordinate Geometry section but enough geometry is done in ‚the first chapter of the Calculus section to make the succeeding chapters intelligible.

BEDFORD, March 1965   L. H. C.

iv

Contents

ALGEBRA

1 Indices and Surds   3
Positive indices   3
Negative and fractional indices   3
More difficult expressions   5
Solving equations   5
Surds   6
Rationalization   7

2 Logarithms   11
The graph of 10^x 11
Logarithms   11
Logarithmic theory   12
Change of base   14
Solving an equation where the unknown is an index   14

3 Solutions of Equations   17
Simultaneous equations, one linear and one quadratic   17
Particular cases   18
Linear equations in three variables   19

4 The Remainder Theorem   22
Roots of an equation   22
The remainder theorem   22
Solution of equations   24
Values worth testing   24
To find the remainder when $f(x)$ is divided by
$(x - a)(x - b)$ 25

5 Variation   27
Direct proportion   27
Other cases of direct proportion   27
Inverse proportion   29
Other cases of inverse proportion   30
Joint variation   32
Variation as the sum of two parts   32

v

6 Arithmetic and Geometric Progressions   36
Sequences   36
The arithmetic progression   36
The nth term of an A.P.   37
The arithmetic mean   37
The sum to n terms of an A.P.   37
The geometric progression   39
The nth term of a G.P.   39
The geometric mean   39
The sum of n terms of a G.P.   40
Worked examples   41

7 The Quadratic Function   44
Factors of an expression and roots of an equation   44
Sum and product of the roots of a quadratic equation   44
Discriminant of a quadratic equation   44
Graph of the quadratic function   45
Symmetrical function of the roots   47
Finding an equation with given roots   47

8 Permutations and Combinations   50
Permutations   50
Factorial notation   50
Objects not all different   50
Combinations   52
Two important formulae   53

9 The Binomial Theorem   57
The binomial theorem   57
Pascal's triangle   58
The binomial theorem for other values of n 60

10 Probability   63
Use of symmetry   63
Definition   63
Importance of equiprobable outcomes   64
Dependent and independent events   64
Mutually exclusive events   65
Addition law   65
Conditional probability: multiplication law   67
Independent events   68
Use of tree diagram   68
Use of set notation   69
Binomial distribution   72
Geometric distribution   76

MATRICES AND VECTORS

11 Matrices 81
Definition and laws of association 81
Transformations and mappings 82
Reflection in a coordinate axis 83
Rotation about the origin 84
Enlargement 85
Shear 86
Mapping described by a singular matrix 86
To find the matrix describing a given transformation 87
Change of area in a transformation 89

12 Vectors 92
Definition 92
Free vectors, localized vectors 92
Addition of vectors 92
Subtraction of vectors 93
Scalar multiple of a vector 94
The unit vector 94
Components of a vector 94
Magnitude of a vector 95
Position vectors 96
Addition of vectors, given their magnitude and directions 99

13 Geometrical Applications of Vectors 102
Section theorem 102
The diagonals of a parallelogram bisect each other 103
Midpoint theorem 103
To prove that the medians of a triangle are concurrent 104

14 Vectors in Three Dimensions 109
Magnitude 109
Direction 109
Direction cosines 110

15 The Scalar Product 116
Definition 116
The distributive law 116
Unit vectors 117
Parallel and perpendicular vectors 117
Angle between two vectors 117
Proof of trigonometric formulae 118

Equation of a straight line   119
Geometrical applications   119
Cosine formula   120
The altitudes of a triangle are concurrent   120
Scalar products in three dimensions   121
Scalar product   121
To find the angle between two vectors   122
To find the equation of the plane through a given point,
   perpendicular to a given vector   122
To find the angle between two planes   123

CALCULUS

16 Gradient   129
Gradient of a line   129
Gradient of a curve   130
The distance–time graph   131
Velocity at a point   132
The gradient of $y = x^2$ at any point   134
The gradient of the line $y = k$ 134
The gradient of the line $y = mx$ 134
The gradient of the line $y = mx + c$ 135
The line $y = mx + c$ 136
Equation of a line of given gradient through a given point   136
Equation of a tangent at a given point to a curve   137

17 The Derived Function   139
The derived function   139
The derived function of a constant   139
The derived function of $mx + c$ 139
The derived function of $ax^2 + bx + c$ 139
The derived function of x^3 140
The derived function of $1/x$ 141
The derived function of $1/x^2$ 141
The general rule   142
The derived function of x^n 142
Notation   143
The increment notation   143
The differential coefficient   144

18 Applications of the Derived Function   149
Maxima and minima   149
Velocity and acceleration   155
Rate of change   156

19 Integration
The inverse of differentiation   160
Notation   161
Velocity and acceleration   163
Area   164
The definite integral   165
Area under a velocity–time graph   166
Area under an acceleration–time graph   167
Sign of the area   169
Solid of revolution   170
Centre of gravity   174

20 More Differentiation and Integration   179
The limit of sin x/x 179
The differential coefficient of sin x 180
The differential coefficient of cos x 180
The integrals of sin x and cos x 181
Products   181
Quotients   182
Function of function   184
Implicit functions   188
Integration of a function of a function   190
Integration by substitution   191
Definite integrals by substitution   192

21 Approximate Methods of Integration   196
The trapezoidal rule   197
Simpson's rule   198

COORDINATE GEOMETRY

22 Distances, Mid-points and Gradients   207
Coordinates   207
The distance between two points   207
The mid-point of a straight line   209
The area of a triangle   212
Gradient   213
The equation of a line   214
Equation of the line of gradient m through (x_1, y_1) 215
Determination of a linear law   216
The relation $y = ax^n$ 218

23 The Straight Line   221
The point dividing AB in a given ratio   221
The centre of gravity of a triangle   222
The line joining (x_1, y_1) to (x_2, y_2) 223
Intercept form   224
(p, α) form   224
(r, θ) form   226
Length of perpendicular from a point to a line   228
The sign of the perpendicular   230
The angle between two lines   232
Lines through the intersection of two given lines   234

24 Loci   237

25 The Circle   242
The equation of a circle   242
The form $x^2 + y^2 + 2gx + 2fy + c = 0$ 242
The equation of a tangent   244
Condition for a line to be a tangent   245
Equations of tangents of gradient m 246
Length of tangent from a point to a circle   247
The intersections of two circles   248

26 Parameters   252
Parametric equations of a circle   252
To find the parametric equations of a curve through the
   origin   253
Parametric forms of well-known curves   254
Curve tracing   255
Equations of tangent and normal   257

TRIGONOMETRY

27 Circular Measure   263
Definition of a radian   263
Length of arc   264
Area of sector   265
Proof that x must lie between $\sin x$ and $\tan x$ 265

28 The General Angle   270
Negative angles   270
Ratios of $(90° + \theta)$ 272
Ratios of $(180° - \theta)$ 273
Ratios of $(180° + \theta)$ 273
Ratios of $(270° - \theta)$ 273

Ratios of $(270° + \theta)$ 274
Ratios of $(360° - \theta)$ 274
Ratios of any angle   274

29 Graphs of the Circular Functions   278
The graph of sin x 278
The graph of cos x 279
The graph of tan x 280
Solution of equations by graphs   281

30 Identities and Equations; the Sine and Cosine Formulae   283
Identities   283
Elimination   285
Solution of equations   286
The sine formula   287
The cosine formula   289
Identities using the sine and cosine formulae   291

31 Compound Angles   294
The formula for cos $(A - B)$ 294
The tangent formulae   296
The subsidiary angle   296
Summary   298

32 Multiple Angles   301
Double angle formulae   301
Two useful formulae   303
cos $3A$ and sin $3A$ 303
tan $(A + B + C)$ 304
cos $2A$ and sin $2A$ in terms of tan A 305

33 Sums and Products   310
Products as sums or differences   310
Sums as products   311

34 Half Angle Formulae: Area of Triangle   316
The half angle formulae   316
Two sides and the included angle   318
The area of a triangle   320
Heron's formula   321
The in-radius   321
Three-dimensional problems   322

Answers   331

Index   354

ALGEBRA

1 Indices and Surds

Positive indices

It is assumed that the three formulae

   (i) $a^m \times a^n = a^{m+n}$,
   (ii) $a^m \div a^n = a^{m-n}$,
   (iii) $(a^m)^n = a^{mn}$,

where m and n are positive integers and $m > n$ in (ii) are familiar to the reader. These formulae may easily be proved from first principles and examples are given below.

$$a^2 \times a^3 = (a \times a) \times (a \times a \times a) = a^5.$$

$$a^5 \div a^2 = \frac{a \times a \times a \times a \times a}{a \times a} = a^3.$$

$$(a^2)^3 = (a \times a) \times (a \times a) \times (a \times a) = a^6.$$

Negative and fractional indices

In order to define expressions containing negative and fractional indices, it is necessary to *assume* that these three equations hold for all values of m and n, positive, negative and fractional.

For example, $\dfrac{a^2}{a^2} = a^0$ (by (ii)).

But $\dfrac{a^2}{a^2} = 1.$

$$\therefore a^0 = 1 \qquad\qquad\qquad\text{(iv)}$$

So any number raised to the power of zero equals 1.

For example, $4^0 = 1; \quad (\tfrac{2}{3})^0 = 1.$

Now consider the case of a negative index.

$$7^{-2} = \frac{7^0}{7^2} \quad \text{(by (ii)).}$$

$$= \frac{1}{7^2} \quad \text{(by (iv)).}$$

$$\therefore 7^{-2} = \frac{1}{7^2} = \frac{1}{49}.$$

In general
$$a^{-n} = \frac{a^0}{a^n}$$

$$= \frac{1}{a^n}.$$

So
$$a^{-n} = \frac{1}{a^n} \qquad\qquad \text{(v)}$$

Now consider a fractional index as in $5^{\frac{1}{2}}$.

By (iii),
$$(5^{\frac{1}{2}})^2 = 5^1$$
$$= 5.$$
$$\therefore 5^{\frac{1}{2}} = \sqrt{5}.$$

In general,
$$(a^{p/q})^q = a^p.$$
$$\therefore a^{p/q} \text{ is a value of } \sqrt[q]{a^p} \text{ or } (\sqrt[q]{a})^p \qquad\qquad \text{(vi)}$$

Since a number may have more than one qth root, we say that $a^{p/q}$ is *a* value of the expression.

Example 1. $8^{\frac{2}{3}} = (\sqrt[3]{8})^2 = 2^2 = 4.$

Example 2. $8^{-\frac{2}{3}} = \frac{1}{8^{\frac{2}{3}}} = \frac{1}{4}.$

To evaluate a number raised to a negative index, first equate it to the reciprocal of the same number raised to the corresponding positive index.

Exercise 1a

Write down the values of the following:

1. $2^{-3}.$	**2.** $3^2.$	**3.** $5^0.$	**4.** $9^{\frac{1}{2}}.$	**5.** $9^{-\frac{1}{2}}.$
6. $16^{\frac{1}{2}}.$	**7.** $16^{-\frac{1}{2}}.$	**8.** $25^{\frac{1}{2}}.$	**9.** $25^{-\frac{1}{2}}.$	**10.** $16^{\frac{1}{4}}.$
11. $16^{-\frac{1}{4}}.$	**12.** $49^{\frac{1}{2}}.$	**13.** $49^{-\frac{1}{2}}.$	**14.** $81^{\frac{1}{4}}.$	**15.** $81^{-\frac{1}{4}}.$
16. $81^{\frac{1}{2}}.$	**17.** $81^{-\frac{1}{2}}.$	**18.** $81^{\frac{3}{4}}.$	**19.** $81^{-\frac{3}{4}}.$	**20.** $10^{-2}.$
21. $10^{-3}.$	**22.** $10^0.$	**23.** $10^{-1}.$	**24.** $\left(\frac{3}{4}\right)^{-1}.$	**25.** $\left(\frac{4}{5}\right)^{-2}.$
26. $\left(\frac{4}{9}\right)^{\frac{1}{2}}.$	**27.** $\left(\frac{4}{9}\right)^{-\frac{1}{2}}.$	**28.** $\left(\frac{4}{9}\right)^{-1}.$	**29.** $\left(\frac{4}{9}\right)^{-2}.$	**30.** $\left(\frac{121}{144}\right)^{-1}.$
31. $\left(\frac{121}{144}\right)^{\frac{1}{2}}.$	**32.** $\left(\frac{121}{144}\right)^{-\frac{1}{2}}.$	**33.** $64^{\frac{1}{2}}.$	**34.** $64^{-\frac{1}{2}}.$	**35.** $64^{\frac{1}{3}}.$
36. $64^{-\frac{1}{3}}.$	**37.** $64^{\frac{2}{3}}.$	**38.** $64^{-\frac{2}{3}}.$	**39.** $64^{\frac{1}{6}}.$	**40.** $64^{-\frac{1}{6}}.$
41. $64^{\frac{5}{6}}.$	**42.** $64^{-\frac{5}{6}}.$	**43.** $\left(\frac{5}{7}\right)^0.$	**44.** $100^{-\frac{1}{2}}.$	**45.** $125^{\frac{1}{3}}.$
46. $125^{-\frac{1}{3}}.$	**47.** $125^{\frac{2}{3}}.$	**48.** $125^{-\frac{2}{3}}.$	**49.** $11^{-2}.$	**50.** $3^{-3}.$

More difficult expressions

Remember that $a^m \times a^n = a^{m+n}$ is true only if the numbers raised to the powers of m and n are equal. The formula cannot be used to evaluate an expression such as $4^{\frac{1}{2}} \times 8^{\frac{1}{3}}$. This must be evaluated by finding the value of each of the factors.

$$4^{\frac{1}{2}} \times 8^{\frac{1}{3}} = 2 \times 2 = 4.$$

To simplify a more complicated expression, it is generally best to put each of the numbers in its prime factors.

Example. *Evaluate* $(24)^{\frac{1}{2}} \times (32)^{\frac{2}{3}} \times 6^{\frac{1}{6}} \times 3^{\frac{1}{3}}$.

$$24 = 2^3 \times 3; \quad 32 = 2^5; \quad 6 = 2 \times 3.$$
$$\therefore \text{ the expression } = (2^3 \times 3)^{\frac{1}{2}} \times (2^5)^{\frac{2}{3}} \times (2 \times 3)^{\frac{1}{6}} \times 3^{\frac{1}{3}}$$
$$= 2^{\frac{3}{2}} \times 3^{\frac{1}{2}} \times 2^{\frac{10}{3}} \times 2^{\frac{1}{6}} \times 3^{\frac{1}{6}} \times 3^{\frac{1}{3}}$$
$$= 2^{(\frac{3}{2} + \frac{10}{3} + \frac{1}{6})} \times 3^{(\frac{1}{2} + \frac{1}{6} + \frac{1}{3})}$$
$$= 2^5 \times 3^1$$
$$= 32 \times 3 = 96.$$

Solving equations

1. If the unknown in an equation is the index, try putting each of the numbers in its prime factors.

Example. *Solve the equation* $32^n = 4$.
$$32 = 2^5 \quad \text{and} \quad 4 = 2^2.$$
$$\therefore (2^5)^n = 2^2.$$
$$\therefore 2^{5n} = 2^2$$
and
$$5n = 2 \quad \text{or} \quad n = \tfrac{2}{5}.$$

2. The equation may reduce to a quadratic as in the following example.

Example. *Solve the equation* $4^x - 3.2^x + 2 = 0$.
$$4^x = (2^2)^x = 2^{2x} = (2^x)^2.$$
Put
$$2^x = y$$
Then
$$y^2 - 3y + 2 = 0.$$
$$\therefore (y - 1)(y - 2) = 0 \quad \text{and} \quad y = 1 \text{ or } 2.$$
$$\therefore 2^x = 1 \quad \text{or} \quad 2^x = 2.$$
If $2^x = 1$, $x = 0$; if $2^x = 2$, $x = 1$.
$$\therefore x = 0 \text{ or } 1.$$

3. The equation may be such that the unknown occurs as a negative or fractional power.

Example. *Find x in terms of y and z given that* $(xy)^{\frac{1}{3}} = z^2$.

Here
$$x^{\frac{1}{3}}y^{\frac{1}{3}} = z^2.$$
$$\therefore xy = (z^2)^3 = z^6.$$
$$\therefore x = \frac{z^6}{y}.$$

Exercise 1b

Evaluate

1. $4^0 \times 2^2$. **2.** $4^{\frac{1}{2}} \times 16^{\frac{1}{2}}$. **3.** $2^{\frac{1}{2}} \times 32^{\frac{1}{2}}$. **4.** $18^{\frac{1}{2}} \times 2^{\frac{1}{2}}$.

5. $8^{\frac{1}{3}} \times 27^{-\frac{1}{3}}$. **6.** $2^{\frac{1}{3}} \times 4^{\frac{1}{3}}$. **7.** $16^{-\frac{1}{4}} \times 81^{\frac{1}{4}}$. **8.** $3^{\frac{1}{3}} \times 27^{\frac{1}{3}}$.

9. $27^{\frac{1}{2}} \times 3^{-\frac{1}{2}}$. **10.** $5^{\frac{1}{3}} \times 25^{\frac{1}{3}}$.

Solve the equations

11. $4^n = 8$. **12.** $3^x = 27$. **13.** $5^x = 125$. **14.** $5^y = 25$.

15. $3^y = \frac{1}{9}$. **16.** $9^n = \frac{1}{3}$. **17.** $8^n = 2$. **18.** $7^x = \frac{1}{7}$.

19. $8^x = 4$. **20.** $8^x = \frac{1}{4}$. **21.** $4^x - 5.2^x + 2^2 = 0$.

22. $4^x - 6.2^x + 2^3 = 0$. **23.** $9^x - 4.3^x + 3 = 0$.

24. $9^x - 10.3^x + 3^2 = 0$. **25.** $9^x - 4.3^{x+1} + 3^3 = 0$.

Find x from the equations

26. $(xy)^{\frac{1}{2}} = 1$. **27.** $x^{\frac{2}{3}}y^{\frac{1}{3}} = 1$. **28.** $(xy)^{\frac{1}{3}} = z$.

29. $x^2y = z^3$. **30.** $x^3y^2 = z$. **31.** $x^{-\frac{1}{3}}y = z$.

32. $x^{-\frac{1}{2}} = yz$. **33.** $x^{\frac{1}{2}}y^2 = z^3$. **34.** $(xy)^{\frac{1}{3}} = z^{\frac{1}{2}}$.

35. $x^{\frac{1}{3}}y^{\frac{1}{2}} = z$.

Simplify

36. $\dfrac{(a + b)}{a^{-1} + b^{-1}}$. **37.** $100^x \div 10^x$. **38.** $\dfrac{x - x^{-1}}{x - 1}$.

39. $\dfrac{1 - x^{-1}}{x - 1}$. **40.** $4^n \div 2^n$. **41.** $9^{\frac{1}{2}} + 4^{-\frac{1}{2}}$.

42. $16^{\frac{1}{2}} + 9^{\frac{1}{2}}$. **43.** $8^{\frac{1}{3}} + 27^{\frac{2}{3}}$. **44.** $2x^{\frac{1}{2}} \times 5x^{\frac{1}{2}}$.

45. $(4^n + 16^n) \div 2^n$. **46.** $(16 + 9)^{\frac{1}{2}}$. **47.** $\dfrac{1 + y^{-1}}{1 + y}$.

48. $\dfrac{y - 1}{y^{\frac{1}{2}} - 1}$. **49.** $125^n \div 5^n$. **50.** $(x^{-2})^{-3}$.

Surds

Some numbers are **perfect squares,** i.e. they have exact square roots. For example, $\sqrt{9} = \pm 3$, $\sqrt{\frac{16}{9}} = \pm\frac{4}{3}$. The square roots of most numbers, however, cannot be expressed as fractions. Suppose for example that $\sqrt{2}$ can be put in the form $\dfrac{p}{q}$ where p and q are integers having no common factor.

Then $$\sqrt{2} = \frac{p}{q}$$

and so $$2 = \frac{p^2}{q^2}.$$

$$\therefore p^2 = 2q^2.$$

Since the right-hand side contains the factor 2, p^2 cannot be odd and so p must be even. Since p is even, p^2 has the factor 4 and so q^2 must contain the factor 2. The square of an odd number cannot have 2 as a factor and so q must be even, which means that p and q must have a common factor. This is contrary to our assumption and so the equation cannot be satisfied by integral values of p and q.

A number which cannot be expressed as a fraction is called **irrational**. Examples of irrational numbers are π and $\sqrt{2}$. Irrational numbers in the form of roots are called **surds**. Examples of surds are $\sqrt{3}$, $\sqrt[3]{7}$ and $\sqrt[4]{10}$.

Surds may often be expressed in terms of other surds as in the following examples:

$$\sqrt{18} = \sqrt{2 \times 3^2} = 3\sqrt{2}.$$

$$\sqrt[3]{81} = \sqrt[3]{3^3 \times 3} = 3\sqrt[3]{3}.$$

$$\sqrt{\frac{4}{5}} = \sqrt{\frac{4 \times 5}{25}} = \frac{2}{5}\sqrt{5}.$$

Rationalization

To evaluate expressions such as $\dfrac{1}{\sqrt{5}}$ and $\dfrac{1}{\sqrt{5}-1}$, the arithmetic is simplified if the surd can be removed from the denominator of the expression. This process is called **rationalizing** the denominator.

Example. *Given that* $\sqrt{5} = 2.236$ *to 4 s.f., evaluate* $\dfrac{1}{\sqrt{5}}$ *and* $\dfrac{1}{\sqrt{5}-1}$, *each to 3 s.f.*

$$\frac{1}{\sqrt{5}} = \frac{\sqrt{5}}{\sqrt{5} \times \sqrt{5}} = \frac{\sqrt{5}}{5} = \frac{2.236}{5} = 0.447.$$

$$\frac{1}{\sqrt{5}-1} = \frac{\sqrt{5}+1}{(\sqrt{5}-1)(\sqrt{5}+1)} = \frac{\sqrt{5}+1}{5-1} = \frac{3.236}{4} = 0.809.$$

The general method is to multiply both numerator and denominator of the fraction by the denominator with the sign of its surd changed.

Exercise 1c

Express in terms of simpler surds

1. $\sqrt{8}$. **2.** $\sqrt{32}$. **3.** $\sqrt[3]{16}$. **4.** $\sqrt[3]{32}$. **5.** $\sqrt[5]{64}$.

6. $\sqrt{500}$. **7.** $\sqrt{1000}$. **8.** $\sqrt[3]{250}$. **9.** $\sqrt[3]{128}$. **10.** $\sqrt[3]{54}$.

Express without roots in the denominator

11. $\dfrac{1}{\sqrt{2}}$. **12.** $\dfrac{3}{\sqrt{8}}$. **13.** $\dfrac{2}{\sqrt{32}}$. **14.** $\dfrac{2}{5\sqrt{2}}$. **15.** $\dfrac{10}{\sqrt{50}}$.

16. $\dfrac{4}{\sqrt{128}}$. **17.** $\dfrac{1}{\sqrt{2}-1}$. **18.** $\dfrac{1}{\sqrt{2}+1}$. **19.** $\dfrac{1}{2-\sqrt{3}}$. **20.** $\dfrac{1}{2+\sqrt{3}}$.

Remove brackets from

21. $\sqrt{2}(\sqrt{2}-1)$. **22.** $\sqrt{3}(\sqrt{3}+1)$. **23.** $(\sqrt{2}-1)(\sqrt{2}+1)$.

24. $(2+\sqrt{3})(2-\sqrt{3})$. **25.** $(1+2\sqrt{3})(1+\sqrt{3})$.

26. $(\sqrt{3}-1)(2-\sqrt{3})$. **27.** $(\sqrt{2}-1)(2\sqrt{2}-1)$.

28. $(\sqrt{3}+1)(\sqrt{3}+2)$. **29.** $(\sqrt{3}+\sqrt{2})(\sqrt{3}-\sqrt{2})$.

30. $(\sqrt{a}+1)(\sqrt{a}-1)$.

Rationalize

31. $\dfrac{1}{\sqrt{3}-1}$. **32.** $\dfrac{1}{2\sqrt{3}-1}$. **33.** $\dfrac{1}{\sqrt{7}-\sqrt{5}}$. **34.** $\dfrac{1}{\sqrt{7}+\sqrt{5}}$.

35. $\dfrac{1}{\sqrt{3}+\sqrt{2}}$. **36.** $\dfrac{1}{2\sqrt{3}+\sqrt{2}}$. **37.** $\dfrac{1}{\sqrt{7}-2}$. **38.** $\dfrac{1}{\sqrt{6}+1}$.

39. $\dfrac{1}{2\sqrt{5}-\sqrt{3}}$. **40.** $\dfrac{1}{\sqrt{5}-\sqrt{2}}$.

Exercise 1d: Miscellaneous

1. Evaluate $(\sqrt{7} + \sqrt{3})(\sqrt{7} - \sqrt{3})$.
2. Simplify $2xy \times 3z^2 \times (xz)^{-1}$.
3. Simplify $2^x \times 4^{-x} \times 8^{x+1}$.
4. Find x given that $2 \times 4^{x+1} = 8^x$.
5. If $y = a^2$ and $a^3b = 1$, express y in the form b^n.
6. If $b = 64a^{-\frac{2}{3}}$, find the value of a when $b = 25$.
7. Evaluate $(18)^{\frac{3}{2}} \times (27)^{-\frac{1}{3}} \times \dfrac{1}{\sqrt{6}}$.
8. If $x = y^{\frac{1}{3}}$ and $y = (xc)^{-\frac{1}{2}}$, express c as a power of x.
9. Solve the equation $25^x - 6.5^x + 5 = 0$.
10. Find x given that $3^{1/x} = \frac{1}{243}$.
11. Evaluate $10^{1.7} \times 10^{1.3}$.
12. Evaluate $10^{1.2} \times \sqrt{10^{1.6}}$.
13. Express 0.03125 as a power of 2.
14. Multiply $x^{\frac{2}{3}} + y^{\frac{2}{3}}$ by $x^{\frac{2}{3}} - x^{\frac{1}{3}}y^{\frac{1}{3}} + y^{\frac{2}{3}}$.
15. Evaluate $\dfrac{1}{2 + \sqrt{3}} + \dfrac{1}{2 - \sqrt{3}}$.
16. Express as a surd the length of the side of a square whose diagonal is 8 cm long.
17. Express as a surd the area of an equilateral triangle of side 5 cm.
18. Express as a surd the length of the hypotenuse of a right-angled triangle whose other sides are $(\sqrt{2} + 1)$ cm and $(\sqrt{2} - 1)$ cm long.
19. Solve the equation $3^{2x+1} - 4.3^x + 1 = 0$.
20. Square $(\sqrt{3} - \sqrt{2})$.
21. Square $(\sqrt{a} - \sqrt{b})$.
22. Using the result of question 21, find the positive square root of $8 - 2\sqrt{15}$.
23. Find the positive square root of $14 - 6\sqrt{5}$.
24. Given that $\sqrt{3} = 1.732$, evaluate $\dfrac{1}{\sqrt{3} + 1}$.
25. Given that $\sqrt{2} = 1.414$, evaluate $\dfrac{\sqrt{2} + 1}{\sqrt{2} - 1}$.
26. Evaluate $\dfrac{1}{3 - \sqrt{5}} + \dfrac{1}{3 + \sqrt{5}}$.
27. The sides of a rectangle are $(2 - \sqrt{3})$ cm and $(2 + \sqrt{3})$ cm. Express the length of the diagonal as a surd.

28. Solve the equation $\sqrt{x + 5} = 5 - \sqrt{x}$ by squaring both sides. Check that your answer satisfies the equation.

29. Solve the equation $\sqrt{x} + \sqrt{x + 7} = 7$.

30. Two circles of radii $(\sqrt{3} + 1)$ cm and $(\sqrt{3} - 1)$ cm touch externally. Find the length of a common tangent in the form of a surd.

2 Logarithms

The graph of 10ˣ

To draw the graph of 10^x, we need to find values of 10^x for a number of values of x. We already know that

$$10^{-1} = 0.1; \quad 10^0 = 1; \quad 10^1 = 10.$$

Other values are found as follows:

$$10^{\frac{1}{2}} = \sqrt{10} \simeq 3.162.$$
$$10^{\frac{1}{4}} = \sqrt{10^{\frac{1}{2}}} \simeq \sqrt{3.162} \simeq 1.779.$$
$$10^{\frac{3}{4}} = 10^{\frac{1}{2}} \times 10^{\frac{1}{4}} \simeq 5.62 \quad \text{(to 3 s.f.)}.$$

$$10^{-\frac{1}{2}} = \frac{1}{10^{\frac{1}{2}}} = \frac{\sqrt{10}}{10} \simeq 0.3162$$

We can now tabulate the values of 10^x which are given to 3 s.f.

x	-1	$-\frac{1}{2}$	0	$\frac{1}{4}$	$\frac{1}{2}$	$\frac{3}{4}$	1
10^x	0.1	0.316	1.0	1.78	3.16	5.62	10.0

Fig. 2.1 shows the graph of 10^x between $x = -1$ and $x = +1$. From this graph we can express any number between 0.1 and 10 as a power of x. For example,

$$4 = 10^{0.6} \text{ approx.}$$

Logarithms

We have seen that $4 = 10^{0.6}$ approx. 0.6 is called the logarithm of 4 to the base 10. Generally, if $y = 10^x$, x is called the logarithm of y to the base 10.

The logarithm of a number is the power to which 10 must be raised to give the number.

The graph may be used to find logarithms of numbers greater than 10 in the following way.

$$40 = 4 \times 10 = 10^{0.6} \times 10 = 10^{1.6} \text{ approx.}$$

So $\qquad \log 40 = 1.6$ approx.

Similarly, $\qquad \log 400 = 2.6$.

Again, $\qquad 0.4 = \frac{4}{10} = \frac{10^{0.6}}{10} = 10^{\bar{1}.6}$

and $\qquad \log 0.4 = \bar{1}.6$.

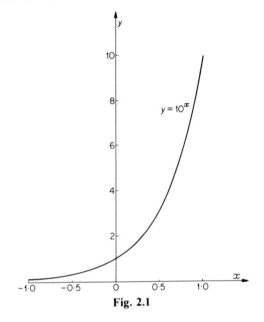

Fig. 2.1

Exercise 2a

Draw the graph of $y = 10^x$ between $x = -1$ and $x = +1$ and from it write down the values of:

1. $10^{\frac{1}{3}}$.   **2.** $\sqrt[3]{10}$.   **3.** $10^{-\frac{1}{3}}$.   **4.** $10^{0.6}$.   **5.** $10^{0.7}$.

From your graph deduce the values of:

6. $10^{1.2}$.   **7.** $10^{1.4}$ **8.** $10^{2.2}$.   **9.** $10^{1.7}$.   **10.** $10^{2.4}$.

From your graph write down the logarithms of:

11. 2.   **12.** 2.5.   **13.** 3.   **14.** 5.   **15.** 7.2.

Deduce the logarithms of:

16. 20.   **17.** 0.25.   **18.** 0.03.   **19.** 500.   **20.** 7.2×10^5.

Logarithmic theory

The logarithm of a number need not be to base 10; in fact, it may be to any base. The logarithm of a number to base b is the power to which b must be raised to give the number, i.e. if

$$y = b^x \quad \text{then} \quad x = \log_b y.$$

If the base is not stated, it may be assumed that it is 10. In all other cases, the base should be clearly stated.

There are three important formulae connecting logarithms which correspond to the three index equations. To prove any formula connecting logarithms, make use of the corresponding index equation as shown below.

1.
$$\log_b x + \log_b y = \log_b xy.$$

Suppose that $\log_b x = u$ and that $\log_b y = v$.

Then $b^u = x$ and $b^v = y$.

$\therefore xy = b^u \times b^v = b^{u+v}.$

From definition, $\log_b xy = u + v = \log_b x + \log_b y$.

2.
$$\log_b x - \log_b y = \log_b \frac{x}{y}.$$

Again supposing that $\log_b x = u$ and $\log_b y = v$,

$$\frac{x}{y} = \frac{b^u}{b^v} = b^{u-v}.$$

From definition, $\log_b \dfrac{x}{y} = u - v = \log_b x - \log_b y$.

3.
$$\log_b x^n = n \log_b x.$$

If $\log_b x = u, b^u = x.$

$\therefore x^n = (b^u)^n = b^{un}.$

From definition, $\log_b x^n = nu = n \log_b x$.

The first of these formulae tells us how to add logarithms.

For example, $\log 4 + \log 25 = \log(4 \times 25) = \log 100 = 2$.

The second tells us how to subtract logarithms.

For example, $\log 80 - \log 8 = \log \dfrac{80}{8} = \log 10 = 1$.

The third tells us how to find the logarithm of a power.

For example, $\log 27 = \log 3^3 = 3 \log 3.$ $\therefore \dfrac{\log 27}{\log 3} = 3.$

Exercise 2b

Evaluate or simplify the following without using tables:

1. $\log 30 - \log 3$.

2. $\log_2 16 - \log_2 8$.

3. $\log_3 2.7 + \log_3 10$.

4. $\dfrac{\log 8}{\log 2}$.

5. $\log 4 + 2 \log 5$.

6. $\dfrac{\log 32}{\log 2}$.

7. $\dfrac{\log x^4}{\log x}$.

8. $\dfrac{\log \sqrt{6}}{\log 6}$.

9. $\log_a a\sqrt{a}$.

10. $\log 7 + 2 \log 5 - \log \frac{7}{4}$.

11. $\log 18 + \log 5 - \log 9$.

12. $\log_a a^2 \div \log_a \sqrt{a}$.

13. $\log_6 42 - \log_6 7$.

14. $\log_9 3 + \log_9 27$.

15. $\log x^4 + \log x$.

16. $(\log a^2 + \log b^2) \div \log ab$.

17. $\log 3 + \log 15 - \log 4.5$.

18. $\log_4 1.6 + \log_4 40$.

19. $\log_8 72 - \log_8 \frac{9}{8}$.

20. $\log_7 \sqrt[3]{7}$.

Change of base

The formula which tells us how to change from one base to another is

$$\log_b x = \frac{\log_a x}{\log_a b}.$$

Suppose that $\log_b x = u$, so that $b^u = x$.

Take logarithms to base a.

$$\log_a b^u = \log_a x$$

or

$$u \log_a b = \log_a x.$$

$$\therefore u = \frac{\log_a x}{\log_a b}$$

and so

$$\log_b x = \frac{\log_a x}{\log_a b}.$$

Example. *Calculate $\log_3 8$.*

$$\log_3 8 = \frac{\log_{10} 8}{\log_{10} 3}$$

$$= \frac{0.9031}{0.4771}$$

$$= 1.892 \text{ or } 1.89 \text{ to 3 s.f.}$$

No.	Log
0.9031	$\bar{1}.9557$
0.4771	$\bar{1}.6786$
1.892	0.2771

Solving an equation where the unknown is an index

To solve an equation such as $3^x = 8$, where the unknown is an index, take logs.

If

$$3^x = 8,$$

$$x \log 3 = \log 8.$$

$$\therefore x = \frac{\log 8}{\log 3}$$

$$= 1.89 \quad \text{from the previous example.}$$

Note the similarity between this example and the last. In fact if $3^x = 8$, we can say immediately that $x = \log_3 8$ and so $x = 1.89$ from the example in the last paragraph.

Exercise 2c

Evaluate

1. $\log_3 4$.　　**2.** $\log_7 17$.　　**3.** $\log_5 14$.　　**4.** $\log_6 28$.　　**5.** $\log_{12} 41$.

Find x from the equations

6. $4^x = 3$.　　**7.** $7^x = 11$.　　**8.** $11^x = 5$.　　**9.** $10^x = 7$.　　**10.** $12^x = 19$.

Exercise 2d: Miscellaneous

1. If $\log y = 3 - 2 \log x$, express y in the form ax^n.

2. If $x = y^{1.6}$ find x when $y = 17$.

3. Evaluate without tables $\log (\frac{1}{3} + \frac{1}{4}) + 2 \log 2 + \log \frac{3}{7}$.

4. Solve the equation $2^{2x} - 9(2^x) + 20 = 0$.

5. Find the value of x which satisfies the equation $\log_x 10 = -3$.

6. Find x given that $e^x = 4$ and $e = 2.718$.

7. Write down the value of $\log (\sqrt{10} \times \sqrt[3]{10})$.

8. Write down the value of $\log \dfrac{100}{\sqrt{10}}$.

9. Write down the value of $\dfrac{\log 81}{\log 27}$.

10. Evaluate without using tables $\log 2.88 - \log 360 + \log 0.125$.

11. Evaluate $\log_{0.25} 8$.

12. Find the value of $\log 75 + \log 112 - \log 0.84$.

13. Show that $\log a + \log ax + \log ax^2 = 3(\log a + \log x)$.

14. Express x in terms of y, given that $\log x + \log y = 2$.

15. Solve the equation $5^x = 8$.

16. Evaluate $\dfrac{\log 16 - \log 1}{\log 2 - \log 1}$.

17. The amount £A of a sum of money £P invested at $r\%$ per annum compound interest for n years is given by $A = P\left(1 + \dfrac{r}{100}\right)^n$. Find the number of years in which a sum of money invested at 5% per annum doubles itself.

18. Find the value of x^{-y} when $x = 4.2$ and $y = 0.4$.

19. Write down the values of (i) $\log_3 9$, (ii) $\log_{81} 9$, (iii) $\log_{\sqrt{3}} 9$.

20. Given that log 2 = 0.3010 and log 3 = 0.4771, find the values of (i) log 12, (ii) log 5.

21. Find x given that $3^x = \frac{1}{5}$.

22. Express log 108 in terms of log 2 and log 3.

23. Express y in terms of x from the equation $\log_6 x + \log_6 y^2 = 1$.

24. Given that $2^{x+1} = 3^x$, find x.

25. Given that $3^{y+1} = 4^{y-1}$, find y.

3 Solutions of Equations

Simultaneous equations, one linear and one quadratic

Two simultaneous equations, one linear and one quadratic, have in general two pairs of solutions. Graphically, the equations describe a straight line and a curve of the second degree as shown in Fig. 3.1. The values of x and y at the points of intersection are the solutions of the equations.

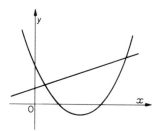

Fig. 3.1

The method of solution is as follows. Express either x or y in terms of the other from the linear equation. The choice of x or y is immaterial—choose whichever is the easier. Substitute this value in the other equation and a quadratic in one variable will result. The solution of this equation gives the pair of solutions for one variable and the corresponding values of the other may be obtained from the linear equation.

Example. *Solve the equations $2x + y = 4$, $x^2 - xy + 1 = 0$.*

From the linear equation, $\qquad\qquad\qquad y = 4 - 2x.$

Substituting, $\qquad\qquad x^2 - x(4 - 2x) + 1 = 0.$

$\qquad\qquad\qquad\therefore x^2 - 4x + 2x^2 + 1 = 0.$

or $\qquad\qquad\qquad\qquad 3x^2 - 4x + 1 = 0.$

$\qquad\qquad\qquad\therefore (3x - 1)(x - 1) = 0.$

Either $\qquad\qquad\qquad x = \tfrac{1}{3}$ or $x = 1.$

When $x = \tfrac{1}{3}$, $\qquad\qquad y = 4 - \tfrac{2}{3} = 3\tfrac{1}{3}.$

When $x = 1$, $\qquad\qquad y = 4 - 2 = 2.$

N.B. Make it clear in your answer that $x = 1$ must pair with $y = 2$.
Do not write $x = 1$ or $\tfrac{1}{3}$, $y = 2$ or $3\tfrac{1}{3}$.

Particular cases

In some cases other methods of solution may be quicker and neater. If, for example, the quadratic equation is

$$x^2 + 2xy + y^2 = 4,$$

it is simpler to write $(x + y)^2 = 4$, from which follows that $x + y = \pm 2$ and the equations may be considered as two pairs of simultaneous linear equations.

Again, if one of the equations is $xy = 2$, it follows that $x = \dfrac{2}{y}$ and this substitution may be easier than one found from the linear equation. In an example of this sort, it may be possible to solve the equations even if the second equation is also quadratic.

Example 1. *Solve the equations $x + y = 6$, $x^2 - 2xy + y^2 = 4$.*
   The quadratic equation gives $x - y = \pm 2$.
   The solution of $x + y = 6$ and $x - y = 2$ is $x = 4$, $y = 2$.
   The solution of $x + y = 6$ and $x - y = -2$ is $x = 2$, $y = 4$.

Example 2. *Solve the equations $xy = 6$, $x^2 + y^2 = 13$.*

From the first equation, $\qquad\qquad x = \dfrac{6}{y}.$

Substituting, $\qquad\qquad\qquad \dfrac{36}{y^2} + y^2 = 13.$

$$\therefore y^4 - 13y^2 + 36 = 0.$$
$$(y^2 - 4)(y^2 - 9) = 0.$$
$$\therefore y^2 = 4 \text{ or } 9, \quad \text{and} \quad y = \pm 2, \pm 3.$$

When $y = 2$, $x = 3$; when $y = -2$, $x = -3$; when $y = 3$, $x = 2$; when $y = -3$, $x = -2$.

Exercise 3a

Solve the equations

1. $x + y = 3$, $xy = 2$.
2. $x - y = 1$, $xy = 6$.
3. $2x + y = 3$, $2x^2 - xy = 1$.
4. $x + 3y = 5$, $x^2 + y^2 = 5$.
5. $x + y = 7$, $y^2 - xy + x^2 = 13$.
6. $3x = 2y$, $x^2 - xy + y^2 = 7$.
7. $x + \dfrac{1}{y} = 3$, $2xy = 1$.
8. $x^2 + xy + y^2 = 3$, $2x + y = 3$,
9. $x^2 + 2xy + y^2 = 9$, $x - y = 5$.
10. $xy = 4$, $x^2 - y^2 = 15$.

Linear equations in three variables

Three linear equations in three variables such as

$$x + y + 2z = 4 \qquad \text{(i)}$$
$$2x + y - z = 1 \qquad \text{(ii)}$$
$$3x + 2y - z = 3 \qquad \text{(iii)}$$

may be solved by reducing the equations to a pair of simultaneous equations in two variables.

Multiply (ii) by 2 and add to (i) to eliminate z.

$$5x + 3y = 6 \qquad \text{(iv)}$$

Subtract (ii) from (iii) in order to eliminate z.

$$x + y = 2 \qquad \text{(v)}$$

Solving (iv) and (v) gives

$$x = 0, \quad y = 2.$$

From (i) $\qquad 2 + 2z = 4 \quad \text{and} \quad z = 1.$

The solution is $\qquad x = 0, \quad y = 2 \quad \text{and} \quad z = 1.$

In graphical work, we should need three mutually perpendicular axes to represent the equations. Each of the equations represents a plane and three planes in general meet in a point. The graphical interpretation does suggest that in particular cases a unique solution will not occur. Two or more of the planes may be parallel or the third plane may pass through the line of intersection of the first two, in which case any point on the line would satisfy.

The equations may in fact give an infinite number of solutions or no solutions at all.

Example 1. *Consider the equations*

$$x + y + z = 5,$$
$$2x + y + 3z = 7,$$
$$3x + 2y + 4z = 12.$$

Here the first two equations when added give the third and so the third gives us no new information. In such a case the equations are said to be **dependent** and any trio of values satisfying the first two equations is a solution.

Example 2. *Consider the equations*

$$x + y + z = 5,$$
$$2x + y + 3z = 7,$$
$$3x + 2y + 4z = 10.$$

The addition of the first two equations gives

$$3x + 2y + 4z = 12,$$

which contradicts the third equation. The equations are said to be **inconsistent** and there is no solution.

Exercise 3b

Solve the equations

1. $x + y + z = 3$, $2x + y - z = 3$, $x + 2y + 3z = 6$.
2. $2x + y + z = 2$, $x + 2y + z = 0$, $x + 3y + 3z = 1$.
3. $x + y + 2z = 3$, $2x + y - z = 5$, $3x + 2y + 5z = 8$.
4. $x + 2y - z = 2$, $2x + y - z = 1$, $x + 2y + 3z = 14$.
5. $2x - y - z = 1$, $x + 3y + z = 6$, $x + y + 2z = 1$.

State whether the following equations are dependent or inconsistent:

6. $x + y + z = 3$, $2x + y + 5z = 4$, $3x + 2y + 6z = 7$.
7. $x + y + z = 3$, $2x + y + 5z = 4$, $3x + 2y + 6z = 1$.
8. $2x - y - z = 5$, $x + y + 3z = 2$, $x - 2y - 4z = 3$.
9. $2x - y - z = 5$, $x + y + 3z = 1$, $4x + y + 5z = 1$.
10. $3x + 2y + z = 7$, $x + y + z = 4$, $6x + 5y + 4z = 10$.

Exercise 3c: Miscellaneous

1. Solve the equations $x + 2y = 7$, $2x^2 + xy = 5$.
2. Solve the equations $x + y + z = 6$, $x - y + 2z = 3$, $2x - 3y + z = 1$.
3. Solve the equations $2x + 3y = 5$, $2x^2 + 3xy = 15$.
4. Are the following equations consistent:
$$x + y - z = 7,$$
$$2x + y - 3z = 4,$$
$$5x + 3y - 7z = 1 ?$$
5. Solve the equations $\dfrac{x}{2} = \dfrac{y}{3}$, $x^2 + y^2 = 52$.
6. If $x + y = a + b$ and $x^2 + xy = b^2 + ab$, find x and y in terms of a and b.
7. Solve the equations $xy = 4$, $x + 3y = 8$.
8. Solve the equations $4x^2 + 4xy + y^2 = 9$, $2x - y = 5$.
9. Solve the equations $\dfrac{1}{x} + \dfrac{1}{y} = 5$, $xy = \dfrac{1}{6}$.
10. If $x:y:z = 1:2:3$ and $xyz = 48$, find x, y and z.
11. A rectangular box has a square base. The sum of the lengths of the twelve edges is 40 cm and the sum of the areas of the six faces is 66 cm^2. Find the volume of the box.
12. Three chemical products A, B and C are made from constituents x, y and z in the following proportions:

	x	y	z
A	4	3	1
B	2	2	4
C	1	1	6

If 35 000 kg of x, 30 000 kg of y and 47 000 kg of z are available, find how many kilograms of A, B and C may be made if all supplies are used.

13. A number consists of three digits whose sum is 9. When the digits are reversed, the number is decreased by 198. The sum of the number and the number formed by reversing the digits is 666. Find the number.

14. The sum of the lengths of the three edges of a rectangular box in which the length is twice the breadth is 13 cm. The sum of the areas of the six faces is 108 cm². Find the volume of the box.

15. A, B and C are three towns at which the drivers of a firm call from time to time. One driver does the round trip from A to B to C to A and covers 75 km. A second driver goes from A to B and then to C and returns to A by the same route. He covers 90 km. A third driver goes from A to C and then to B and covers 55 km. Find the distance from A to B.

4 The Remainder Theorem

Roots of an equation

The roots of the equation $f(x) = 0$ are the values of x which satisfy the equation. If $x = a$ is a root of the equation, $f(a) = 0$, and conversely if $f(a) = 0$, $x = a$ is a root.

The quadratic equation $x^2 - 2x - 3 = 0$ has roots -1 and 3. This is obvious because the factors of the left-hand side are $(x + 1)$ and $(x - 3)$. This suggests that if $x = a$ is a root of $f(x) = 0$ then $(x - a)$ is a factor of $f(x)$. This is in fact true and is a consequence of the remainder theorem which we shall now prove.

The remainder theorem

The remainder theorem tells us how to find the remainder when a polynomial such as $x^3 - 5x^2 - 2x + 1$ is divided by a linear function such as $(x - 2)$. The remainder is found by putting $x = 2$ in the polynomial and is $2^3 - 5(2)^2 - 2(2) + 1$ or -15.

The value assigned to x is the value which makes the divisor $(x - 2)$ zero and the result may be checked by long division.

To prove the remainder theorem, suppose that when $f(x)$ is divided by $(x - a)$, the quotient is $Q(x)$ and the remainder R. In an algebraic division, the process is continued until the remainder is of lower degree than the divisor. Since the divisor is of the first degree in this case, the remainder cannot contain x and so R is a number which is indicated by the omission of the brackets containing x.

When 49 is divided by 8, the quotient is 6 and the remainder 1. This is expressed by the equation $49 = 8(6) + 1$. In the general case, therefore,

$$f(x) = (x - a)Q(x) + R.$$

This is an identity, i.e. both sides of the equation are exactly the same when simplified and an identity is satisfied by all values of x.

Choose to put x equal to a. Then

$$f(a) = R.$$

So the remainder is $f(a)$.

Example 1. *Find the remainder when $x^3 + 5x^2 - 2x - 1$ is divided by $(x + 3)$.*

Give x the value which makes $(x + 3)$ zero, i.e. -3. The remainder

is
$$(-3)^3 + 5(-3)^2 - 2(-3) - 1$$

or
$$-27 + 45 + 6 - 1$$

or
$$23.$$

Example 2. *Find the value of k if* $(x + 2)$ *is a factor of* $x^3 + kx^2 - 4x + 4$. *Find the other factors of the expression when k has the value found.*

Put $x = -2$ in the expression.

The remainder is zero and so

$$(-2)^3 + k(-2)^2 - 4(-2) + 4 = 0.$$
$$\therefore -8 + 4k + 8 + 4 = 0.$$
$$\therefore k = -1.$$

The expression is $x^3 - x^2 - 4x + 4$.

Divide by $(x + 2)$ which we know is a factor.

$$
\begin{array}{r}
(x + 2)\, x^3 - x^2 - 4x + 4 \,(x^2 - 3x + 2) \\
\underline{x^3 + 2x^2} \qquad\qquad\qquad \\
- 3x^2 - 4x \qquad\quad \\
\underline{- 3x^2 - 6x} \qquad\quad \\
2x + 4 \\
\underline{2x + 4}
\end{array}
$$

The factors are $(x + 2)(x^2 - 3x + 2)$ or $(x + 2)(x - 1)(x - 2)$.

Exercise 4a

1. If $f(x) = x^3 - 3x^2 + 6x - 7$, write down the value of $f(0)$.
2. If $f(x) = a^x$, write down the value of $f(0)$.
3. If $f(x) = 6x^3 - 18$, write down the value of $f(2)$.
4. If $f(y) = 7y^3 - 4y^2 + 3y - 1$, write down the value of $f(-1)$.
5. If $f(x) = x^3 - 3x$, write down the expression representing $f(a)$.
6. Find the remainder when $x^3 - 6x^2 + 1$ is divided by $(x - 1)$.
7. Find the remainder when $x^3 - 2x^2 + 3x + 1$ is divided by $(x + 1)$.
8. Find the remainder when $x^4 - 6x^2 + 1$ is divided by $(x + 2)$.
9. Find the remainder when $x^5 - 3x^3 - 1$ is divided by $(x - 3)$.
10. Find the remainder when $x^{23} - 1$ is divided by $(x + 1)$.
11. Find the value of k if $(x + 2)$ is a factor of $x^3 + kx^2 - 10x - 8$.
12. Find the value of k if $(x - 1)$ is a factor of $x^3 + kx^2 - 5x - 2$.
13. Find the values of a and b if $(x + 1)$ and $(x + 2)$ are both factors of $x^3 + ax^2 + bx + 4$.
14. Find the values of a and b if $(x - 2)$ and $(x - 3)$ are both factors of $x^3 + ax^2 + bx + 6$.
15. Given that $(x + 2y)$ is a factor of $x^3 + 6x^2y + 11xy^2 + 6y^3$, factorize the expression completely.
16. The expression $ax^2 + bx - 2$ has $(x + 2)$ as a factor. When the expression is divided by $(x - 1)$, the remainder is 6. Find the values of a and b.
17. Find the value of k if $(x + 1)$ is a factor of $x^3 - 4x^2 + kx + 6$. Find also the other factors of the expression.
18. Find the values of a and b if the expression $x^3 + ax^2 + bx - 4$ is exactly divisible by $(x^2 - 4)$.

19. What value of k will make $x^3 - 2x^2 + kx + 6$ vanish when $x = 1$? For what other values of x does the expression vanish?

20. The expression $ax^2 + 5x + b$ leaves remainder -4 when divided by $(x + 1)$ and remainder 0 when divided by $(x + 2)$. Find a and b.

Solution of equations

If $f(a) = 0$, the remainder when $f(x)$ is divided by $(x - a)$ is zero and so $(x - a)$ is a factor of $f(x)$. This fact may be used to factorize a polynomial or to solve an equation.

Example. *Solve the equation* $x^3 - 9x^2 + 26x - 24 = 0$.

Let
$$f(x) = x^3 - 9x^2 + 26x - 24.$$
$$f(1) = 1 - 9 + 26 - 24 = -6.$$

$\therefore (x - 1)$ is not a factor.
$$f(2) = 8 - 36 + 52 - 24 = 0.$$

$\therefore (x - 2)$ is a factor.

By division,
$$x^3 - 9x^2 + 26x - 24 = (x - 2)(x^2 - 7x + 12)$$
$$= (x - 2)(x - 3)(x - 4).$$

The roots of the equation are 2, 3 and 4.

Values worth testing

What values of x are worth trying in the solution of the equation $x^3 - 9x^2 + 26x - 24 = 0$?

Suppose that $x = a$ is a root, then
$$a^3 - 9a^2 + 26a - 24 = 0 \quad \text{or} \quad a^3 - 9a^2 + 26a = 24.$$

Since a is a factor of the left-hand side of the equation, it must also be a factor of 24.

In this example, the list of possible values is a long one, namely ± 1, ± 2, ± 3, ± 4, ± 6, ± 8, ± 12, ± 24, but in other examples the number of values worth testing may be small.

For example, to factorize $x^3 + 2x^2 - x - 2$, the only values which can work are ± 1, ± 2.

If the coefficient of x^3 is not unity, the number of factors is still limited. If $(ax + b)$ is a factor of $3x^3 + 8x^2 + 3x - 2$, then a must be a factor of 3 and b a factor of 2. The only possible factors are $(x \pm 1)$, $(x \pm 2)$, $(3x \pm 1)$ and $(3x \pm 2)$.

If
$$f(x) = 3x^3 + 8x^2 + 3x - 2,$$
$$f(1) = 3 + 8 + 3 - 2 = 12.$$
$$f(-1) = -3 + 8 - 3 - 2 = 0.$$

$\therefore (x + 1)$ is a factor.

By division,
$$3x^3 + 8x^2 + 3x - 2 = (x + 1)(3x^2 + 5x - 2)$$
$$= (x + 1)(3x - 1)(x + 2).$$

To find the remainder when $f(x)$ is divided by $(x-a)(x-b)$

The remainder will be linear in x, so suppose it is $Ax + B$. Then
$$f(x) = (x - a)(x - b)Q(x) + Ax + B,$$
where $Q(x)$ is the quotient. Putting $x = a$ and $x = b$,
$$f(a) = Aa + B$$
and
$$f(b) = Ab + B.$$
From these,
$$f(a) - f(b) = A(a - b)$$
and
$$bf(a) - af(b) = B(b - a).$$
Therefore the remainder is
$$\frac{f(a) - f(b)}{a - b} x + \frac{af(b) - bf(a)}{a - b}.$$

We shall conclude this chapter by factorizing an expression of the fourth degree in a, b and c.

Example. *Factorize* $a^3(b - c) + b^3(c - a) + c^3(a - b)$

If we put $b = a$ in this expression, it becomes
$$a^3(a - c) + a^3(c - a) + c^3(0) = 0.$$
Therefore $(a - b)$ must be a factor of the expression and similarly $(b - c)$ and $(c - a)$ must be factors. But the expression is of the fourth degree in a, b and c and so the only other factor possible is one of the first degree. It must also be symmetrical in a, b and c and therefore can only be $(a + b + c)$. No other factor except a constant is possible.
$$a^3(b - c) + b^3(c - a) + c^3(a - b) = \lambda(a - b)(b - c)(c - a)(a + b + c),$$
where λ is a constant. λ may be found by equating coefficients (of a^3b for example) or by giving a, b and c suitable values. Be careful not to give any two equal values, otherwise the equation will become $0 = 0$ which is not very helpful. Try $a = 0$, $b = 1$, $c = 2$ and we have
$$0 + 1(2) + 8(-1) = \lambda(-1)(-1)(2)(3).$$
$$\therefore -6 = 6\lambda$$
and
$$\lambda = -1.$$
The factors are $-(a - b)(b - c)(c - a)(a + b + c)$.

Exercise 4b: Miscellaneous

1. Solve the equation $x^3 - 3x^2 - 4x + 12 = 0$.
2. Solve the equation $x^3 - 2x^2 + 1 = 0$.
3. Find the remainder when $x^{19} + x^{17}$ is divided by $(x + 1)$.

4. Solve the equation $x^4 - x^2 + 4x - 4 = 0$.

5. Solve the equation $x^3 - 3x^2 - x + 6 = 0$.

6. Find the value of k if $(x + 2)$ is a factor of $x^3 + kx^2 + 6x - 4$.

7. Show that $(x - a)$ is a factor of $(x - b)^7 + (b - a)^7$.

8. For what values of n is $(x + 1)$ a factor of $(x^n + 1)$?

9. Find the values of a and b if $2x^3 + 15x^2 + ax + b$ is exactly divisible by $(x + 4)$ and also by $(2x + 1)$.

10. When $ax^2 + bx + c$ is divided by $(x - 1)$ the remainder is 8; when divided by $(x + 1)$ the remainder is -6 and when divided by $(x + 2)$ the remainder is -4. Find a, b and c.

11. Show that $(a + b + c)$ is a factor of $a^3 + b^3 + c^3 - 3abc$.

12. When the expression $x^3 + kx^2 + 2$ is divided by $(x + 2)$, the remainder is 1 less than when divided by $(x + 1)$. Find k.

13. When the expression $x^3 + ax^2 + 2x + 1$ is divided by $(x - 2)$, the remainder is three times as great as when the expression is divided by $(x - 1)$. Find a.

14. Show that $(x + 2)$ cannot be a factor of $x^{2n} + 4$.

15. For what values of n is $(x - 1)$ a factor of
$$x^n - x^{n-1} + x^{n-2} - \ldots + (-1)^n?$$

16. Show that $(x - 1)^2$ is a factor of $x^3 - 4x^2 + 5x - 2$.

17. Solve the equation $x^4 + 3x^3 - 2x^2 - 12x = 8$.

18. Find the remainder when $x^{23} - x^{19} - 1$ is divided by $(x + 1)$.

19. Factorize $bc(b - c) + ca(c - a) + ab(a - b)$.

20. Factorize $(a - b)^3 + (b - c)^3 + (c - a)^3$.

21. For what values of n is $x^2 + 1$ a factor of $x^n + 1$?

22. Find a factor of $(x - b - c)(x - a - c)(x - a - b) - abc$.

23. Factorize $(a^2 + b^2)(a - b) + (b^2 + c^2)(b - c) + (c^2 + a^2)(c - a)$.

24. Find an expression for the remainder when $f(x)$ is divided by $(x^2 - a^2)$.

25. Factorize $a^2(b - c) + b^2(c - a) + c^2(a - b)$.

5 Variation

Direct proportion

If the ratio of y to x is always constant, then y is said to vary directly as x. The equation connecting the two quantities is $\dfrac{y}{x} = k$ or $y = kx$. The relationship may be written $y \propto x$ which is read as 'y is proportional to x' or 'y varies directly as x'. The graph of y plotted against x is a straight line through the origin and conversely any straight line through the origin represents a case of direct proportion.

If x is doubled, so is y; if x is trebled, so is y; if x is halved, so is y, and so on.

There are many illustrations of direct proportion in everyday life and three are given below.

1. The distance travelled by a train moving at constant speed is proportional to the time taken.

2. The perimeter of a square is directly proportional to its length of side.

3. The extension of a spring is directly proportional to its tension.

Direct proportion is, in fact, so common that there is a tendency to assume that any two quantities are in direct proportion. The height of a child, for instance, is not proportional to its weight.

Other cases of direct proportion

The volume of a cylinder is given by the equation $V = \pi r^2 h$. The volume is a function of two variables r and h. If r is kept constant, V varies directly as h. If h is kept constant, V does not vary directly as r. It, in fact, varies directly as the square of r. This example is one of joint variation, in which V is said to vary jointly as h and the square of r.

In general, if y varies as the square of x, then $\dfrac{y}{x^2}$ is constant. The equation connecting x and y is $y = kx^2$, which may be written $y \propto x^2$. In such a case if x is doubled, y is quadrupled; if x is trebled, the value of y is 9 times greater than its previous value.

The volume of a sphere is given by $V = \frac{4}{3}\pi r^3$. Here V varies directly as the cube of r.

The time of swing of a simple pendulum is given by

$$T = 2\pi \sqrt{\frac{l}{g}}.$$

Here T varies directly as the square root of l.

Example 1. *The extension of a stretched spring is directly proportional to its tension. A tension of 6 newtons produces an extension of 4 cm. Find the extension produced by a tension of 8 newtons.*

First method. $\dfrac{\text{Tension}}{\text{Extension}}$ is constant.

Let x cm be the extension produced by a tension of 8 newtons.

Then
$$\frac{6}{4} = \frac{8}{x}.$$

$$\therefore x = 5\tfrac{1}{3}.$$

Second method. Let $T = kx$, where T is the tension in newtons and x the extension in cm. A different choice of units merely alters the value of the constant, k.

When $T = 6$, $x = 4$.
$$\therefore 6 = 4k \quad \text{and} \quad k = 1\tfrac{1}{2}.$$

$\therefore T = 1\tfrac{1}{2}x$, and when $T = 8$,
$$x = \tfrac{16}{3} = 5\tfrac{1}{3}.$$

Example 2. *The volume of a cone is directly proportional to its height and to the square of its base radius. Two cones are such that the height of the first is twice that of the second and the radius of the first is half that of the second. Find the ratio of their volumes.*

The formula for the volume of the cone is $V = kr^2h$. Suppose that the radius of the first cone is r', its height h' and its volume V'. Then $V' = kr'^2h'$. The radius of the second cone is $2r'$ and its height is $\tfrac{1}{2}h'$. Suppose its volume is V''. Then
$$V'' = k(2r')^2 \cdot \tfrac{1}{2}h'$$
$$= 2kr'^2h'.$$

$$\therefore \frac{V''}{V'} = \frac{2kr'^2h'}{kr'^2h'} = 2$$

The volume of the second cone is twice that of the first.

Exercise 5a

1. Write down a connection between x and y given that x electric bulbs cost y p and each bulb cost 10 p.
2. The area of a circle varies as r^n. Write down the value of n.
3. Complete the statements:
   (i) If A varies as the square of r, then r varies as ...
   (ii) If $V \propto r^3$, then r ...
4. A rectangle of area $100 \, \text{cm}^2$ is x cm long and y cm wide. Is x directly proportional to y?
5. The area of a rectangle in which one side is double the other is A cm². How does A vary with the shorter side?

6. The resistance R newtons to the motion of a car varies directly as the square of the speed, V km/h. Express this as an equation.

7. If $y \propto x$ and $y = 8$ when $x = 4$, find x when $y = 17$.

8. If $v \propto u^2$ and $u = 3$ when $v = 8$, find v when $u = 6$.

9. Compare the surface of two spheres whose radii are in the ratio $3:1$.

10. Compare the volumes of two spheres whose radii are in the ratio $2:1$.

11. Two bottles are of the same shape. The smaller holds 500 cm^3 and the larger has each of its linear dimensions double the corresponding dimensions of the other. How much does the larger hold?

12. If $(y + 2)$ is directly proportional to x, and $y = 4$ when $x = 3$, find y when $x = 4$.

13. A solid sphere of radius 2 cm weighs 1 kg. Find the weight of a sphere of the same material of radius 3 cm.

14. The distance to the horizon on a clear day varies directly as the square root of the height above sea level. At a height of 2 m above sea level, it is possible to see 5 km. At what height does the view extend for 20 km?

15. In what ratio must the radius of a sphere be increased to double its surface area?

16. How does V, the volume of a sphere, vary with S, its surface area?

17. A model of a car is made on a scale of $1:50$. What ratio has the area of the bonnet of the model to that of the car?

18. The radius of a sphere is double that of a second and the density of the first is half that of the second. Compare their weights.

19. If y varies directly as x and $x = 4$ when $y = 3$, plot y against x.

20. The formula for simple interest is $I = \dfrac{PRT}{100}$. Express this as a statement saying how I varies with P, R and T.

Inverse proportion

If y varies as $\dfrac{1}{x}$, y is said to be inversely proportional to x.

In symbols, if $y \propto \dfrac{1}{x}$, then $y = \dfrac{k}{x}$ or $xy = k$.

If y is plotted against x, the resulting graph is a curve called a rectangular hyperbola, but if y is plotted against $\dfrac{1}{x}$, the graph is a straight line.

Three examples of quantities in inverse proportion are given:

1. The time taken by a train to cover a certain stretch of line is inversely proportional to its speed.

2. The volume of a gas at constant temperature is inversely proportional to its pressure (Boyle's Law).

3. If the area of a rectangle is constant, its length is inversely proportional to its breadth.

Other cases of inverse proportion

One quantity may vary inversely as the power of another. For example, the gravitational pull between two bodies is inversely proportional to the square of the distance between them. In symbols, if P is the force and d the distance between them,

$$P = \frac{k}{d^2}.$$

Another example is the electrical resistance of a wire which varies inversely as the square of its radius.

If y varies inversely as the cube of x, then $y = \dfrac{k}{x^3}$.

If y varies inversely as the square root of x, then $y = \dfrac{k}{\sqrt{x}}$.

Example. *The electrical resistance of a wire varies inversely as the square of its radius. Given that the resistance is 0.8 ohm when the radius is 0.2 cm, find the resistance when the radius is 0.3 cm.*

If R is the resistance in ohms and r the radius in centimetres,

$$R = \frac{k}{r^2}.$$

When $r = 0.2$, $R = 0.8$.

$$\therefore 0.8 = \frac{k}{(0.2)^2} \quad \text{and so} \quad k = 0.032.$$

The equation connecting R and r is $R = \dfrac{0.032}{r^2}$.

When $r = 0.3$, $\qquad R = \dfrac{0.032}{0.09} = 0.36$ ohm (to 2 s.f.).

Exercise 5b

1. If y is inversely proportional to x and $y = 8$ when $x = 6$, find y when $x = 4$.
2. If y is inversely proportional to the square of x and $y = 4$ when $x = 3$, find y when $x = 2$.
3. If y is inversely proportional to the square root of x and $y = 8$ when $x = 4$, find x when $y = 2$.

4. If $(y + 2)$ varies inversely as x and $y = 3$ when $x = 2$, find y when $x = 5$.

5. Complete the statements:

   (i) If V varies inversely as the square of h, then h varies ...

   (ii) If $A \propto \dfrac{1}{\sqrt{r}}$, then $r \propto \cdots$

6. If x varies inversely as y and y varies inversely as z, how does x vary with z?

7. If x varies inversely as y and y varies inversely as the square of z, how does x vary with z?

8. If a gas at constant temperature has a volume of 500 cm^3, find its volume when the pressure is doubled.

9. The kinetic energy of a body is proportional to the square of its velocity. When the velocity is 8 m/s, the kinetic energy is 12 joules. Find the velocity when the kinetic energy is 48 joules.

10. Rectangular tanks with square bases are made so that they all have the same capacity. How does the height vary with the edge of the square base?

11. If $(y - 2)$ varies inversely as the square of x and $y = 6$ when $x = 3$, find y when $x = 2$.

12. Corresponding values of u and v are given in the table:

v	3	4	6	12
u	48	36	24	12

How does v vary with u?

13. Corresponding values of x and y are given in the table:

y	2	8	32	72
x	12	6	3	2

How does y vary with x?

14. If y varies inversely as the nth power of x, write down an equation between x and y.

15. If 12 men working 6 hours a day finish a job of work in 10 days, how long will the same job take 4 men working 8 hours a day?

16. If x varies directly as y and inversely as the square of z and $x = 4$ when $y = 3$ and $z = 2$, find the equation connecting x, y and z.

17. Given that y varies inversely as x and $y = 3$ when $x = 2$, draw the graph of y against x.

18. With the information of question 17, plot y against $\dfrac{1}{x}$.

19. If $x \propto \dfrac{1}{y}$ and if $x \propto z^3$, what is the effect on y of doubling z?

20. If $(y - 3) \propto \dfrac{1}{x^2}$ and $x \propto \dfrac{1}{z}$, find y when $z = 3$ given that $y = 7$ when $z = 2$.

Joint variation

We have already met cases of joint variation. If a function V is the product of powers (positive or negative) of other variables x, y, z then V is said to vary jointly as the individual powers of x, y and z.

Example 1. The time of swing of a pendulum is given by the formula $T = 2\pi\sqrt{\dfrac{l}{g}}$. The time of swing varies directly as the square root of l and inversely as the square root of g.

Example 2. The height of a cylinder can be expressed in terms of its volume and base radius by the equation $h = \dfrac{V}{\pi r^2}$. The height varies directly as the volume and inversely as the square of the radius.

Example 3. The height (h) which the outer rail of a curve is raised above the inner rail is directly proportional to the gauge (g) and the square of the maximum velocity (v) and inversely proportional to the radius of the curve (r). Express this in the form of an equation.

$$h = k\frac{gv^2}{r}.$$

Variation as the sum of two parts

The function $(ax + bx^2)$ obviously is a function of x but is neither directly proportional to x nor to the square of x. It is in fact the sum of two quantities ax, which varies directly as x, and bx^2, which varies directly as the square of x. This is an example of the variation as the sum of two parts or variation by parts. The function is one which varies partly as x and partly as x^2. It is not always made sufficiently clear in a question whether the example is one of joint variation or variation by parts and thought should be given to clarify this before a formula is written down.

Example 1. *The resistance to the motion of a car is partly constant and partly varies as the square of the velocity. Write down an equation for the resistance R in terms of the velocity v.*

$$R = a + bv^2.$$

Notice that the two constants a and b are in general not equal. A common mistake is to write $R = k + kv^2$.

Example 2. *The distance travelled by a particle moving under constant acceleration varies partly as the time and partly as the square of the time. Express this as an equation.*

If s is the distance and t the time,

$$s = at + bt^2.$$

Example 3. *The cost of catering per head is partly constant and partly inversely proportional to the number of people present. If the cost per head for 10 people is £0.35 and the cost per head for 20 people is £0.30, find the cost per head for 40 people.*

Suppose the cost per head is £C and the number of people n.
Then

$$C = a + \frac{b}{n}.$$

When $n = 10$, $C = 0.35$ $\therefore 0.35 = a + \frac{b}{10}.$

When $n = 20$, $C = 0.30$ $\therefore 0.30 = a + \frac{b}{20}.$

Subtracting the equations,

$$0.05 = \frac{b}{10} - \frac{b}{20} = \frac{b}{20}.$$

$\therefore b = 1$ and, by substitution, $a = 0.25$.

The formula connecting C and n is

$$C = 0.25 + \frac{1}{n}$$

When $n = 40$, $C = 0.275$.
The cost per head for 40 people is $27\frac{1}{2}$ p.

Exercise 5c: Miscellaneous

1. If y is inversely proportional to x^2 and $y = 18$ when $x = 4$, find y when $x = 3$.
2. If x varies directly as the square of y and y varies inversely as the cube of z, prove that xz^6 is constant.
3. The safe speed for a train rounding a corner is proportional to the square root of the radius. If the safe speed for a curve of radius 49 metres is 28 km/h, find the safe speed for a curve of radius 64 metres.
4. A motorist estimates that his annual expenditure is partly constant and partly varies as the distance travelled. The cost when he travels 4000 km is £100 and when he travels 6000 km is £125. Find the cost when he travels 6400 km.
5. The illumination of a bulb varies inversely as the square of the distance. If the illumination is 6 lux at a distance of 6 metres, what is the illumination at a distance of 3 metres?
6. The attraction between two bodies is directly proportional to the product of their masses and inversely proportional to the square of the distance between them. If each mass is trebled, how must the distance between them be altered to give the same attraction?

7. The heat generated by a current in a wire varies directly as the time, directly as the square of the voltage and inversely as the resistance. If the voltage is 40 volts and the resistance 20 ohms, the heat generated is 80 joules per s. Find the heat generated in 1 minute if the voltage is 60 volts and the resistance 120 ohms.

8. A solid cone of height 6 cm weighs 10.8 kg. Find the weight of a similar cone of height 5 cm.

9. A solid sphere of radius 6 cm weighs 2.7 kg. Find the weight of a shell of the same material whose internal and external radii are 4 cm and 8 cm respectively.

10. If $V \propto \dfrac{x^3}{y}$ and $y \propto xt$, find how V varies with x and t.

11. The resistance to the motion of a car is partly constant and partly varies as the square of the velocity. When the velocity is 10 km/h, the resistance is 420 newtons. When the velocity is 20 km/h, the resistance is 480 newtons. Find the resistance when the velocity is 30 km/h.

12. The effort required to raise a load is partly constant and partly proportional to the load. The effort necessary for a load of 6 newtons is 4 newtons and for a load of 9 newtons is 5 newtons. Find the effort necessary for a load of 18 newtons.

13. A model is made of a ship. If the ratio of the displacement of the ship to that of the model is 27 000:1, find the ratio of the areas of deck space.

14. If z varies directly as x and inversely as the square of y, find the percentage change in z when each of x and y is increased by 10%.

15. Find the increase in the volume of a cone when the base radius is increased by 10% and the height by 5%.

16. The periodic time (T) of a planet and semi-major axis (a) of its orbit are such that T^2 is directly proportional to the cube of a (Kepler's third law). If two orbits are such that the major axis of one is double that of the other, find the ratio of their periodic times.

17. It is given that W varies directly as the fourth power of r and inversely as the square of h. If r is doubled, in what ratio must h be altered to leave W unchanged?

18. The pressure of a gas is inversely proportional to its volume and directly proportional to its absolute temperature. When the pressure is 10^6 N m^{-2} and the temperature is 300 K, the volume is 1000 cm^3. Find the volume when the pressure is 1.8×10^6 N m^{-2} and the temperature 324 K.

19. The deflection at the centre of a girder of given material with fixed ends under a uniformly distributed load varies directly as W, the load, directly as the cube of the length, L, and inversely as the moment of inertia of the girder, I. If the load is doubled, the length of the girder halved and its moment of inertia halved, how is the deflection changed?

20. In motion under gravity, the velocity at any moment is directly proportional to the time from the moment of release. The distance fallen is directly proportional to the square of that time. How does velocity vary with distance?

21. The distance in which a car may be brought to rest by application of the

brakes is partly proportional to the velocity and partly proportional to the square of the velocity. If a car travelling at 20 km h^{-1} can be brought to rest in 8 m and the same car travelling at 40 km h^{-1} can be brought to rest in 20 m, find the distance in which it can be brought to rest when travelling at 30 km h^{-1}.

22. A quantity z is the sum of two terms, the first of which varies as x and the second of which varies as y. Write down an equation connecting x, y and z. When $x = 1$ and $y = 2$, $z = 8$; when $x = 2$ and $y = 1$, $z = 7$. Find the value of z when $x = y = 1$.

23. A particle is fastened to a string and made to describe a horizontal circle about the fixed end of the string. The tension in the string is directly proportional to the square of the velocity of the particle and inversely proportional to the length of the string. Given that the tension is 8 newtons when the velocity is 12 m s^{-1} and the length of string is 4 m, express v, the velocity in m s^{-1}, in terms of T, the tension in newtons, and l, the length in metres.

24. If V varies as x when y is constant and V varies as y^2 when x is constant, express V in terms of x and y given that $V = 8$ when $x = 2$ and $y = \frac{1}{2}$.

25. In order to produce a given note, the tension in a wire must vary as the square of the length of the wire. Two wires produce the same note. The length of the first wire is 2 metres and it is under a tension of 50 newtons. If the second wire is under a tension of 72 newtons, find its length.

6 Arithmetic and Geometric Progressions

Sequences

Most readers will have heard or seen programmes on radio or television which ask for the next term in a certain sequence.

In the sequence $\quad$ 1, 4, 9, 16, 25 ...

it is fairly obvious that the next term is 6^2 or 36.

In the sequence $\quad$ 1, 5, 7, 17, 31, 65 ...

it is not so easy to spot the next term. Each term is, in fact, found by adding the preceding term to twice the term before that.

$$65 = 31 + 2(17); \quad 31 = 17 + 2(7).$$

The next term of the sequence is therefore

$$65 + 2(31) = 127.$$

The nth term of the sequence may be expressed as $2^n + (-1)^n$.

For example, the third term is $2^3 - 1$ or 7; the sixth term is $2^6 + 1$ or 65. The term asked for is the 7th which equals $2^7 - 1$ or 127.

A television programme would be more likely to give a sequence very difficult to spot and with some trick incorporated, such as

$$12, 51, 81, 12, 42, 72 \ldots$$

This is formed from the sequence

$$12, 15, 18, 21, 24, 27 \ldots$$

in which each term is three greater than the preceding. The digits of each term in this sequence (except the first) are then reversed to produce the given sequence. The next term is 03 or 3.

The arithmetic progression

The terms of a sequence such as

$$12, 15, 18, 21 \ldots$$

are said to form an **arithmetic progression** (A.P.). The difference between any term and the preceding is constant and is called the **common difference.** Examples are:

(i) 7, 13, 19, 25 ...; $\quad$ common difference 6.
(ii) 8, 5, 2, -1 ...; $\quad$ common difference -3.
(iii) $a, a + d, a + 2d$ ...; $\quad$ common difference d.

The nth term of an A.P.

Consider the sequence 12, 15, 18, 21 ...

To find the second term, $3(1)$ is added to 12;
to find the third term, $3(2)$ is added to 12;
to find the fourth term, $3(3)$ is added to 12;
to find the nth term, $3(n-1)$ is added to 12.

The nth term is therefore $3n - 3 + 12$ or $3n + 9$.

If any value is given to n, we shall find the corresponding term of the series. For example, the sixth term is $3(6) + 9$ or 27.

In the general sequence whose first term is a and whose common difference is d, the nth term is $a + (n-1)d$.

Example. *Find which term 84 is in the sequence 12, 15, 18, 21, ...*

The difference between 12 and 84 is 72. So the common difference 3 has been added 24 times to the first term 12. Therefore 84 is the 25th term of the sequence.

$$\text{The } n\text{th term} = a + (n-1)d.$$
$$84 = 12 + (n-1)3.$$
$$72 = 3(n-1) \quad \text{and} \quad n = 25.$$

The arithmetic mean

The arithmetic mean of two quantities x and y is their 'average' $\frac{1}{2}(x + y)$. This is the quantity which placed between x and y gives three terms of an A.P.

If the terms a, b and c are three consecutive terms of an A.P., then

$$b - a = c - b$$

or

$$a + c = 2b.$$

Example. The arithmetic mean of 7 and 29 is $\frac{1}{2}(7 + 29)$ or 18.

N.B. If a question concerns three terms in A.P., it is usually easier to consider $(a - x)$, a and $(a + x)$ rather than a, $(a + x)$ and $(a + 2x)$.

The sum of n terms of an A.P.

To find the sum of 100 terms of the series

$$1 + 2 + 3 + 4 \ldots$$

would be a tedious job by addition. However, if the series is written down backwards and added to itself, something helpful occurs.

Let $S = 1 + 2 + 3 + \ldots + 100.$
Then $S = 100 + 99 + 98 + 97 + \ldots + 1.$
Adding, $2S = 101 + 101 + 101 + \ldots + 101.$

The terms of this sequence are all the same and there are 100 terms.

$$\therefore 2S = 100 \times 101 = 10\,100.$$
$$\therefore \quad S = 5050.$$

Let us try the same device with the general series. Call the nth term l; the term before the nth is $(l - d)$. So

$$S = a + (a + d) + (a + 2d)\ldots + (l - d) + l.$$

Also
$$S = l + (l - d) + (l - 2d)\ldots + (a + d) + a.$$

Adding,
$$2S = (a + l) + (a + l) + \ldots + (a + l) + (a + l).$$

$$\therefore 2S = n(a + l), \quad \text{or} \quad S = \frac{n}{2}(a + l).$$

l is the nth term of this series and therefore

$$l = a + (n - 1)\,d.$$

Substituting for l,

$$S = \frac{n}{2}\{2a + (n - 1)\,d\}.$$

Example. *Find the least number of terms of the A.P.*

$$28 + 25 + 22 + \ldots$$

needed to give a negative sum.

Here $a = 28$ and $d = -3$.

The sum to n terms $= \dfrac{n}{2}\{56 - 3(n - 1)\}$.

For this to be negative $3(n - 1) > 56$.
$$\therefore 3n > 59.$$
$$\therefore n > 19\tfrac{2}{3}.$$

So 20 terms must be taken to give a negative sum.

Exercise 6a

1. Find the 15th term of the A.P. 1, 8, 15, 22 ...
2. Find the 40th term of the A.P. 8, 6, 4 ...
3. Find an expression for the nth odd number.
4. Find the sum of the first 20 odd numbers.
5. Is 243 a term of the A.P. 4, 7, 10 ...?
6. Find the value of n given that 49 is the nth term of the A.P. 1, 7, 13 ...
7. If 5, x, y, 32 are in A.P., find x and y.
8. The 4th term of an A.P. is 18 and the 7th term is 30. Find the first term.
9. Find the sum of 15 terms of the A.P. $3 + 7 + 11$ ...
10. Find the sum of 20 terms of the A.P. $8 + 5 + 2$ ...
11. What is the nth term of the A.P. 4, 4.1, 4.2, 4.3 ...?

12. Find the sum of all even numbers less than 100.

13. How many terms are there in the A.P. 2, 5, 8 ... $(6n + 2)$?

14. How many terms are there in the A.P. 15, 12, 9 ... $(21 - 3n)$?

15. The nth term of a sequence is $(6n + 5)$. Find the third term.

16. The nth term of a sequence is $(2n + 3)$. Find the sum to n terms.

17. The sum to n terms of a sequence is $n(n + 5)$. Find the first three terms.

18. The sum to n terms of a sequence is $n(n - 6)$. Find the nth term.

19. Find the value of the first term of the A.P. 100, 94, 88 ... to be negative.

20. Find the least number of terms of the A.P. $100 + 94 + 88 + ...$ needed to give a negative sum.

The geometric progression

If a sequence of terms is such that each term is a constant multiple of the preceding, the sequence is called a **geometric progression** (G.P.). The constant multiple is called the **common ratio.**

Examples of term in G.P. are:

(i) 5, 15, 45, 135 ...;   common ratio 3.
(ii) 18, 6, 2 ...;   common ratio $\frac{1}{3}$.
(iii) 2, -6, 18, -54 ...;   common ratio -3.
(iv) $a, ar, ar^2, ar^3 \cdots$;   common ratio r.

The nth term of a G.P.

Consider the sequence 4, 12, 36 ...

The second term is $4(3)$.
The third term is $4(3^2)$.
The fourth term is $4(3^3)$.
The nth term is $4(3^{n-1})$.

Similarly the nth term of a G.P. whose first term is a and whose common ratio is r is ar^{n-1}.

The geometric mean

If x, y, z are three consecutive terms of a G.P., the common ratio is equal to $\frac{y}{x}$ or to $\frac{z}{y}$.

$$\therefore x = \frac{z}{y} \quad \text{or} \quad y^2 = xz.$$

This is the condition for x, y and z to be consecutive terms of a G.P. The middle term, y, is called the geometric mean of x and z; the geometric mean of two numbers is, therefore, the square root of their product.

For example, the geometric mean of 4 and 16 is $\sqrt{4 \times 16}$ or 8.

The sum of *n* terms of a G.P.

Suppose that S denotes the sum of n terms of the G.P. whose first term is a and whose common ratio is r.

Then $\quad S = a + ar + ar^2 + \ldots + ar^{n-2} + ar^{n-1}$ \hfill (i)

Multiply both sides by r:

$$rS = ar + ar^2 + \ldots \quad + ar^{n-1} + ar^n \qquad \text{(ii)}$$

Subtracting (ii) from (i):

$$S - rS = a - ar^n.$$
$$\therefore S(1 - r) = a(1 - r^n),$$

or $$S = a\frac{1 - r^n}{1 - r}.$$

If $r < 1$, both numerator and denominator of this fraction are positive and this is the most convenient form for S. If, however, $r > 1$, both numerator and denominator are negative and a more convenient form is

$$S = a\frac{r^n - 1}{r - 1}.$$

Example. *Find the sum of the first ten terms of the G.P.*
$$3, 6, 12, 24 \ldots$$

Here $a = 3$ and $r = 2$.

$$S = a\frac{r^n - 1}{r - 1} = 3 \cdot \frac{2^{10} - 1}{2 - 1} = 3(2^{10} - 1) = 3 \times 1023 = 3069.$$

Exercise 6b

(Answers may be left in index form.)

1. Write down the 20th term of the G.P. 2, 10, 50, 250 ...
2. Write down the 17th term of the G.P. 3, 6, 12 ...
3. What is the 18th term of the G.P. 81, 27, 9 ...?
4. What is the 10th term of the G.P. 1, -7, 49 ...?
5. Find n given that 3^{17} is the nth term of 9, 27, 81 ...
6. Find n given that $\dfrac{1}{2^{14}}$ is the nth term of 8, 4, 2 ...
7. Find the geometric mean of 18 and 50.
8. If 5, x, y, 320 are in G.P., find x and y.
9. Find the sum of 12 terms of the G.P. 2, 6, 18 ...
10. Find the sum of 20 terms of the G.P. 1, 5, 25 ...
11. Find the sum of $2n$ terms of the G.P. 3, 6, 12 ...
12. What is the nth term of the G.P. 12, 6, 3 ...?
13. What is the nth term of the G.P. $5a$, $10a^2$, $20a^3$ ...?

14. A number expressed in the base two has n digits each unity. Find its value in base ten.

15. How many terms of the G.P. $2 + 8 + 32 \ldots$ are needed to give a sum greater than 1000?

16. The 3rd term of a G.P. is 18 and the 5th 162. Find the 1st term.

17. The arithmetic mean of two numbers is $6\frac{1}{2}$ and their geometric mean is 6. Find the numbers.

18. Find the sum of all the numbers less than 100 which are powers of 2 (exclude unity).

19. Insert three geometric means between 3 and 1875.

20. The nth term of a series is 3.2^{n-1}. Find the sum of the first n terms.

Worked examples

Three typical examples of problems on sequences are worked below.

Example 1. *If the first, second and fifth terms of an A.P. are three consecutive terms of a G.P., find the common ratio.*

Suppose the terms are a, ar and ar^2. If d is the common difference of the A.P.,

$$ar - a = d$$
and
$$ar^2 - a = 4d.$$
$$\therefore ar^2 - a = 4(ar - a)$$
or
$$r^2 - 4r + 3 = 0.$$
$$\therefore (r - 1)(r - 3) = 0$$
and
$$r = 1 \quad \text{or} \quad r = 3.$$

If $r = 1$, the three terms are identical, which may be discounted, and so the common ratio is 3.

Example 2. *The sum of the first n terms of a sequence is $n \times 2^n$ for all values of n. Find the first three terms.*

The first term is found by putting n equal to 1 and is 1×2^1 or 2.

The sum of the first two terms is 2×2^2 or 8. The second term is therefore 6.

The sum of the first three terms is 3×2^3 or 24. The third term is therefore 16.

The first three terms are 2, 6, 16.

Example 3. *Find approximately how many grains of corn are needed to place one on the first square of a chess-board, two on the second square, four on the third square and so on.*

There are 64 squares. So the required number is $1 + 2 + 2^2 + 2^3 \cdots$ to 64 terms.

The sum is
$$\frac{2^{64} - 1}{2 - 1} = 2^{64} - 1.$$

$$\log 2 = 0.3010.$$
$$\log 2^{64} = 64 \times 0.3010 = 19.264.$$
$$\therefore 2^{64} = 1.837 \times 10^{19} \quad \text{or} \quad 1.84 \times 10^{19} \qquad \text{(to 3 s. f.)}.$$

Subtracting 1 from 2^{64} will not affect this answer, which is true to 3 significant figures only.

Exercise 6c: Miscellaneous

1. Find a formula for the sum of the first n odd numbers.

2. Show that the sum of the odd numbers from 1 to 125 inclusive is equal to the sum of the odd numbers from 169 to 209 inclusive.

3. How many terms of the A.P. $1 + 3 + 5 \ldots$ are needed to give a sum greater than 500?

4. Find a formula for the sum of n terms of the A.P. $1 + 4 + 7 \ldots$

5. A man saves £60 of his salary in a certain year and in each succeeding year saves £20 more than the previous year. How much does he save in 20 years?

6. The sum of the first n terms of a series is $n(n + 5)$. Find the first three terms of the series.

7. The sum of the first n terms of a series is 2.5^n. Find the first three terms of the series.

8. If the pth term of an A.P. is x and the $2p$th term is y, find the $3p$th term.

9. Find the sum of all the positive terms of the A.P. $14, 12\frac{1}{2}, 11 \ldots$

10. Find the sum of all odd numbers less than 100 which are multiples of 3.

11. Given that x, y, z are in G.P., prove that $\log x$, $\log y$ and $\log z$ are in A.P.

12. Find the sum of 15 terms of the series
$$\log 2 + \log + \log 18 + \ldots$$

13. The 2nd, 4th and 7th terms of an A.P. are consecutive terms of a G.P. Find the common ratio.

14. The 3rd, 4th and 7th terms of an A.P. are consecutive terms of a G.P. Show that the next term of the G.P. is the 16th term of the A.P.

15. The nth term of an A.P. is x and the sum of the first n terms is S. Find an expression for the first term.

16. A ball allowed to drop from a height of h m on to horizontal ground rebounds to a height of $\frac{4}{5}h$ m. If the ball is dropped from a height of 10 m, find, to the nearest metre, the total distance the ball will have travelled when it hits the ground for the ninth time.

17. Find the sum of n terms of the G.P.
$$\tfrac{1}{2} + \tfrac{1}{6} + \tfrac{1}{18} + \ldots$$
Find the least number of terms which must be taken to give a sum greater than $\frac{29}{40}$.

18. Find the sum of all the integers between 200 and 500 which are divisible by 9.

19. The lengths of the sides of a triangle are in G.P. The length of the shortest side is 6 cm and the perimeter of the triangle is $28\frac{1}{2}$ cm. Find the lengths of the other sides.

20. The nth term of a certain series is $a + bn + c \times 3^n$, where a, b and c are constants. Given that the first three terms are 6, 14 and 34 respectively, find the 4th term.

21. Given that the 4th term of an A.P. is x and that the 10th term is y, find an expression for the sum of the first 10 terms.

22. A lorry delivers gravel from a base on a road to dumps on the road which are 20 metres apart. The first dump is 40 metres from the base and the second 60

metres from base. The lorry starts from the base, delivers the load to the first dump, returns to base, delivers that second load to the second dump and so on. Find how far the lorry has travelled when it has just returned to base after delivering the 20th load.

23. The 3rd term of an A.P. is 17 and the 7th is 33. Find a formula for the sum of the first n terms.

24. Find the values of x for which $(x - 6)$, $2x$ and $(8x + 20)$ are consecutive terms of a G.P.

25. Find, to 3 significant figures, the sum of 20 terms of the G.P.
$$1 + 1.05 + 1.05^2 + 1.05^3 + \ldots$$

7 The Quadratic Function

Factors of an expression and roots of an equation

Since $2x^2 - 10x + 12 = 2(x^2 - 5x + 6)$
$$= 2(x - 2)(x - 3),$$
the roots of the equation $2x^2 - 10x + 12 = 0$ are $x = 2$ or $x = 3$. The **factors** of the expression $2x^2 - 10x + 12$ are $2(x - 2)(x - 3)$; the **roots** of the equation $2x^2 - 10x + 12 = 0$ are $x = 2$ or $x = 3$. Similarly the factors of the expression $3x^2 - 3x - 18$ are $3(x + 2)(x - 3)$; the roots of $3x^2 - 3x - 18 = 0$ are $x = -2$ or $x = 3$.

Sum and product of the roots of a quadratic equation

Suppose that the roots of the quadratic equation $ax^2 + bx + c = 0$ are denoted by α and β. Then the factors of $ax^2 + bx + c$ are $a(x - \alpha)(x - \beta)$,

i.e.
$$a\left(x^2 + \frac{b}{a}x + \frac{c}{a}\right) = a(x - \alpha)(x - \beta)$$

$$= a(x - (\alpha + \beta)x + \alpha\beta)$$

Since this is an identity, $b/a = -(\alpha + \beta)$, $c/a = \alpha\beta$,

i.e. $\qquad\qquad \alpha + \beta = -b/a, \ \alpha\beta = c/a$

The sum of the roots of the quadratic equation is $-b/a$; the product of the roots is c/a.

A more awkward method, which cannot be generalized to equations of higher degree than two, is to use the formula for the roots of a quadratic equation. Since the roots are $\dfrac{-b + \sqrt{b^2 - 4ac}}{2a}$ and $\dfrac{-b - \sqrt{b^2 - 4ac}}{2a}$ the sum of the roots is $-2b/2a$, i.e., $-b/a$. The product of the roots is

$$\left(\frac{-b + \sqrt{b^2 - 4ac}}{2a}\right)\left(\frac{-b - \sqrt{b^2 - 4ac}}{2a}\right)$$

$$= \frac{b^2 - (b^2 - 4ac)}{4a^2}$$

$$= \frac{c}{a}$$

Discriminant of a quadratic

The formula for the solution of a quadratic equation was obtained by completing the square, i.e. by making the left-hand side of the equation a perfect square,

viz.
$$ax^2 + bx + c = 0$$

$$x^2 + \frac{b}{a}x + \frac{b^2}{4a^2} = \frac{b^2}{4a^2} - \frac{c}{a},$$

i.e.
$$\left(x + \frac{b}{2a}\right)^2 = \frac{b^2 - 4ac}{4a^2}$$

If the expression on the right-hand side of the equation, $\dfrac{b^2 - 4ac}{4a^2}$, is positive, then the quadratic equation has real roots. If it is zero, the equation has two equal roots, and if it is negative, then there are no real roots. Since the denominator, $4a^2$, is always positive, the sign is determined by $b^2 - 4ac$. This is called the **discriminant** of the quadratic. These results can be summarized:

$$b^2 - 4ac > 0: \text{two real distinct roots.}$$
$$b^2 - 4ac = 0: \text{two equal roots}$$
$$b^2 - 4ac < 0: \text{no real roots.}$$

In the last case, we shall wish to say later that there are two complex roots.

Graph of the quadratic functions

The graph of $y = ax^2 + bx + c$ is a parabola.

If $a > 0$, for large values of x, the value of y is large and positive.

If $a < 0$, for large values of x, the value of y is large and negative.

If $b^2 \geqslant 4ac$, the graph cuts the axis of x in real points (roots real).

If $b^2 < 4ac$, the graph does not cut the axis of x in real points.

The graph of $y = ax^2 + bx + c$ is one of those shown in Fig. 7.1.

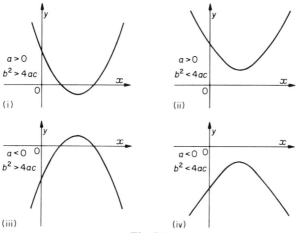

Fig. 7.1

Example. *For what values of x is* $x^2 - 2x < 1$.

If $x^2 - 2x < 1$, $x^2 - 2x + 1 < 2$ (completing the square).

$$\therefore (x - 1)^2 < 2.$$

$$\therefore (x - 1)^2 - 2 < 0 \quad \text{or} \quad (x - 1 - \sqrt{2})(x - 1 + \sqrt{2}) < 0.$$

$$\therefore x \text{ must lie between } 1 \pm \sqrt{2}.$$

Examples of this type depend upon one or other of the following facts:

(i) If $(x - \alpha)(x - \beta) < 0$, then x must lie between α and β.

(ii) If $(x - \alpha)(x - \beta) > 0$, then x cannot lie between α and β.

These follow because if x is greater than both α and β, $(x - \alpha)$ and $(x - \beta)$ are both positive; of x lies between α and β, one factor is positive and the other negative; if x is less than both α and β, $(x - \alpha)$ and $(x - \beta)$ are both negative.

Exercise 7a

1. Find the greatest value of a for which $x^2 - 3x + a = 0$ has real roots.

2. For what range of values of x is $x^2 - 3x > 2$?

3. For what range of values of x is $3x - x^2 > 2$?

4. For what range of values of x is $\dfrac{x - 1}{x - 2} > 0$?

5. Sketch the graph of $y = x^2 - 4x + 3$.

6. Sketch the graph of $y = 5 + 4x - x^2$.

7. For what values of k is $9x^2 + kx + 4$ a perfect square?

8. Prove that the roots of $(x - a)(x - b) = 1$ are real.

9. Prove that the equation $\dfrac{1}{x} + \dfrac{1}{x + a} + \dfrac{1}{x + b} = 0$ has real roots.

10. Find k if $k(x + y)^2 + (x - y)^2$ is a perfect square.

11. Find k if $k(x + y)^2 + x^2 + y^2$ is a perfect square.

12. For what values of c is $x^2 + 6x + c$ positive for all values of x?

13. For what values of b is $x^2 + bx + 9$ positive for all values of x?

14. One root of $x^2 + 5x + k = 0$ is -2. What is the other?

15. One root of $x^2 + ax + 6 = 0$ is 3. What is the other?

16. Find k if $3x^2 + kx + 2 = 0$ has equal roots.

17. One root of $x^2 - 3x + k = 0$ is double the other. Find the roots and the value of k.

18. One root of $x^2 + ax + 3 = 0$ is three times the other. Find the possible values of the roots and the corresponding values of a.

19. Prove that the roots of $ax^2 - (a + c)x + c = 0$ are real.

20. Show that $(a + b - 1)^2 + 4(a + b)$ is a perfect square.

Symmetrical functions of the roots

Using the equations

$$\alpha + \beta = -\frac{b}{a} \quad \text{and} \quad \alpha\beta = \frac{c}{a},$$

any symmetrical function of the roots α and β of the equation $ax^2 + bx + c = 0$ can be expressed in terms of a, b and c.

Example. *Given that α and β are the roots of the equation*
$$ax^2 + bx + c = 0,$$

express in terms of a, b and c: (i) $\alpha^2 + \beta^2$; (ii) $\alpha^3 + \beta^3$; (iii) $\dfrac{\alpha^2}{\beta} + \dfrac{\beta^2}{\alpha}$.

$$\alpha + \beta = -\frac{b}{a}; \quad \alpha\beta = \frac{c}{a}.$$

(i) $\qquad \alpha^2 + \beta^2 = (\alpha + \beta)^2 - 2\alpha\beta$

$$= \frac{b^2}{a^2} - 2\frac{c}{a} = \frac{b^2 - 2ac}{a^2}.$$

(ii) $\qquad \alpha^3 + \beta^3 = (\alpha + \beta)(\alpha^2 - \alpha\beta + \beta^2)$

$$= \left(-\frac{b}{a}\right)\left(\frac{b^2 - 2ac}{a^2} - \frac{c}{a}\right), \quad \text{using (i)}$$

$$= \left(-\frac{b}{a}\right)\left(\frac{b^2 - 3ac}{a^2}\right) = \frac{b(3ac - b^2)}{a^3}.$$

(iii) $\qquad \dfrac{\alpha^2}{\beta} + \dfrac{\beta^2}{\alpha} = \dfrac{\alpha^3 + \beta^3}{\alpha\beta}$

$$= \frac{a}{c} \cdot \frac{b}{a^3} (3ac - b^2) \quad \text{using (ii)}$$

$$= \frac{b}{ca^2} (3ac - b^2).$$

Finding an equation with given roots

To find an equation with given roots, the usual method is to calculate the sum and product of the roots. If S is the sum and P the product, the equation required is

$$x^2 - Sx + P = 0.$$

Example 1. *Given that α and β are the roots of $x^2 + 6x + 7 = 0$, find the equation whose roots are $\dfrac{1}{\alpha}$ and $\dfrac{1}{\beta}$.*

$$\alpha + \beta = -6 \quad \text{and} \quad \alpha\beta = 7.$$

The sum of the new roots $= \dfrac{1}{\alpha} + \dfrac{1}{\beta} = \dfrac{\alpha + \beta}{\alpha\beta} = -\dfrac{6}{7}.$

The product of the new roots $= \dfrac{1}{\alpha\beta} = \dfrac{1}{7}.$

The equation required is $x^2 + \frac{6}{7}x + \frac{1}{7} = 0$

or $\qquad\qquad\qquad 7x^2 + 6x + 1 = 0.$

Example 2. *Given that α and β are the roots of $ax^2 + bx + c = 0$, form the equation whose roots are $\dfrac{\alpha}{\beta}$ and $\dfrac{\beta}{\alpha}$.*

$$\alpha + \beta = -\dfrac{b}{a} \quad \text{and} \quad \alpha\beta = \dfrac{c}{a}.$$

The sum of the new roots $= \dfrac{\alpha}{\beta} + \dfrac{\beta}{\alpha}$

$$= \dfrac{\alpha^2 + \beta^2}{\alpha\beta} = \dfrac{(\alpha + \beta)^2 - 2\alpha\beta}{\alpha\beta}$$

$$= \dfrac{\dfrac{b^2}{a^2} - 2\dfrac{c}{a}}{\dfrac{c}{a}} = \dfrac{(b^2 - 2ac)}{a^2} \cdot \dfrac{a}{c}$$

$$= \dfrac{b^2 - 2ac}{ac}$$

The product of the new roots $= \dfrac{\alpha}{\beta} \cdot \dfrac{\beta}{\alpha} = 1.$

The equation is $\qquad x^2 - \dfrac{b^2 - 2ac}{ac}x + 1 = 0$

or $\qquad\qquad acx^2 - (b^2 - 2ac)x + ac = 0.$

Exercise 7b: Miscellaneous

1. If α, β are the roots of the equation $ax^2 + bx + c = 0$, find the condition that $\alpha = 2\beta$.
2. Find the condition that one of the roots of the equation $ax^2 + bx + c = 0$ is treble the other.

3. If α and β are the roots of the equation $3x^2 - 4x + 5 = 0$, find the value of $\dfrac{\alpha}{\beta} + \dfrac{\beta}{\alpha}$.

4. If α and β are the roots of the equation $2x^2 + x + 5 = 0$, find the value of $\dfrac{1}{\alpha^2} + \dfrac{1}{\beta^2}$. What can you deduce about the roots of the equation?

5. Form the equation whose roots are treble those of the equation $z^2 - 3z + 1 = 0$.

6. Form the equation whose roots are the reciprocals of those of the equation $4x^2 - 2x - 7 = 0$.

7. Form the equation whose roots are 2 less than those of the equation $x^2 - 5x - 1 = 0$.

8. Find the condition that the roots of the equation $x^2 + bx + c = 0$ differ by b.

9. If α, β are the roots of $ax^2 + bx + c = 0$, form the equation whose roots are $(1 + \alpha)$ and $(1 + \beta)$.

10. If α, β are the roots of the equation $ax^2 + bx + c = 0$, form the equation whose roots are $\alpha + 2\beta$ and $\beta + 2\alpha$.

11. Prove that the inequality $x^2 - 2px + q > 0$ holds for all values of x if and only if $q > p^2$.

12. If α and β are the roots of the equation $2x^2 - 5x - 1 = 0$, find the value of $\alpha^3 + \beta^3$.

13. Sketch the graph of $y = 1 - x - x^2$.

14. Show that the roots of the equation
$$(x - b)(x - c) + (x - c)(x - a) + (x - a)(x - b) = 0$$
are real.

15. If α and β are the roots of $x^2 - 5x + 2 = 0$, find the value of $\alpha^4 + \beta^4$.

16. For what values of k does the equation
$$x^2 + (3k - 4)x + (2k + 8) = 0$$
have real roots?

17. Find the quadratic equation whose roots are the squares of the roots of the equation $ax^2 + bx + c = 0$.

18. Show that the equation $x(x - 2) = k(x - 1)$ has real roots for all values of k.

19. Show that the equation $(b^2 - 4ac)x^2 + 4(a + c)x - 4 = 0$ always has real roots.

20. If α and β are the roots of $x^2 - 3x + 1 = 0$ find the value of $\alpha^6 + \beta^6$.

21. Show that the roots of $x(3 - x) + k(x - 2) = 0$ are real for all values of k.

22. Find the equation whose roots are double those of the equation $x(x - 1) = 2$.

23. Form the equation whose roots are ± 2, ± 3.

24. Show that the sum of the roots of the equation $ax^4 + bx^2 + c = 0$ is zero.

25. Show that the sum of the squares of the roots of the equation $ax^4 + bx^2 + c = 0$ is $-\dfrac{2b}{a}$.

8 Permutations and Combinations

Permutations

A woman has three indoor plants, an azalea (A), a cactus (C) and a hyacinth (H). In how many ways can she arrange them on a shelf? The first plant she can choose in three ways, the second in two ways, and the third must be the only one left, so that there are $3 \times 2 \times 1$, i.e. 6, different arrangements. We can tabulate them to check.

A	C	H
A	H	C
C	A	H
C	H	A
H	A	C
H	C	A

Notice when tabulating that, if we have some order in the way we list them, it helps ensure that we have not omitted any possible arrangements: here we wrote first A, then the two ways of arranging C and H.

If the woman was choosing three plants out of ten different plants in her greenhouse to bring in and arrange on a shelf, then she could choose the first in 10 different ways, the second in 9 different ways, and the third in 8 ways, so that there are $10 \times 9 \times 8$, i.e. 720, different ways. The number of ways of arranging 3 objects out of 10 different objects is written $_{10}P_3$ or $^{10}P_3$. Generally, the number of ways of arranging r objects chosen from n is $_nP_r$ where

$$_nP_r = n(n - 1)(n - 2) \ldots (n - r + 1)$$

Factorial notation

Clearly, in studying arrangements, we shall have examples in which we need the product of many consecutive integers. A convenient abbreviation for the product of all integers up to and including n is $n!$,

e.g.
$$5! = 5 \times 4 \times 3 \times 2 \times 1 = 120,$$
$$8! = 8 \times 7 \times 6 \times 5 \times 4 \times 3 \times 2 \times 1 = 40\,320,$$
$$n! = n(n - 1)(n - 2) \ldots 3 \times 2 \times 1.$$

Many calculators have factorial buttons, and many tables of logarithms have the logarithms of factorials, to help with calculations.

Objects not all different

Suppose that the woman buys another hyacinth. How many ways can she now find in which to arrange her plants? If she labels the hyacinths H_1 and H_2 to make sure she can distinguish between them, then there are $4 \times 3 \times 2 \times 1$, i.e. 24, different arrangements. But if she removes the labels, and the hyacinths really are indistinguishable, then to any one arrangement, say ACH_1H_2, there corresponds another arrangement ACH_2H_1 that cannot be distinguished from the first, once the labels have been removed. So the number of ways of arranging 4 objects, 2 of which are identical and the other two different, is $\dfrac{4 \times 3 \times 2 \times 1}{1 \times 2}$ i.e. 12. Similarly, if she had three identical hyacinths and two other plants, there would be $\dfrac{5 \times 4 \times 3 \times 2 \times 1}{1 \times 2 \times 3}$, i.e. 20, ways of arranging these, since any one arrangement of the letters $ACH_1H_2H_3$ occurs in a group of 3!, the number of different ways of arranging the three letters $H_1H_2H_3$.

Example 1. *Find the number of different arrangements that can be made using the letters* (i), *PALMERSTON,* (ii) *ADDITIONAL.*

(i) Here we have 10 letters, all different, so that there are 10!, i.e. 3 628 800 different ways of arranging these.

(ii) Although there are again 10 letters, there are two letters A, two letters D and two letters I, so there are $\dfrac{10!}{2!\,2!\,2!}$, i.e. 453 600, different ways of arranging them.

Exercise 8a

1. Find the value of the following factorials:
   (i) 3! (ii) 6! (iii) 11!

2. Find the value of the following expressions:
   (i) $\dfrac{13!}{10!}$ (ii) $\dfrac{13!}{10!\,3!}$ (iii) $\dfrac{8!}{(4!)^2}$.

3. Write the following products in terms of factorials:
   (i) $8 \times 7 \times 6$ (ii) $12 \times 11 \times 10 \times 9$
   (iii) $\dfrac{20 \times 19 \times 18 \times 17}{1 \times 2 \times 3 \times 4}$ (iv) $n(n-1)(n-2)$.

4. Tabulate the number of different arrangements that can be made of the four letters
   (i) LAMP (ii) FOOD (iii) POOP.

5. Tabulate the number of arrangements that can be made of two letters chosen from the five letters TAKEN. Write down the value of $_5P_2$.

6. How many different arrangements can be made of each of the following letters?
   (i) PRIZE (ii) TANDEM (iii) FORMED

7. How many different arrangements can be made of each of the following letters?
   (i) SETTEE (ii) CALCULUS (iii) OPPOSITE.

8. In how many ways can a first, a second, and a third prize be awarded in a class of 10 pupils?

9. In how many ways can a French, a Latin, and a Greek prize be awarded in a class of 10 pupils?

10. How many numbers greater than 600 can be formed from the digits 2, 3, 4 and 6, if no digit may be used twice?

Combinations

In how many ways can a traveller choose three ties to take on a journey, if he owns ten ties? The first tie—denote it by T_a—can be chosen in 10 ways, the next—T_b—can be chosen in 9 ways, the next—T_c—can be chosen in 8 ways, as there are now only 8 ties left from which he can choose. So there might seem to be $10 \times 9 \times 8$ different selections he can make. But the orderings

$$
\begin{array}{ccc}
T_a & T_b & T_c \\
T_a & T_c & T_b \\
T_b & T_a & T_c \\
T_b & T_c & T_a \\
T_c & T_a & T_b \\
T_c & T_b & T_a
\end{array}
$$

all give him the same selection of ties, so that there are only $\dfrac{10 \times 9 \times 8}{1 \times 2 \times 3}$ different selections.

Similarly, if we are choosing 5 postcards from a selection of 50 offered for sale, there are $\dfrac{50 \times 49 \times 48 \times 47 \times 46}{1 \times 2 \times 3 \times 4 \times 5}$ different selections we can make. This is often written $_{50}C_5$ or $^{50}C_5$ or sometimes $\begin{pmatrix} 50 \\ 5 \end{pmatrix}$, though the last can be confused with a matrix. It is sometimes convenient to abbreviate the evaluation of this by saying

$$
\frac{50 \times 49 \times 48 \times 47 \times 46}{1 \times 2 \times 3 \times 4 \times 5} = \frac{50!}{5! \, 45!}
$$

Generally, $_nC_r$, the number of ways of choosing r objects from n, is

$$
\frac{n!}{r!(n-r)!}.
$$

Example 2. *Find the number of different selections that can be made of three letters from SCHOOL.*

Notice that there are two letters O. If we choose both, we have four ways in which the other letter can be chosen, so there are four different selections containing two Os, i.e. OOS, OOC, OOH, OOL. Now consider the number of different selections that can be made from the letters SCHOL. There are $_5C_3$, i.e. $\dfrac{5 \times 4 \times 3}{1 \times 2 \times 3}$, i.e. 10 different selections. Altogether there is a total of 4 + 10, or 14 different selections of three letters from SCHOOL.

Exercise 8b

1. Find the value of the following
   (i) $_6C_3$ (ii) $_7C_4$ (iii) $_7C_3$

2. Tabulate the different selections of two letters that can be made from the letters TAKEN. Deduce the value of $_5C_2$.

3. How many different selections of two letters can be made from
   (i) LAMP (ii) FOOD (iii) POOP?

4. Eight people are to be divided into two groups, one group containing three people and the other containing five. In how many different ways can this be done?

5. In how many ways can a football team of eleven players be chosen from a class of 15? In how many ways can the four spectators be chosen from the class of 15?

6. In how many ways can a football team of eleven players be chosen from a class of 15, if one boy owns the ball and insists on playing?

7. Given five points in a plane, no three of which are collinear, how many straight lines can be formed by joining the points?

8. There are 5 points in a plane, no 4 of which are concyclic. How many circles can be drawn to pass through three of these points?

9. Given a polygon with n vertices, how many diagonals can be drawn? (A diagonal is any straight line joining two not-adjacent vertices.)

10. A tennis four is chosen from 6 people. In how many ways can this be done? How many different matches can be arranged among the 6 people?

Two important formulae

(a) $\qquad\qquad _nC_r = {_n}C_{n-r}$

Suppose that we wish to choose five persons from a group of eight. Then each choice of 5 excludes 3 persons, so the number of ways of choosing 5 persons from 8 must be the same as the number of ways of choosing 3 persons from 8. More generally, if we choose r persons from n, we exclude $(n - r)$ persons from n.

This can also be proved algebraically.

$$_nC_r = \frac{n!}{r!(n-r)!} \; ; \; {_n}C_{n-r} = \frac{n!}{(n-r)! \, n!}$$

$$\therefore \; {_n}C_r = {_n}C_{n-r}.$$

(b) $$_{n+1}C_{r+1} = {_nC_r} + {_nC_{r+1}}$$

Consider 8 boys from whom we want to choose 5. Call one of the boys Bill. Then our selections either include Bill or do not include Bill. If our selection includes Bill, then we have to choose the other 4 boys from 7. If our selection does not include Bill, then we have to choose all 5 boys from the remaining 7. Thus

$$_8C_5 = {_7C_4} + {_7C_5}.$$

Generalizing, choosing $r + 1$ boys from $n + 1$, if we do include Bill, we have to choose the remaining r boys from n; if we do not include Bill, we have to choose all $r + 1$ boys from the remaining n, thus

$$_{n+1}C_{r+1} = {_nC_r} + {_nC_{r+1}}.$$

To prove this result algebraically,

$$_nC_r = \frac{n!}{r!(n-r)!}; \quad {_nC_{r+1}} = \frac{n!}{(r+1)!(n-r-1)!}.$$

$$\therefore {_nC_r} + {_nC_{r+1}} = \frac{n!}{r!(n-r)!} + \frac{n!}{(r+1)!\,(n-r-1)!}$$

$$= \frac{n!(r+1)}{(r+1)!(n-r)!} + \frac{n!(n-r)}{(r+1)!(n-r)!}$$

$$= \frac{n!(r+1+n-r)}{(r+1)!(n-r)!}$$

$$= \frac{n!(n+1)}{(r+1)!(n-r)!}$$

$$= \frac{(n+1)!}{(r+1)!(n-r)!}$$

$$= {_{n+1}C_{r+1}}.$$

This formula is used in building Pascal's triangle, as shown on page 58.

The next example illustrates how two different methods can be used to solve the same problem.

Example 3. *How many ways are there of choosing any number of objects out of n different objects?*

We can choose one object in $_nC_1$ ways; we can choose two objects in $_nC_2$ ways; we can choose three objects in $_nC_3$ ways, and so the total number of ways in which any number of objects can be chosen from n is $_nC_1 + {_nC_2} + {_nC_3} + \ldots + {_nC_n}$.

Alternatively, consider each object in turn. The first can be either chosen or rejected. The second can be either chosen or rejected, and so on for all n. Thus there are 2^n ways of choosing or rejecting any number of objects. But in one of these all n objects will have been rejected so the number of ways in which any number of objects can be chosen from n is

$$2^n - 1$$

The result

$$2^n = {}_nC_0 + {}_nC_1 + {}_nC_2 + \ldots + {}_nC_n$$

can be obtained by putting $x = 1$ in the expansion of $(1 + x)^n$, by the Binomial Theorem (page 60), for

$$(1 + x)^n = 1 + {}_nC_1 x + {}_nC_2 x^2 \ldots + x^n$$

Exercise 8c

1. Find the number of arrangements of the letters NECROPOLIS.
2. Find the numbers of arrangements of the letters ARREARS.
3. Find the number of ways 5 boys can be seated (i) on a straight bench, (ii) round a circular table.
4. If a cricket team is chosen from 13 men, in how many ways can the batting order be made?
5. A committee consists of 8 men and 4 women. In how many ways can a sub-committee of 3 men and 1 woman be chosen?
6. In how many ways can a committee of 4 men and 3 women be chosen from 7 men and 8 women? In how many ways can it be chosen if Mr X refuses to serve if Mrs Y is on the committee?
7. From a committee of 4 women and 6 men, in how many ways can a sub-committee of 4 be chosen to include one particular man?
8. A railway carriage has 8 seats. In how many ways can 6 people seat themselves in the carriage?
9. Find the number of ways of arranging 5 identical white balls and 4 identical red balls in line.
10. Find the number of combinations of 8 things taken 4 at a time if 2 of the things are alike.
11. In how many ways can 6 books be arranged on a shelf if 2 particular books are to be next to each other?
12. In how many ways can a jury of 12 persons be chosen from 8 men and 8 women? In how many of these ways will all the women have been chosen?
13. In how many ways can 8 different coins be given to Arthur, Anne, and Millicent so that Arthur and Anne receive two each and Millicent has four?
14. How many triangles can be drawn by joining 10 points, no three of which are collinear?
15. What is the greatest possible number of points of intersection of 8 straight lines?
16. What is the greatest possible number of points of intersection of 4 straight lines and 3 circles?
17. How many code words of 3 letters can be formed (i) if no letter is repeated, (ii) if a letter can be used more than once?

18. A car registration number is formed by 2 letters and 4 digits, of which the first cannot be zero. How many different registration numbers can be formed?

19. How many even numbers can be formed from the digits 2, 3, 4, and 5 if no digit is used twice?

20. How many arrangements can be made of the 5 letters SEEDY. In how many of these are the two letters E adjacent?

21. How many whole numbers are factors of 2310, not counting 1 and the number itself? (*Hint:* use Example 3.)

22. How many whole numbers are factors of 840, not counting 1 and the number itself?

23. Given 2n objects of which n are alike and the other n are all different, how many different selections can be made if any number is to be taken?

24. How many different results are possible for 13 football matches, if any one match can end in one of three ways?

9 The Binomial Theorem

Consider the continued product $(x + a)(x + b)(x + c)$. Since each term of one bracket will be multiplied by each term of the other brackets, the number of terms in the product is $2 \times 2 \times 2$ or 8.

If the x term is chosen from each bracket, the product is x^3.

If the x term is chosen from two brackets and the other letter from the third bracket, we get three terms, namely $x^2(a + b + c)$.

If the x term is chosen from one bracket and the other letter from the other brackets, we again get three terms, namely
$$x(ab + bc + ca).$$
The product of the last three terms in the brackets is abc.
$$\therefore (x + a)(x + b)(x + c) \equiv x^3 + x^2(a + b + c)$$
$$+ x(ab + bc + ca) + abc.$$

The number of terms in the coefficient of x^2 is the number of ways of choosing 1 letter from 3, i.e. $_3C_1$ or 3.

The number of terms in the coefficient of x is the number of ways of choosing 2 letters from 3, i.e. $_3C_2$ or 3.

If now b and c are made equal to a, the identity becomes
$$(x + a)^3 \equiv x^3 + 3x^2a + 3xa^2 + a^3.$$

The binomial theorem

Let us now consider the continued product
$$(x + a)(x + b)(x + c) \dots (x + k),$$
in which there are n factors.

The product starts with x^n.

The coefficient of x^{n-1} is $(a + b + c + \dots)$. The number of such terms is $_nC_1$.

The coefficient of x^{n-2} is $(ab + bc + ca \dots)$. The number of such terms is $_nC_2$.

The coefficient of x^{n-3} is $(abc + bcd + \dots)$. The number of such terms is $_nC_3$.

Finally the term independent of x is $abc \dots k$.
$$\therefore (x + a)(x + b) \dots (x + k)$$
$$= x^n + x^{n-1}(a + b + \dots) + x^{n-2}(ab + bc + \dots)$$
$$+ x^{n-3}(abc + bcd + \dots) + \dots + abc \dots k.$$

If now $b, c, d \dots k$ are all made equal to a, we get the identity
$$(x + a)^n \equiv x^n + {}_nC_1x^{n-1}a + {}_nC_2x^{n-2}a^2 + {}_nC_3x^{n-3}a^3 +$$
$$\dots + {}_nC_{n-1}xa^{n-1} + a^n.$$

This is called the binomial theorem and is true for all values of a and x provided that n is a positive integer.

Pascal's triangle

The coefficients in the expansion of $(x + a)^n$ may be built up using Pascal's triangle.

				1		1				$n = 1$
			1		2		1			$n = 2$
		1		3		3		1		$n = 3$
	1		4		6		4		1	$n = 4$
1		5		10		10		5	1	$n = 5$
1	6	15	20	15	6	1				$n = 6$

A triangle is built as shown in the diagram with each term of the two outer diagonals unity. The inside numbers are formed by adding the two adjacent numbers in the line above. For example, in the line for $n = 5$, the coefficient 10 is found by adding 4 and 6 which are immediately above and to either side.

Thus, reading from Pascal's triangle,

$(x + a)^6 = x^6 + 6x^5a + 15x^4a^2 + 20x^3a^3 + 15x^2a^4 + 6xa^5 + a^6.$

The reason that the triangle gives the correct coefficients is seen by writing down the triangle with the numbers in the form of combinations.

$$
\begin{array}{ccccccccc}
& & & & 1 & & 1 & & \\
& & & 1 & & {}_2C_1 & & 1 & \\
& & 1 & & {}_3C_1 & & {}_3C_2 & & 1 \\
& 1 & & {}_4C_1 & & {}_4C_2 & & {}_4C_3 & & 1 \\
1 & & {}_5C_1 & & {}_5C_2 & & {}_5C_3 & & {}_5C_4 & & 1
\end{array}
$$

Considering ${}_5C_3$, for example, we see that it is formed by adding ${}_4C_2$ and ${}_4C_3$. Using the formula

$$_{n+1}C_{r+1} = {}_nC_r + {}_nC_{r+1}$$

proved in the last chapter it follows that ${}_5C_3$ does in fact equal the sum of ${}_4C_2$ and ${}_4C_3$.

Example 1. *Expand* $(2x - 3y)^5$.

The expansion is

$(2x)^5 + 5(2x)^4(-3y) + 10(2x)^3(-3y)^2$
$\qquad\qquad\qquad + 10(2x)^2(-3y)^3 + 5(2x)(-3y)^4 + (-3y)^5$

or $\qquad 32x^5 - 240x^4y + 720x^3y^2 - 1080x^2y^3 + 810xy^4 - 243y^5.$

Example 2. *Expand* $(1 + x + x^2)^6$ *in ascending powers of* x *as far as the term in* x^3.

$$(1 + x + x^2)^6 = (1 + x(1 + x))^6$$
$$= 1 + 6x(1 + x) + 15x^2(1 + x)^2 + 20x^3(1 + x)^3 \cdots$$

Terms after this will contain a factor of x^4 at least and so are not needed.

To the order required,

$$(1 + x + x^2)^6 = 1 + 6x(1 + x) + 15x^2(1 + 2x) + 20x^3(1), \quad \text{powers greater}$$

than x^3 having been neglected.

$$\therefore (1 + x + x^2)^6 = 1 + 6x + 6x^2 + 15x^2 + 30x^3 + 20x^3 + \ldots$$
$$= 1 + 6x + 21x^2 + 50x^3 + \ldots$$

Example 3. *Use the binomial theorem to evaluate* $(1.02)^4$, *correct to 5 significant figures.*

$$(1 + .02)^4 = 1 + 4(.02) + 6(.02)^2 + 4(.02)^3 + (.02)^4$$
$$= 1 + 4(.02) + 6(.0004) + 4(.000008) + .00000016$$
$$= 1.08243 \quad \text{or} \quad 1.0824 \text{ to 5 s.f.}$$

Exercise 9a

1. Write down the expansion of $(x + a)(x + b)(x + c)(x + d)$.
2. Expand $(1 + x)^5$.
3. Expand $(2 - x)^4$.
4. Expand $\left(x + \dfrac{1}{x}\right)^3$.
5. Find the coefficient of x^3 in $(2x - 1)^7$.
6. Find the coefficient of x^4 in $(1 - 3x)^5$.
7. What is the term independent of x in the expansion of $\left(2x + \dfrac{1}{x}\right)^6$?
8. How many terms are there in the expansion of $(m + n)^{12}$?
9. Find the coefficient of x^2 in the expansion of $(1 - x + x^2)^5$.
10. Find the coefficient of x^3 in the expansion of $(1 + x)(1 - x)^3$.
11. Write down the term containing x^r in the expansion of $(x + a)^n$.
12. Find the coefficient of x^r in the expression of $(x - 2)^n$.
13. Evaluate $(\sqrt{3} + 1)^4 + (\sqrt{3} - 1)^4$.
14. Evaluate $(\sqrt{2} + 1)^5 - (\sqrt{2} - 1)^5$.
15. Use the binomial theorem to evaluate 101^4.
16. Use the binomial theorem to evaluate 99^3.
17. Find the value of $(1.01)^7$ correct to 5 significant figures.
18. Find the value of $(0.99)^4$ correct to 5 significant figures.
19. Expand $(1 - x + x^2)^n$ as far as the term in x^2.
20. Find the coefficient of x^n in the expansion of $(1 + 2x + x^2)^n$.

The binomial theorem for other values of *n*

The expansion for $(a + x)^n$ holds only if n is a positive integer. The corresponding expansion for $(1 + x)^n$ written in the form

$$(1 + x)^n = 1 + nx + \frac{n(n - 1)}{1 \cdot 2}x^2 + \frac{n(n - 1)(n - 2)}{1 \cdot 2 \cdot 3}x^3 + \ldots$$

holds for all values of n, positive, negative and fractional, provided that x lies between ± 1. Note that for n negative or fractional the expansion holds only when the first term is unity.

The truth of these statements is not proved but the result may be assumed.

If n is a positive integer, the series terminates and has $(n + 1)$ terms. If n is not a positive integer, the coefficient

$$\frac{n(n - 1)(n - 2)\ldots}{1 \cdot 2 \cdot 3 \ldots}$$

never becomes zero and the expansion is in the form of an infinite series. The chief use of the expansion is when x is small and the terms get progressively smaller. If x is small enough, an approximation for $(1 + x)^n$ is $1 + nx$.

Example 1. *Find approximations for* (i) $(1.01)^3$; (ii) $(0.99)^4$; (iii) $\dfrac{(1.02)^3 \times (1.04)^2}{(0.99)^4}$

$(1 + x)^n = 1 + nx$ approximately.

(i) $(1.01)^3 = 1 + 3(0.01) \simeq 1.03$.

(ii) $(0.99)^4 = (1 - 0.01)^4 \simeq 1 - 0.04 = 0.96$.

(iii) $(1.02)^3 \simeq 1.06$ and $(1.04)^2 \simeq 1.08$.

$\therefore \quad (1.02)^3(1.04)^2 \simeq 1.06 \times 1.08$

$\qquad\qquad\qquad = 1.14$, neglecting 0.06×0.08

Now $\qquad \dfrac{1}{(0.99)^4} = (1 - 0.01)^{-4}$

$\qquad\qquad\qquad \simeq 1 + 0.04 = 1.04$,

$\therefore \quad \dfrac{(1.02)^3(1.04)^2}{(0.99)^4} \simeq 1.14 \times 1.04$

$\qquad\qquad\qquad = 1.18$, neglecting 0.14×0.04.

Example 2. *Write down the first four terms in the expansion of* $(1 - 2x)^{\frac{3}{2}}$ *when* x *is positive and less than* $\frac{1}{2}$.

Since $0 < x < \frac{1}{2}$, $0 < 2x < 1$ and the binomial expansion is valid,

$$(1 - 2x)^{\frac{3}{2}} = 1 + \tfrac{3}{2}(-2x) + \frac{(\tfrac{3}{2})(\tfrac{1}{2})}{1 \cdot 2}(-2x)^2 + \frac{(\tfrac{3}{2})(\tfrac{1}{2})(-\tfrac{1}{2})}{1 \cdot 2 \cdot 3}(-2x)^3 \cdots$$

$$= 1 - 3x + \tfrac{3}{2}x^2 + \tfrac{1}{2}x^3 \cdots$$

Example 3. *Use the expansion of* $(1 - \frac{1}{50})^{\frac{1}{2}}$ *to find the value of* $\sqrt{2}$ *to 4 significant figures.*

$$(1 - \tfrac{1}{50})^{\frac{1}{2}} = 1 + \tfrac{1}{2}(-\tfrac{1}{50}) + \frac{(\tfrac{1}{2})(-\tfrac{1}{2})}{1 \cdot 2}(-\tfrac{1}{50})^2 + \frac{\tfrac{1}{2}(-\tfrac{1}{2})(-\tfrac{3}{2})}{1 \cdot 2 \cdot 3}(-\tfrac{1}{50})^3 \cdots$$

$$= 1 + \tfrac{1}{2}(-0.02) - \tfrac{1}{8}(0.0004) + \tfrac{1}{16}(-0.000008)$$
$$= 1 - 0.01 - 0.00005 \text{ to 5 s.f.}$$
$$= 0.98995.$$

$$\therefore \sqrt{\tfrac{49}{50}} = 0.98995 \quad \text{or} \quad \sqrt{\tfrac{98}{100}} = 0.98995.$$

$$\therefore \frac{7\sqrt{2}}{10} = 0.98995 \quad \text{and} \quad \sqrt{2} = \frac{9.8995}{7}.$$

$$\therefore \sqrt{2} = 1.4142 \quad \text{or} \quad 1.414 \text{ to 4 s.f.}$$

Exercise 9b: Miscellaneous

1. Write down approximations for $(1.03)^{-2}$ and $(0.98)^3$.

2. Write down the approximate value of $\dfrac{1.03^2 \times 1.01}{1.04}$.

3. Write down the first four terms in the expansion of $(1 - x)^{\frac{1}{3}}$ if $0 < x < 1$.

4. Write down the first four terms in the expansion of $\dfrac{1}{(1 + x)^3}$ if $-1 < x < 1$.

5. Each side of a cube block when heated to a certain temperature increases by 0.5%. Find the percentage increase in the volume.

6. By long division find the first four terms in ascending powers of x when $(1 - x)$ is divided by $(1 + x)$. Check your result using the binomial theorem.

7. Expand $\dfrac{1 + 2x}{(1 - x)^2}$ in ascending powers of x as far as the term in x^3 given that x is small.

8. Write down the binomial expansion for $(x + y)^4$ and use your expansion to evaluate $(2.01)^4$ correct to three decimal places.

9. Find the coefficient of a^5b^4 in the expansion of $(3a - \frac{1}{2}b)^9$.

10. Find the term independent of x in the expansion of $\left(2x^2 - \dfrac{1}{x^2}\right)^6$.

11. Using the expansion of $(1 - \frac{1}{8})^{\frac{1}{3}}$, evaluate $\sqrt[3]{7}$ correct to three decimal places.

12. Find the coefficient of x^2 in the expansion of $(1 + x)^4 (1 - 3x)^5$.

13. Expand $\sqrt{\dfrac{1 + x}{1 - x}}$ as far as the term containing x^2, given that x is small. By taking x as $\frac{1}{10}$, find the value of $\sqrt{11}$ to 3 significant figures.

14. Write down the first three terms in the expansion of $(1 - x)^{-2}$, when x is small, and find the coefficient of x^n in the expansion.

15. Write down the first three terms in the expansion of $\sqrt{1 - 2x}$ when x is small. By putting $x = \frac{1}{20}$, find the value of $\sqrt{10}$ to 4 significant figures.

16. Find the numerical values of a and b if in the expansion of $\dfrac{1 + ax + bx^2}{(1 - x)^2}$ there are no terms in x or x^2.

17. Find the coefficient of x^2 in the expansion of $(1 - x)(1 + x)^n$.

18. Find the coefficient of x^n in the expansion of $\dfrac{1 + x}{(1 - x)^2}$.

19. Expand $(1 - x + 3x^2)^4$ in ascending powers of x as far as the term in x^4.

20. Expand $(1 - 5x)^{\frac{1}{3}}$ as far as the term in x^3. By putting $x = \frac{1}{20}$ find the value of $\sqrt[3]{6}$ to 3 significant figures.

21. Find the values of a and n if the first three terms in the expansion of $(1 + ax)^n$ are $1 + 15x + 90x^2$.

22. Find the term independent of x in the expansion of $\left(x^2 + \dfrac{1}{x}\right)^{3n}$.

23. If x is so large that $\dfrac{1}{x^2}$ and higher powers of $\dfrac{1}{x}$ can be neglected, show that
$$\frac{x + 1}{(x + 2)^2} = \frac{1}{x} - \frac{3}{x^2}.$$

24. Find the first four terms in the expansion of $\sqrt{1 + 2x}$. By putting $x = \frac{1}{100}$, find $\sqrt{102}$ correct to 4 decimal places.

25. Find the first three terms in the expansion of $\dfrac{1}{(1 - x)^2} - \dfrac{1}{1 - 2x}$ in ascending powers of x, given that x is small.

10 Probability

Some events are more likely to happen than others. We may say that 'the odds against a certain horse winning a race are 3 to 1'; we may say that there is a 'fifty-fifty' chance of rain stopping play in a cricket match at Old Trafford; sometimes one even hears that there is 'a 90 % probability' that a certain footballer will score from the penalty spot.

In mathematics we assign numbers between 0 and 1 to measure the relative likelihood of events happening. If an event will definitely not take place, we give it the probability 0: if it is certain to happen, we say that the probability is 1. A fair coin (a mathematical coin that cannot land on its edges) has a probability $\frac{1}{2}$ of landing showing heads, $\frac{1}{2}$ of landing showing tails. A fair die has a probability of $\frac{1}{6}$ of showing '5' when thrown, since each of the six faces is equally likely to be showing.

Use of symmetry

These probabilities were found by studying the symmetry of the problem. With the coin, one or other face must be showing, yet each was as likely to be showing as the other, so the probability of each showing must be $\frac{1}{2}$. Six faces to the fair die, each equally likely to be showing; since one face must be showing, we need six equal numbers, whose sum is 1, so the probability of each face being uppermost is $\frac{1}{6}$. If we have a well-shuffled pack of playing cards, the probability of a card drawn at random belonging to any one of the four suits is $\frac{1}{4}$.

Definition

If we have a number of equiprobable events, such as drawing cards from a well-shuffled pack, the probability of a particular outcome is defined as

$$\frac{\text{the number of equiprobable favourable outcomes}}{\text{the total number of equiprobable outcomes}}$$

Thus the probability of drawing a king from a well-shuffled pack of 52 cards is $\frac{4}{52}$; if one joker is included in the pack the probability is $\frac{4}{53}$; the probability of drawing a red king from a pack containing both jokers is $\frac{2}{54}$, since either the King of Hearts or the King of Diamonds is acceptable, and there are now a total of 54 equiprobable outcomes.

Importance of equiprobable outcomes

A family of two children may consist of two boys, one boy and one girl, or two girls, but unless we know that these three events are equiprobable, we cannot say that the probability of any one of these events is $\frac{1}{3}$. If a baby is equally likely to be a boy or a girl, equiprobable outcomes are

$$\text{B, B};\quad \text{B, G};\quad \text{G, B};\quad \text{G, G}$$

Thus the probability of there being two boys is $\frac{1}{4}$, since one out of the four equiprobable outcomes is favourable. The probability of there being a boy and a girl is $\frac{2}{4}$, and the probability of there being two girls in the family is $\frac{1}{4}$.

To find the probability of an event happening, decide how many equiprobable outcomes are favourable, how many equiprobable outcomes there are, and the probability is the ratio of these. The probability of an event is often abbreviated

$$\text{Pr ( )}$$

e.g. Pr(a 'six' with a fair die) $= \frac{1}{6}$.

Example 1. *In a well-shuffled pack of 52 cards, find the probability that a card drawn at random is a red card.*

There are 52 possible outcomes, all equiprobable, and of these 26 are favourable, so the probability that a card drawn at random is red is $\frac{26}{52}$, i.e. $\frac{1}{2}$.

Example 2. *From a well-shuffled pack of cards, the three and four of spades are removed. Find the probability of drawing (i) a club, (ii) a spade.*

(i) There are now 50 cards in the pack, of which 13 are clubs, so the probability of drawing a club is $\frac{13}{50}$.

(ii) There are still 50 cards from which the draw is made, but only 11 of these are favourable, so that the probability of drawing a spade is $\frac{11}{50}$.

Example 3. *A bag of marbles contains 4 red marbles and 11 blue marbles. One marble is drawn at random, what is the probability that it is red?*

Since there are 15 marbles in the bag, 4 of which are red,

$$\text{Pr (marble drawn at random is red)} = \frac{4}{15}.$$

Dependent and independent events

If the marble drawn in Example 3 is replaced, and another draw made, the probability of drawing a red marble on the second draw is still $\frac{4}{15}$. Whatever the marble drawn and replaced, the outcome of the first draw does not affect the probabilities of the second draw. These are **independent** events. But if the first marble is not replaced, then the probabilities are altered. If the first marble is red, the probability of the second marble being red is $\frac{3}{14}$, whereas if the first marble is blue, the probability of the second marble being red is $\frac{4}{14}$. The two draws are now **dependent**.

Mutually exclusive events

Some events are such that one automatically excludes the other. A number cannot be odd and even; a coin cannot show heads and tails simultaneously. Such events are called **mutually exclusive.**

Where only two outcomes are possible, the sum of the probabilities must be 1. Thus if A and B are two such outcomes,

$$\text{Pr (A)} + \text{Pr (B)} = 1$$

Addition Law

The probability that a card drawn at random from a pack of 52 cards is a club is $\frac{13}{52}$, i.e. $\frac{1}{4}$; the probability that it is a diamond is also $\frac{1}{4}$. To find the probability that the card is either a club or a diamond, we note that we now have $13 + 13$ favourable outcomes, so the probability that the card is a club or a diamond is $\frac{26}{52}$, i.e. $\frac{1}{2}$.

To find the probability that the card drawn is a club or a king, though, we note that we have 13 clubs, but only an *extra* three kings, as we have already counted the King of Clubs. Thus there are only 16 favourable outcomes, and so the probability is $\frac{16}{52}$, i.e. $\frac{4}{13}$. These results can be illustrated by Venn diagrams, as in Figure 10.1.

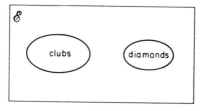

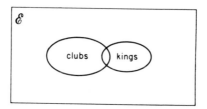

Fig. 10.1

The Addition Law can be summarized thus:
  If A and B are mutually exclusive events,

$$\text{Pr (A or B)} = \text{Pr (A)} + \text{Pr (B)},$$

e.g. Pr (club) = $\frac{1}{4}$, Pr (diamond) = $\frac{1}{4}$, Pr (club or diamond) = $\frac{1}{2}$, but if A and B are not mutually exclusive,

Pr (A or B or both) = Pr (A) + Pr (B) − Pr (both A and B),

e.g. Pr (club) = $\frac{1}{4}$, Pr (king) = $\frac{1}{13}$, Pr (King of Clubs) = $\frac{1}{52}$, Pr (club or a king) = $\frac{1}{4} + \frac{1}{13} - \frac{1}{52} = \frac{16}{52} = \frac{4}{13}$.

Exercise 10a

1. From a well-shuffled pack of 54 cards, including two jokers, one card is drawn. Find the probability that it is (i) a joker, (ii) not a joker, (iii) a king, (iv) a diamond, (v) either a king or a diamond, (vi) neither a king nor a diamond, (vii) the Queen of Hearts, (viii) any heart except the Queen, (ix) a two, three, or four, (x) not a two, three or four.

2. Fifty identical cards are black on one side, and the other side are numbered from 1 up to 50. They are then placed, number downwards, on a table, and one card is drawn. Find the probability that the card has a number that is (i) even, (ii) a multiple of three, (iii) not a multiple of three, (iv) a multiple of seven, (v) a multiple of twenty-one.

   The cards numbered 31 to 50 are now removed, and one card is drawn at random as before. Find the probability that the card has a number on it that is (vi) a multiple of three, (vii) a prime, (viii) larger than 13, (ix) smaller than 13, (x) 32.

3. A box of 20 Christmas tree lights contains 8 that are defective, but these are indistinguishable in appearance from the rest.
   (i) Find the probability that a light drawn at random is defective.
   (ii) A light is drawn and found to be defective and not replaced. Find the probability that the next light is defective.
   (iii) Two lights are drawn and found to be defective, and not replaced. Find the probability that the next light is not defective.
   (iv) Six lights are drawn, of these two are found to be defective and are not replaced. Find the probability that the next light drawn is defective.
   (v) Eight lights are drawn, all are found defective and not replaced. What is the probability that the next light drawn is not defective?

4. All the chocolates in a certain brand of chocolates are identical in appearance, but some are hard-centred and some soft-centred. One bag of these chocolates is known to contain 20 soft centres and 25 hard centres.
   (i) Find the probability that one chocolate drawn at random is soft-centred.
   (ii) After five hard centres have been eaten, what is the probability that the next chocolate is soft-centred?
   Six chocolates are eaten, four having hard centres. What is now the probability that the next chocolate eaten is (iii) soft-centred, (iv) hard centred?
   Twenty-five chocolates have been eaten of which 5 are hard-centred.
   (v) What is the probability that the next eaten is soft-centred?

5. In order to use the labels in a competition, a woman removed the labels

from 8 tins of cat-food and 12 tins of dog-food. Without their labels the tins are now identical.

(i) What is the probability that a tin chosen at random contains dog-food?

(ii) After the dog has eaten two tins of cat food, what is the probability that the next tin opened contains dog-food?

(iii) When two tins of dog-food and one of cat-food have been removed, what is the probability that the next tin contains dog-food?

(iv) What is the least number of tins of dog-food that the animals have to eat for the cat to be more likely than not to have cat-food when the next tin is opened?

6. List the possible outcomes of throwing a pair of dice, e.g. (1,1) to denote both dice showing ones, (1,2) to denote the first die showing one and the second showing two. There should be 36 such outcomes, all equiprobable.

(i) Show that the probability that the total shown on the two dice is 2 is $\frac{1}{36}$.

(ii) List the probability of all totals from 2 to 12, and verify that these add up to 1.

(iii) What is the most likely total obtained from the throwing of two fair dice?

(It is commonly believed that all totals are equiprobable; this enables 'craps' to be played as a gambling game, in which, after an initial round, the banker wins on '7'.)

Conditional probability: multiplication law

Suppose that we have two fair dice, one numbered 1 to 6 and the other with faces coloured blue (B), green (G), orange (O), red (R), white (W), and yellow (Y). When they are thrown together, what is the probability that they show yellow and an even number?

The set of all equiprobable outcomes can be represented by crosses on a grid:

	1	2	3	4	5	6
Y	×	⊗	×	⊗	×	⊗
W	×	×	×	×	×	×
R	×	×	×	×	×	×
O	×	×	×	×	×	×
G	×	×	×	×	×	×
B	×	×	×	×	×	×

The favourable outcomes, yellow and an even number, we can ring; there are 36 equiprobable outcomes, three of which are favourable, so that

Pr (one die shows yellow and the other an even number) $= \frac{3}{36}$
$$= \frac{1}{12}$$

It is often tedious—or impossible—to list all possible outcomes so we devise a shorter method. We know that the probability that an even number is shown is $\frac{1}{2}$, so that when a coloured die is thrown the probability that it is accompanied by an even number is equal to the probability that it is accompanied by an odd number,

i.e. $\text{Pr (even and yellow)} = \frac{1}{2} \text{Pr (yellow)}$
$$= \frac{1}{2} \times \frac{1}{6}$$
$$= \frac{1}{12}$$

i.e. $\text{Pr (even and yellow)} = \text{Pr (even)} \times \text{Pr (yellow)}$

Independent events

On page 64, we said that events were independent if the outcome of one event did not affect the outcome of the other. Thus the outcomes of these two dice are independent. In mathematical terms, we can define independence by saying that two events, A and B, are independent if

$$\text{Pr (both A and B)} = \text{Pr (A)} \times \text{Pr (B)}$$

Use of tree diagrams

When we have two or more events, the probability of any one outcome can often be found most easily from a tree-diagram. Continuing with the example of the two dice, we can draw the following tree-diagram.

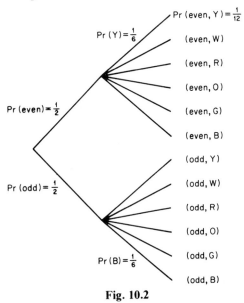

Fig. 10.2

We see that we have twelve equiprobable outcomes, only one of which is acceptable, so that its probability is $\frac{1}{12}$.

If we wish to find the probability that the dice showed an even number with red or with yellow, then there are two acceptable outcomes, so that the probability is $\frac{2}{12}$, i.e. $\frac{1}{6}$.

Use of set notation

We often find it convenient to consider the **set** of all equiprobable outcomes of experiments, and the favourable subset in which we are at any one time interested. In the example with the two dice we have a universal set $\mathscr{E}$ consisting of 36 elements, each an equiprobable result of the experiment. (This set is sometimes called the **outcome** space or the **possibility** space.) If the subset $\{A\}$ is the set favourable for this problem, $\{A\}$ consists of the three elements we can denote by

$$(Y, 2), (Y, 4), (Y, 6)$$

If $n\{A\}$ is the number of elements in $\{A\}$, we can define the probability of an event associated with an element of A as

$$\text{Pr}\,(A) = \frac{n\{A\}}{n\{\mathscr{E}\}}$$

Thus here $n\{A\} = 3$, $n\{\mathscr{E}\} = 36$ and $\text{Pr}\,(A) = \frac{3}{36}$

The Addition Law can be expressed clearly in set notation. If A is the set of all outcomes favourable for one event, and B the set of all outcomes favourable for another, $A \cap B$ is the set of all outcomes favourable for both, e.g. if A is the set of all clubs, B the set of all kings, then $A \cap B$ will contain just the King of Clubs. $A \cup B$ is the set of all events favourable for A or B or both, so that

$$\text{Pr}(A \cup B) = \text{Pr}(A) + \text{Pr}(B) - \text{Pr}(A \cap B).$$

Example 1. *A bag contains 8 marbles, 5 white and 3 red. One is drawn at random and not replaced. Another draw is made. What is the probability that both marbles are red?*

First draw a tree-diagram. From it we can see that the probability that both marbles are red is $(\frac{3}{8})(\frac{2}{7}) = \frac{3}{28}$. We can also see that the probability that both are white is $(\frac{5}{8})(\frac{4}{7}) = \frac{5}{14}$. To find the probability that the marbles are not the same colour, we can either say

$$\text{Pr(not the same colour)} = 1 - \text{Pr(they are the same colour)}$$
$$= 1 - \frac{3}{28} - \frac{5}{14}$$
$$= \frac{15}{28}$$

or we can say that acceptable outcomes would be 'red then white' or 'white then red'; from our probability tree the probability of each of these outcomes

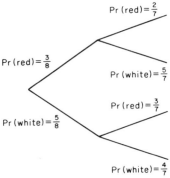

$\Pr(\text{red}) = \frac{2}{7}$

$\Pr(\text{red}) = \frac{3}{8}$

$\Pr(\text{white}) = \frac{5}{7}$

$\Pr(\text{red}) = \frac{3}{7}$

$\Pr(\text{white}) = \frac{5}{8}$

$\Pr(\text{white}) = \frac{4}{7}$

Fig. 10.3

is $(\frac{5}{8})(\frac{3}{7}) = \frac{15}{56}$, so the probability of one or other is $\frac{15}{28}$. If we use this second method, we can check our answer is correct, since $\frac{3}{28} + \frac{5}{14} + \frac{15}{28} = 1$.

Example 2. *A fair die is thrown four times. Find the probability that* (i) *four 'sixes' are shown;* (ii) *no sixes are shown;* (iii) *at least one six is shown.*

(i) Since it is a fair die, $\Pr(\text{six}) = \frac{1}{6}$.

∴ $\qquad\qquad$ $\Pr(\text{first throw shows six}) = \frac{1}{6}$
$\qquad\qquad\qquad$ $\Pr(\text{second throw shows six}) = \frac{1}{6}$
$\qquad\qquad\qquad$ $\Pr(\text{third throw shows six}) = \frac{1}{6}$
$\qquad\qquad\qquad$ $\Pr(\text{fourth throw shows six}) = \frac{1}{6}$
and $\qquad\qquad$ $\Pr(\text{all four throws show sixes}) = (\frac{1}{6})^4$

(ii) Similarly, $\qquad$ $\Pr(\text{first throw is not a six}) = \frac{5}{6}$,
and so $\qquad\qquad$ $\Pr(\text{none of the throws is six}) = (\frac{5}{6})^4$

(iii) To find the probability that at least one 'six' is shown, we note that we have either no sixes shown or at least one six shown, so

$$\Pr(\text{at least one six shown}) = 1 - \Pr(\text{no sixes shown})$$
$$= 1 - (\tfrac{5}{6})^4$$
$$= \tfrac{671}{1296}$$

Example 3. *If it is fine one day, the probability that it is fine the next day is $\frac{3}{4}$; if it is wet one day, the probability that it is wet the next day is $\frac{2}{3}$. If it is fine to-day (Thursday), find the probability that it is fine on* (i) *Saturday,* (ii) *Sunday.*

Draw a tree-diagram.

The probability that it is fine on Saturday is $\frac{3}{4} \times \frac{3}{4} + \frac{1}{4} \times \frac{1}{3}$, i.e. $\frac{31}{48}$; we can check that the probability that it is wet on Saturday is $\frac{1}{4} \times \frac{2}{3} + \frac{3}{4} \times \frac{1}{4}$, i.e. $\frac{17}{48}$, and $\frac{31}{48} + \frac{17}{48} = 1$.

The probability that it is fine on Sunday can be found by reading along the upper of each pair of branches of the tree, so the probability is

$$\tfrac{3}{4} \times \tfrac{3}{4} \times \tfrac{3}{4} + \tfrac{3}{4} \times \tfrac{1}{4} \times \tfrac{1}{3} + \tfrac{1}{4} \times \tfrac{1}{3} \times \tfrac{3}{4} + \tfrac{1}{4} \times \tfrac{2}{3} \times \tfrac{1}{3}$$

i.e. $\frac{347}{576}$. Again the probability that it is wet on Sunday gives a check. Reading

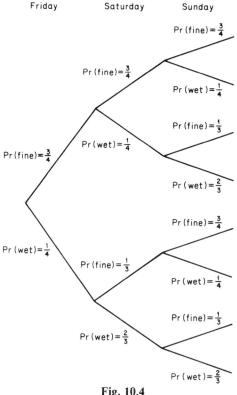

Friday	Saturday	Sunday

Fig. 10.4

along the lower of each pair of branches, the probability that it is wet on Sunday is

$$\tfrac{3}{4} \times \tfrac{3}{4} \times \tfrac{1}{4} + \tfrac{3}{4} \times \tfrac{1}{4} \times \tfrac{2}{3} + \tfrac{1}{4} \times \tfrac{1}{3} \times \tfrac{1}{4} + \tfrac{1}{4} \times \tfrac{2}{3} \times \tfrac{2}{3}$$

i.e. $\tfrac{229}{576}$, or $1 - \tfrac{347}{576}$. Notice that by ordering the way we label the branches of the tree we make it easier to see where to find acceptable outcomes.

Exercise 10b

1. The probability that a certain biased coin shows 'heads' is $\tfrac{2}{3}$. Find the probability that when tossed twice it will show (i) two heads, (ii) two tails, (iii) exactly one head.

2. A certain fair die has one face marked '1', two faces marked '2' and three faces marked '3'. Find the probability that, when thrown twice, (i) both faces show '1', (ii) both faces show '2', (iii) both faces show '3'.

3. When the fair die in question 2 is thrown twice, the numbers showing are added together to give a total, so that totals 2, 3, 4, 5 and 6 can be obtained. Find the probability of each of these totals.

4. The probability that a certain marksman scores a 'bull' is $\frac{1}{3}$. Find the probability that, with three shots, (i) all are 'bulls'; (ii) none are 'bulls'; (iii) at least one is a 'bull'.

5. A card is drawn from a pack of 52 cards and not replaced. A second card is then drawn. Find the probability that (i) both cards are aces; (ii) neither card is an ace; (iii) one card is an ace, but the other not an ace.

6. With the experiment of question 5, find the probability that (i) both cards are hearts; (ii) neither card is a heart; (iii) one card is a heart but the other is not.

7. With the experiment in question 5 find the probability that (i) the first is a heart and the second a spade; (ii) the first is a diamond and the second a spade; (iii) the first is a club and the second a spade; (iv) the second card is a spade, whatever the first card was.

8. A bag contains 5 red marbles, 6 white marbles, and 7 blue marbles. One marble is drawn and not replaced, then a second marble is drawn. Find the probability that (i) the first marble is red and the second white; (ii) the first marble is white and the second white; (iii) the second marble is white, whatever the first marble was.

9. A certain family holiday either at Blackpool or at Benidorm. If they holiday at Blackpool one year, the probability that they will holiday at Benidorm the next year is $\frac{3}{4}$; if they holiday at Benidorm one year, the probability that they will holiday at Blackpool the next is $\frac{5}{6}$. They spent their holiday at Blackpool in 1978. Find the probability that they holiday at Blackpool in (i) 1980, (ii) 1981.

10. If a certain stockbroker carries an umbrella to work one day there is a probability of $\frac{2}{3}$ that he will not carry it the following day; if he does not carry an umbrella one day, the probability that he will carry one the next day is $\frac{3}{4}$. If he carries one to work on Monday of one week, find the probability that he carries one on (i) Wednesday, (ii) Thursday, of that week.

Binomial distribution

When we toss a fair coin twice,

$$Pr(2\ heads) = \frac{1}{4}$$
$$Pr(1\ head) = \frac{2}{4}$$
$$Pr(0\ head) = \frac{1}{4}$$

If we toss the coin three times,

$$Pr(3\ heads) = \frac{1}{8}$$
$$Pr(2\ heads) = \frac{3}{8}$$
$$Pr(1\ head) = \frac{3}{8}$$
$$Pr(0\ head) = \frac{1}{8}$$

and if we toss it four times,

$$\Pr(4 \text{ heads}) = \tfrac{1}{16}$$
$$\Pr(3 \text{ heads}) = \tfrac{4}{16}$$
$$\Pr(2 \text{ heads}) = \tfrac{6}{16}$$
and so on.

We may notice the numerators of the fractions

$$
\begin{array}{ccccccc}
 & & 1 & & 2 & & 1 \\
 & 1 & & 3 & & 3 & & 1 \\
1 & & 4 & & 6 & \ldots
\end{array}
$$

and see that these are the numbers in three rows of Pascal's triangle. Is this just coincidence?

If we wish to find the probability of 3 heads out of 8 tosses of the coin, then there will be 2^8, i.e. 256, equiprobable outcomes, since each toss can show either heads or tails. One acceptable outcome, with the obvious notation, would be

HHHTTTTT and another would be HTHTHTTT

and we can see that there are as many equiprobable favourable outcomes as there are different arrangements of five letter Ts and three letter Hs, i.e. $\dfrac{8!}{5! \, 3!}$. More generally, to find the probability of r heads out of n tosses, the number of equiprobable favourable outcomes will be the number of different arrangements of n letters, r of which are the letter H and $(n - r)$ of which are the letter T. There are

$$\frac{n!}{r! \, (n - r)!}$$

arrangements of these letters, and this is the coefficient of x^r in the expansion of $(x + y)^n$.

Suppose we have a biased coin, for which the probability of showing head is p. If we denote the probability of it showing tails by q, where $q = 1 - p$, and when the coin is tossed twice

$$\Pr(2 \text{ heads}) = p^2$$
$$\Pr(1 \text{ head}) = 2pq$$
$$\Pr(0 \text{ head}) = q^2.$$

When this coin is tossed three times,

$$\Pr(3 \text{ heads}) = p^3,$$
$$\Pr(2 \text{ heads}) = 3p^2q,$$
$$\Pr(1 \text{ head}) = 3pq^2$$
$$\Pr(0 \text{ head}) = q^3,$$

from the tree-diagram overleaf.

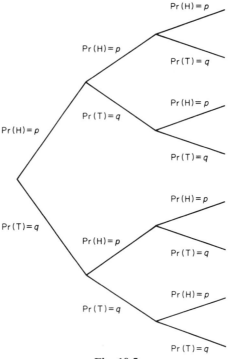

Fig. 10.5

Similarly, when tossed four times,

$$Pr(4 \text{ heads}) = p^4$$
$$Pr(3 \text{ heads}) = 4p^3q$$
$$Pr(2 \text{ heads}) = 6p^2q^2$$

and so on.

We can see that if we have any experiment with just two outcomes, 'success' or 'failure', then if the probability of success in any one trial is p, the probability of r successes and $(n-r)$ failures in n trials is

$$\frac{n!}{r!\,(n-r)!}\, p^r q^{n-r}$$

i.e. $_nC_r\, p^r q^{n-r}$. Such a distribution of probabilities is called a **binomial distribution**.

Example. *The probability that any one seed from a certain packet of cabbage seed germinates is 0.6. If a gardener sows 20 seeds, find the probability of (i) all twenty seeds germinating, (ii) no seeds germinating, (iii) exactly three seeds germinating.*

Since the probability that any one seed germinates is 0.6, the probability that all twenty will germinate is $(0.6)^{20}$, about 3.7×10^{-5}. The probability that any one seed will not germinate is 0.4, so the probability that all twenty will not germinate is $(0.4)^{20}$, about 1.1×10^{-8}. The probability that exactly three seeds will germinate (and so 17 fail to germinate) is

$$\frac{20!}{17!\,3!}(0.6)^3(0.4)^{17},$$

i.e. $\dfrac{20 \times 19 \times 18}{1 \times 2 \times 3}(0.6)^3(0.4)^{17}$, about 4×10^{-5}. *Note* In examinations, it is often acceptable to leave the answers with powers of p and q, e.g. $1140\,(0.6)^3(0.3)^{17}$ for (iii).

Exercise 10c

1. The probability that the seeds of a certain packet of wallflowers produces plants with yellow flowers is 0.2. If 10 seeds are sown find the probability that (i) all 10 plants have yellow flowers; (ii) no plants have yellow flowers; (iii) exactly five plants have yellow flowers.

2. When a certain box of six Christmas crackers became slightly damp, the probability of any one cracker 'cracking' was reduced to 0.8. Find the probability that (i) all six crackers 'cracked'; (ii) no crackers 'cracked'; (iii) exactly three crackers 'cracked'.

3. 'Cutting' a pack of cards is mathematically the same as drawing a card and then replacing it. Find the probability that (i) when a pack is cut 4 times, spades are shown twice, (ii) when a pack is cut 8 times, spades are shown four times.

4. A certain tennis-player serves either an 'ace' or a 'fault'; the probability that he serves an ace is 0.4. Find the probability that he serves (i) four consecutive aces; (ii) eight consecutive faults; (iii) four aces in six attempts.

5. A man throws a fair die 10 times. Find the probability that he obtains (i) 10 sixes, (ii) 9 sixes, (iii) 1 six, (iv) no sixes.
   What is the number of sixes he is most likely to obtain?

6. A certain golfer estimates that the probability that she will sink any one putt is 0.9. Find the probability that she will miss (i) one putt in two attempts; (ii) one putt in four attempts.

7. In a certain stretch of a river, the probability that any one fish caught is a pike is 0.2. Find the probability that (i) of three fish caught, two are pike; (ii) of four fish caught, three are pike; (iii) of five fish caught, four are pike.

8. The probability that a certain Yorkshire opening batsman scores off any one ball is 10^{-2}. Assuming (a) this particular Yorkshire batsman is never out, and (b) $1 - 10^{-2}$ to be sufficiently close to 1 so that the difference can be ignored in answering this question, find the probability that he scores (i) twice in an over of six balls; (ii) three times in an over of six balls.

Geometric distribution

We have seen that the number of favourable outcomes of an 'experiment' with r successes and $n - r$ failures is the number of ways of arranging r letters S and $n - r$ letters F, such as

SSFSFFSF

Certain 'experiments' finish as soon as one failure occurs, so that arrangements like

SSF   SSSSF   SSSSSSSSF

all finish at the first F; such 'experiments' include Russian roulette, playing cricket, when the batsman's innings finishes as soon as he is 'out', and even solving a mathematics problem, for most students cease trying to solve a problem as soon as they have found one solution to it.

If we have an experiment in which the probability of success is p,

the probability of one success in two trials is pq
the probability of two successes in three trials is $p^2 q$
the probability of three successes in four trials is $p^3 q$

and since these are consecutive terms in a geometric series, we call this a **geometric distribution.** Notice that we cannot have only two successes in four trials, since, as soon as there is a failure, the experiment ceases, so we can never have more than one failure.

Example. *Russian roulette is traditionally played with a six-cylinder revolver, a bullet being placed in one of the six cylinders. The player then fires the revolver at his own head. A win is proclaimed if the bullet is not fired when the trigger is pulled. Find the probability that a certain Russian has (i) six wins in six attempts; (ii) alas, only five wins in six attempts.*

The probability of a win at any one attempt is $\frac{5}{6}$, since five of the six chambers are unloaded, so the probability of six consecutive wins is $(\frac{5}{6})^6$, about 0.33. For only five wins in six attempts, we need WWWWWF, probability $(\frac{5}{6})^5(\frac{1}{6})$, about 0.07.

Exercise 10d

1. A certain student is attempting to solve a problem. After each failure he starts again, but as soon as he finds a solution he stops. The probability that he solves the problem at any one attempt is 0.8. Find the probability that (i) he solves the problem at the second attempt; (ii) he solves the problem at the fourth attempt; (iii) the problem is still unsolved after four attempts.

2. The probability that a certain fisherman will catch a shark on any one trip is 0.1; he promises his wife he will stop as soon as he has caught one shark. Find the probability that (i) he catches a shark on his second trip; (ii) he catches a shark on his third trip; (iii) he still has not caught a shark after seven trips.

How many fishing trips should his wife allow him if the probability that he catches a shark on one of these is to exceed 0.995?

3. A salesman is trying to sell a car. He estimates that the probability that any one customer buys the car is 0.1 Find the probability that (i) the car is sold to the second customer; (ii) the car is sold to the fifth customer; (iii) the car is still unsold after being offered to ten customers.

4. Returning to the Yorkshire opening batsman on page 75, find the probability that (i) he scores first off the second ball he receives; (ii) he scores first off the tenth ball he receives; (iii) he does not score in the first six balls he receives.

5. A holiday-maker estimates that the probability of a thunderstorm on any one day of his holiday is 0.3. He arrives on holiday on 1 August when the weather is fine. Find the probability that (i) the first thunderstorm is on 4 August; (ii) the first thunderstorm is on 6 August; (iii) he leaves on the evening of 10 August without having had a thunderstorm.

If every day up to and including 11 August is free from thunderstorms, show that the probability that the first thunderstorm of his holiday is on 14 August is 0.147, and the probability that the first thunderstorm of his holiday is on 16 August is 0.07203. Comment on the relation between these results and the answers to (i) and (ii).

MATRICES AND VECTORS

11 Matrices

Definitions and laws of association

Readers of certain newspapers will have seen an advertisement for ladies silk shirts, with an order form partly reproduced below;

Size

	34	36	38	40
Cream				
Grey				
Peach				

To order, readers had to place numbers indicating how many shirts they required in the appropriate places. If one large family put in zeros to indicate that they did not want some particular shirts, their order may have looked like this:

$$
\begin{matrix}
0 & 2 & 1 & 0 \\
1 & 0 & 1 & 0 \\
0 & 1 & 0 & 1
\end{matrix}
$$

A matrix is a rectangular array of numbers subject to certain laws of association. If we wish to find the total orders received by that firm, we should add together all the corresponding entries for each particular shirt, e.g.

$$
\begin{pmatrix}
0 & 2 & 1 & 0 \\
1 & 0 & 1 & 0 \\
0 & 1 & 0 & 1
\end{pmatrix}
+
\begin{pmatrix}
1 & 0 & 1 & 4 \\
0 & 3 & 2 & 1 \\
0 & 1 & 0 & 2
\end{pmatrix}
=
\begin{pmatrix}
1 & 2 & 2 & 4 \\
1 & 3 & 3 & 1 \\
0 & 2 & 0 & 3
\end{pmatrix}
$$

If we wish to double any one order, we double every element in the matrix describing that order; similarly to find any scalar multiple of the order.

To find the price of the order, suppose that shirts size 34 cost £15; size 36, £16; size 38, £17; and size 40, £18. We can represent these in a matrix:

$$
\begin{pmatrix}
15 \\
16 \\
17 \\
18
\end{pmatrix}
$$

and if we define matrix multiplication in a certain way, the product will give us the cost in £ of the shirts of each colour in the order.

$$\begin{pmatrix} 0 & 2 & 1 & 0 \\ 1 & 0 & 1 & 0 \\ 0 & 1 & 0 & 1 \end{pmatrix} \begin{pmatrix} 15 \\ 16 \\ 17 \\ 18 \end{pmatrix} = \begin{pmatrix} 2 \times 16 + 1 \times 17 \\ 1 \times 15 + 1 \times 17 \\ 1 \times 16 + 1 \times 18 \end{pmatrix} = \begin{pmatrix} 49 \\ 32 \\ 34 \end{pmatrix}$$

The total cost in £ of the order can be found from

$$(1 \quad 1 \quad 1) \begin{pmatrix} 49 \\ 32 \\ 34 \end{pmatrix} = (115),$$

premultiplying by the matrix $(1 \quad 1 \quad 1)$, in the same way that premultiplying the order-matrix by $(1 \quad 1 \quad 1)$ gives the number of shirts of each size.

The reader is probably familiar with matrices*, and this example serves to illustrate that matrices are rectangular arrays of numbers giving information, subject to certain laws of association, addition, and multiplication, defined in the ways illustrated here. One consequence of the manner in which they are defined is that matrix addition is commutative, whereas matrix multiplication is not commutative. Both, it will be remembered, are associative.

We shall be mainly interested in matrices displaying the components of vectors in two or three dimensions, and in matrices describing geometrical transformations in two-dimensional space (often written $\mathbb{R}^2$).

The vertices of a plane figure can be displayed in a matrix, for example, $\begin{pmatrix} 0 & 1 & 1 & 0 \\ 0 & 0 & 1 & 1 \end{pmatrix}$ displays the four vertices of the unit square, two of whose sides are along the coordinate axes and meet at the origin; the position vector of a point relative to the origin can be displayed in a matrix

$$\begin{pmatrix} x \\ y \end{pmatrix}.$$

In studying geometrical transformations, it is sometimes clearer if we consider the effect on a plane figure, rather than just a single straight line.

Transformations and mappings

Since

$$\begin{pmatrix} 2 & 1 \\ -1 & 0 \end{pmatrix} \begin{pmatrix} 0 & 1 & 1 & 0 \\ 0 & 0 & 1 & 1 \end{pmatrix} = \begin{pmatrix} 0 & 2 & 3 & 1 \\ 0 & -1 & -1 & 0 \end{pmatrix}$$

*If not, see chapters 33, 34 of *Ordinary Level Mathematics*.

the matrix giving the position vectors of the four vertices of the unit square S has been mapped, by premultiplication by a matrix $\mathbf{M}$, into a matrix which gives the position vectors of the four vertices of another quadrilateral S'.

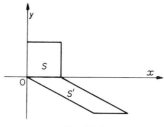

Fig. 11.1

We can say that S has been mapped into S' by the matrix $\mathbf{M}$, or that S has been **transformed** into S' by the matrix $\mathbf{M}$. The matrix $\mathbf{M}$ describes the transformation, which can sometimes be described by words, e.g. reflection in the x-axis, rotation about origin through $90°$ in a clockwise sense. The matrix is almost invariably the shorter and more precise description.

Reflection in a coordinate axis

Let $\mathbf{R}$ denote the matrix $\begin{pmatrix} 0 & 2 & 2 & 0 \\ 0 & 0 & 1 & 1 \end{pmatrix}$ and $\mathbf{T} \begin{pmatrix} 1 & 0 \\ 0 & -1 \end{pmatrix}$. When $\mathbf{R}$ is multiplied by $\mathbf{T}$

$$\begin{pmatrix} 1 & 0 \\ 0 & -1 \end{pmatrix}\begin{pmatrix} 0 & 2 & 2 & 0 \\ 0 & 0 & 1 & 1 \end{pmatrix} = \begin{pmatrix} 0 & 2 & 2 & 0 \\ 0 & 0 & -1 & -1 \end{pmatrix}$$

so that $\mathbf{T}$ is seen to describe a reflection in the x-axis. Similarly $\begin{pmatrix} -1 & 0 \\ 0 & 1 \end{pmatrix}$ describes a reflection in the y-axis.

More generally we can see that if $\begin{pmatrix} x \\ y \end{pmatrix}$ is the position vector relative to the origin of any point (x, y),

$$\begin{pmatrix} -1 & 0 \\ 0 & 1 \end{pmatrix}\begin{pmatrix} x \\ y \end{pmatrix} = \begin{pmatrix} -x \\ y \end{pmatrix}$$

so that every point in $\mathbb{R}^2$ will be reflected in the y-axis.

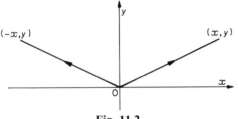

Fig. 11.2

Rotation about the origin

When **R** is multiplied by $\begin{pmatrix} 0 & 1 \\ -1 & 0 \end{pmatrix}$,

$$\begin{pmatrix} 0 & 1 \\ -1 & 0 \end{pmatrix}\begin{pmatrix} 0 & 2 & 2 & 0 \\ 0 & 0 & 1 & 1 \end{pmatrix} = \begin{pmatrix} 0 & 0 & 1 & 1 \\ 0 & -2 & -2 & 0 \end{pmatrix}$$

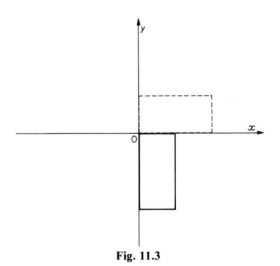

Fig. 11.3

so that $\begin{pmatrix} 0 & 1 \\ -1 & 0 \end{pmatrix}$ is seen to describe a rotation through 90° in a clockwise sense.

Again, we see that if $\begin{pmatrix} x \\ y \end{pmatrix}$ is the position vector relative to the origin of any point (x, y), then $\begin{pmatrix} 0 & 1 \\ -1 & 0 \end{pmatrix}\begin{pmatrix} x \\ y \end{pmatrix} = \begin{pmatrix} y \\ -x \end{pmatrix}$, which Fig. 11.4 shows is a rotation about the origin through 90° in a clockwise sense, whatever the point (x, y).

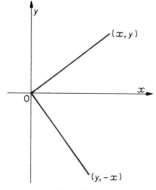

Fig. 11.4

Enlargement

Since

$$\begin{pmatrix} k & 0 \\ 0 & k \end{pmatrix}\begin{pmatrix} x \\ y \end{pmatrix} = \begin{pmatrix} kx \\ ky \end{pmatrix},$$

the matrix $\begin{pmatrix} k & 0 \\ 0 & k \end{pmatrix}$ increases all distances from the coordinate axes by a factor k, so this denotes an increase of all lengths by a factor k.

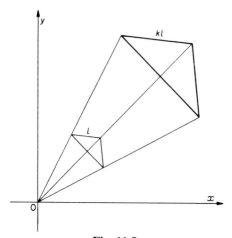

Fig. 11.5

Shear

Since

$$\begin{pmatrix} 1 & k \\ 0 & 1 \end{pmatrix}\begin{pmatrix} 0 & 1 & 1 & 0 \\ 0 & 0 & 1 & 1 \end{pmatrix} = \begin{pmatrix} 0 & 1 & 1+k & k \\ 0 & 0 & 1 & 1 \end{pmatrix}$$

the matrix $\begin{pmatrix} 1 & k \\ 0 & 1 \end{pmatrix}$ describes a shear parallel to the x-axis. Similarly,

$\begin{pmatrix} 1 & 0 \\ k & 1 \end{pmatrix}$ describes a shear parallel to the y-axis.

Mapping described by a singular matrix

The matrix $\begin{pmatrix} a & b \\ ka & kb \end{pmatrix}$ is a singular matrix.*

Since

$$\begin{pmatrix} a & b \\ ka & kb \end{pmatrix}\begin{pmatrix} 0 & 1 & 1 & 0 \\ 0 & 0 & 1 & 1 \end{pmatrix} = \begin{pmatrix} 0 & a & a+b & b \\ 0 & ka & ka+kb & kb \end{pmatrix},$$

the four vertices of the unit square are mapped into four points lying

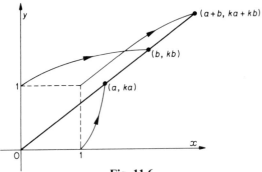

Fig. 11.6

in the straight line $y = kx$. Similarly, when multiplying the position
vector of any point (x, y)

$$\begin{pmatrix} a & b \\ ka & kb \end{pmatrix}\begin{pmatrix} x \\ y \end{pmatrix} = \begin{pmatrix} ax + by \\ kax + kby \end{pmatrix}$$

so that all points in $\mathbb{R}^2$ are mapped into points in the straight line
$y = kx$. Thus all plane figures in $\mathbb{R}^2$ will be mapped into segments of
the straight line $y = kx$.

* The determinant of the 2 × 2 matrix $\begin{pmatrix} a & b \\ c & d \end{pmatrix}$ is $ad - bc$; a matrix whose determinant
is 0 is called a singular matrix.

To find the matrix describing a given transformation in $\mathbb{R}^2$

Suppose that we wish to find the matrix which describes a reflection in $x + y = 0$. From Fig. 11.7 we see that the point (x,y) must be mapped

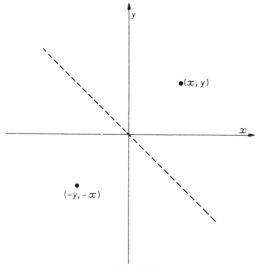

Fig. 11.7

into the point $(-y, -x)$. Let the matrix that describes the transformation be $\begin{pmatrix} a & b \\ c & d \end{pmatrix}$. Then

$$\begin{pmatrix} a & b \\ c & d \end{pmatrix}\begin{pmatrix} x \\ y \end{pmatrix} = \begin{pmatrix} -y \\ -x \end{pmatrix}$$

i.e.

$$\begin{pmatrix} ax + by \\ cx + dy \end{pmatrix} = \begin{pmatrix} -y \\ -x \end{pmatrix}$$

This is an identity, true for all values of x and y, so that $a = 0$, $b = -1$, $c = -1$, and $d = 0$, and the required matrix is $\begin{pmatrix} 0 & -1 \\ -1 & 0 \end{pmatrix}$.

Example 1. *Find the matrix which describes the transformation of the unit square into the parallelogram* $\begin{pmatrix} 0 & 2 & 5 & 3 \\ 0 & 1 & 2 & 1 \end{pmatrix}$, *and the matrix which maps the parallelogram P back into the unit square.*

Let the required matrix be $\begin{pmatrix} a & b \\ c & d \end{pmatrix}$. Then

$$\begin{pmatrix} a & b \\ c & d \end{pmatrix}\begin{pmatrix} 0 & 1 & 1 & 0 \\ 0 & 0 & 1 & 1 \end{pmatrix} = \begin{pmatrix} 0 & 2 & 5 & 3 \\ 0 & 1 & 2 & 1 \end{pmatrix}$$

i.e.
$$\begin{pmatrix} 0 & a & a+b & b \\ 0 & c & c+d & d \end{pmatrix} = \begin{pmatrix} 0 & 2 & 5 & 3 \\ 0 & 1 & 2 & 1 \end{pmatrix}$$

Two matrices are equal if and only if all corresponding elements are equal, so $a = 2$, $b = 3$, $c = 1$, and $d = 1$. Notice that we had six equations for four unknowns; we could only find solutions because the element 5 was the sum of the elements 2 and 3, and the element 2 in the second row was the sum of the two elements 1 and 1.

The unit square of course is determined by the vectors $\begin{pmatrix} 1 \\ 0 \end{pmatrix}$ and $\begin{pmatrix} 0 \\ 1 \end{pmatrix}$, and the parallelogram is determined by the vectors $\begin{pmatrix} 2 \\ 1 \end{pmatrix}$ and $\begin{pmatrix} 3 \\ 1 \end{pmatrix}$. We notice that the matrix of the transformation is $\begin{pmatrix} 2 & 3 \\ 1 & 1 \end{pmatrix}$. This is given by the vectors $\begin{pmatrix} 2 \\ 1 \end{pmatrix}$, $\begin{pmatrix} 3 \\ 1 \end{pmatrix}$ into which the vectors $\begin{pmatrix} 1 \\ 0 \end{pmatrix}$ and $\begin{pmatrix} 0 \\ 1 \end{pmatrix}$ were transformed.

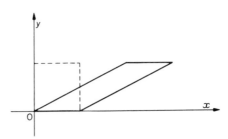

Fig. 11.8

The mapping of P back into the unit square will be by the matrix $\begin{pmatrix} p & q \\ r & s \end{pmatrix}$, where

$$\begin{pmatrix} p & q \\ r & s \end{pmatrix}\begin{pmatrix} 0 & 2 & 5 & 3 \\ 0 & 1 & 2 & 1 \end{pmatrix} = \begin{pmatrix} 0 & 1 & 1 & 0 \\ 0 & 0 & 1 & 1 \end{pmatrix}$$

so that
$$\begin{pmatrix} 0 & 2p+q & 5p+2q & 3p+q \\ 0 & 2r+s & 5r+2s & 3r+s \end{pmatrix} = \begin{pmatrix} 0 & 1 & 1 & 0 \\ 0 & 0 & 1 & 1 \end{pmatrix}$$

i.e. $2p + q = 1$ and $3p + q = 0$, giving $p = -1$ and $q = 3$; and $2r + s = 0$ and $3r + s = 1$, giving $r = 1$ and $s = -2$. Thus the matrix which describes the transformation of P back into the unit square is

$$\begin{pmatrix} -1 & 3 \\ 1 & -2 \end{pmatrix}$$

which is the inverse of the matrix that effected the transformation.

Change of area in a transformation

Fig. 11.9 shows the image of a unit square under the transformation described by the matrix $\begin{pmatrix} a & b \\ c & d \end{pmatrix}$. The coordinates of C' are $(a + b, c + d)$, so the area of the rectangle $OXC'Y$ is $(a + b)(c + d)$. The areas

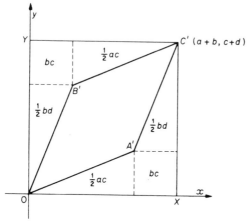

Fig. 11.9

of the triangles and rectangle that are not shaded are marked on the figure, so that the area of the parallelogram $OA'C'B'$ is

$$(a + b)(c + d) - \tfrac{1}{2}ac - bc - \tfrac{1}{2}bd - \tfrac{1}{2}ac - bc - \tfrac{1}{2}bd$$
$$= ad - bc, \text{ the determinant of the matrix.}$$

This can be generalized to show that the factor by which the area of any plane figure is increased under a geometrical transformation that can be described by a matrix is the determinant of that matrix.

Exercise 11: Miscellaneous

1. Describe the transformation of each of the following matrices on the rectangle $\begin{pmatrix} 0 & 2 & 2 & 0 \\ 0 & 0 & 1 & 1 \end{pmatrix}$ and illustrate each by a sketch.

(i) $\begin{pmatrix} 1 & 0 \\ 0 & -1 \end{pmatrix}$

(ii) $\begin{pmatrix} 3 & 0 \\ 0 & 3 \end{pmatrix}$

(iii) $\begin{pmatrix} \tfrac{1}{2} & 0 \\ 0 & \tfrac{1}{2} \end{pmatrix}$

(iv) $\begin{pmatrix} -2 & 0 \\ 0 & 1 \end{pmatrix}$

(v) $\begin{pmatrix} -2 & 0 \\ 0 & -2 \end{pmatrix}$

(vi) $\begin{pmatrix} -2 & 0 \\ 0 & -1 \end{pmatrix}$

(vii) $\begin{pmatrix} 0 & 2 \\ 2 & 0 \end{pmatrix}$ (viii) $\begin{pmatrix} 1 & 2 \\ 0 & 1 \end{pmatrix}$ (ix) $\begin{pmatrix} 1 & \frac{1}{2} \\ 1 & 0 \end{pmatrix}$

(x) $\begin{pmatrix} 1 & 2 \\ 1 & 0 \end{pmatrix}$ (xi) $\begin{pmatrix} 1 & -2 \\ 0 & 1 \end{pmatrix}$ (xii) $\begin{pmatrix} 1 & 0 \\ 2 & 1 \end{pmatrix}$

2. Find the matrices that map the unit square into each of the following, illustrating the mappings by a sketch.

(i) $\begin{pmatrix} 0 & 3 & 3 & 0 \\ 0 & 0 & 3 & 3 \end{pmatrix}$ (ii) $\begin{pmatrix} 0 & 3 & 3 & 0 \\ 0 & 0 & 1 & 1 \end{pmatrix}$

(iii) $\begin{pmatrix} 0 & 1 & 1 & 0 \\ 0 & 0 & -1 & -1 \end{pmatrix}$ (iv) $\begin{pmatrix} 0 & 2 & 2 & 0 \\ 0 & 0 & -2 & -2 \end{pmatrix}$

(v) $\begin{pmatrix} 0 & 2 & -1 & 1 \\ 0 & 1 & 0 & -1 \end{pmatrix}$ (vi) $\begin{pmatrix} 0 & 2 & 1 & -1 \\ 0 & 2 & 0 & -2 \end{pmatrix}$

3. Find the matrix which describes each of the following transformations:
   - (i) reflection in the x-axis;
   - (ii) reflection in the line $y = x$;
   - (iii) enlargement, centre (0,0), factor 5;
   - (iv) rotation through $180°$ about the origin.

4. Find the image of the unit square under each of the following mappings, and the matrix that describes the inverse mapping.

(i) $\begin{pmatrix} 2 & 0 \\ 0 & 2 \end{pmatrix}$ (ii) $\begin{pmatrix} \frac{1}{4} & 0 \\ 0 & \frac{1}{4} \end{pmatrix}$ (iii) $\begin{pmatrix} 0 & -1 \\ 1 & 0 \end{pmatrix}$

(iv) $\begin{pmatrix} 2 & 0 \\ 0 & -2 \end{pmatrix}$ (v) $\begin{pmatrix} 2 & 0 \\ 0 & 3 \end{pmatrix}$ (vi) $\begin{pmatrix} -1 & -1 \\ -1 & 1 \end{pmatrix}$

5. Find the image of the unit square under the mapping described by the matrix $\begin{pmatrix} 2 & 1 \\ 1 & 3 \end{pmatrix}$. Draw the parallelogram into which it is mapped and show, by finding the areas of certain rectangles and triangles, that the area of the parallelogram is 5 square units.

6. Find the area of the region into which the unit square is mapped by each of the following matrices.

(i) $\begin{pmatrix} 2 & 1 \\ 1 & 1 \end{pmatrix}$ (ii) $\begin{pmatrix} 3 & 2 \\ 1 & 1 \end{pmatrix}$ (iii) $\begin{pmatrix} 4 & 2 \\ 1 & 3 \end{pmatrix}$

(iv) $\begin{pmatrix} 1 & 1 \\ -1 & 1 \end{pmatrix}$ (v) $\begin{pmatrix} 1 & 1 \\ 1 & 1 \end{pmatrix}$ (vi) $\begin{pmatrix} 1 & -1 \\ 1 & 0 \end{pmatrix}$

7. If $A = \begin{pmatrix} 1/\sqrt{2} & -1/\sqrt{2} \\ 1/\sqrt{2} & 1/\sqrt{2} \end{pmatrix}$, find A^2.
   - (i) What transformation is described by A^2?

(ii) What transformation is described by **A**?

(iii) What transformation is described by $\mathbf{A}^{-1}$?

8. If $\mathbf{A} = \begin{pmatrix} 1/2 & -\frac{1}{2}\sqrt{3} \\ \frac{1}{2}\sqrt{3} & 1/2 \end{pmatrix}$, find $\mathbf{A}^2$ and $\mathbf{A}^3$. What transformations are described by the following matrices?

(i) $\mathbf{A}^3$,   (ii) $\mathbf{A}$,   (iii) $\mathbf{A}^2$,   (iv) $\mathbf{A}^6$,   (v) $\mathbf{A}^5$,   (vi) $\mathbf{A}^{-1}$.

9. Find the product $\begin{pmatrix} \cos\theta & -\sin\theta \\ \sin\theta & \cos\theta \end{pmatrix} \begin{pmatrix} x \\ y \end{pmatrix}$.

(i) Draw a diagram to show that this matrix describes a rotation about the origin through an angle θ in an anti-clockwise sense.

(ii) Find the matrix which describes a rotation about the origin in a clockwise sense through an angle θ.

(iii) Find the matrix which rotates about the origin through an angle 2θ in an anti-clockwise sense, and compare it with the square of a certain matrix. Show that $\cos 2\theta = \cos^2\theta - \sin^2\theta$, and suggest an expression equal to $\sin 2\theta$.

10. By considering the product of the matrices

$$\begin{pmatrix} \cos\alpha & -\sin\alpha \\ \sin\alpha & \cos\alpha \end{pmatrix}, \begin{pmatrix} \cos\beta & -\sin\beta \\ \sin\beta & \cos\beta \end{pmatrix},$$

show that $\cos(\alpha + \beta) = \cos\alpha \cos\beta - \sin\alpha \sin\beta$

and $\sin(\alpha + \beta) = \sin\alpha \cos\beta + \sin\alpha \cos\beta.$

12 Vectors

Definition

A **vector** can be defined as a physical quantity having both magnitude and direction, any two vectors being combined in addition according to the parallelogram law. This is in contrast to **scalars,** which are quantities that are determined completely by a single number. Examples of scalars are mass, speed, and temperature; examples of vectors are displacement, velocity, acceleration, force, and momentum. We shall be concerned in pure mathematics almost entirely with position vectors, which show the displacement of one point relative to another.

Free vectors, localized vectors

A free vector has magnitude and direction, but has no particular position associated with it. The three vectors represented by the three equal parallel straight lines in Fig. 12.1 are all equal free vectors, though they have different lines of action.

Fig. 12.1

A localized vector can be located along a straight line, i.e. in a particular direction, as the force acting on a body due to the gravitational attraction, or can be localized at a point.

Addition of vectors

Any two vectors are added using the parallelogram law; this is the definition of vector addition.

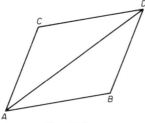

Fig. 12.2

If $\overrightarrow{AB}$ and $\overrightarrow{AC}$ are any two vectors acting at the point A, then their sum is found by completing the parallelogram $ABDC$, the sum being represented by the diagonal AD, shown in Fig. 12.2.

Subtraction of vectors

If $\overrightarrow{OA}$ and $\overrightarrow{OB}$ are two vectors, $\overrightarrow{OA} - \overrightarrow{OB}$ is defined as $\overrightarrow{OA} + (-\overrightarrow{OB})$, where $-\overrightarrow{OB}$ is equal in magnitude to $\overrightarrow{OB}$ but opposite in the direction in which it acts. $\overrightarrow{OA} - \overrightarrow{OB}$ is illustrated in Fig. 12.3.

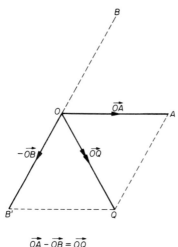

$$\overrightarrow{OA} - \overrightarrow{OB} = \overrightarrow{OQ}$$

Fig. 12.3

It is easily proved that $OQAB$ is a parallelogram and therefore that OQ is equal and parallel to BA. Provided that we are concerned with magnitude and direction only, and not with the position vector,

$$\overrightarrow{BA} = \overrightarrow{OA} - \overrightarrow{OB}$$

whereas $\overrightarrow{OQ} = \overrightarrow{OA} - \overrightarrow{OB}$ in magnitude, direction, and line of action. If $\overrightarrow{OA} = \mathbf{a}$, $\overrightarrow{OB} = \mathbf{b}$, then $\overrightarrow{BA} = \mathbf{a} - \mathbf{b}$.

Scalar multiple of a vector

If we add the vector $\mathbf{a}$ to the vector $\mathbf{b}$, we write the sum as $\mathbf{a} + \mathbf{b}$. If we add the vector $\mathbf{a}$ to the vector $\mathbf{a}$, we write the sum as $2\mathbf{a}$. We can extend this notation to any scalar multiple of $\mathbf{a}$, i.e.

$$n\mathbf{a} = \mathbf{a} + \ldots + \mathbf{a},$$

to n terms, where n is an integer. Similar we define $\left(\dfrac{1}{n}\right)\mathbf{a}$ to be such a vector that

$$\mathbf{a} = \frac{1}{n}\mathbf{a} + \frac{1}{n}\mathbf{a} + \frac{1}{n}\mathbf{a} + \ldots + \frac{1}{n}\mathbf{a}, \text{ to } n \text{ terms.}$$

The extension to a rational multiple of the form $(m/n)\,\mathbf{a}$ follows immediately.

The unit vector

A unit vector in a given direction is a vector with unit magnitude in that direction. If we have a vector $\mathbf{a}$, its magnitude is often written $|\mathbf{a}|$; thus

$$\mathbf{a} = |\mathbf{a}|\hat{\mathbf{a}}$$

where $\hat{\mathbf{a}}$ is the unit vector in the direction of $\mathbf{a}$.

We shall represent the unit vector along the x-axis by $\mathbf{i}$, the unit vector along the y-axis by $\mathbf{j}$, and, later on, the unit vector along an axis Oz, perpendicular to the plane of Ox and Oy, by $\mathbf{k}$. Any point on the x-axis has position vector relative to the origin of $\lambda\mathbf{i}$, where λ is a scalar; any point on the y-axis has position vector relative to O of $\mu\mathbf{j}$, where μ is a scalar. For example, the position vector of the point with coordinates $(2,0)$ is $2\mathbf{i}$; the position vector of the point with coordinates $(0,-3)$ is $-3\mathbf{j}$.

Components of a vector

Any vector $\overrightarrow{OP}$ can be expressed as the sum of two vectors in an infinite number of ways (see Fig. 12.4 for a few of them). But if we take any line through O and call it Ox, draw the line perpendicular to Ox through P, then we have a unique pair of perpendicular vectors equal to $\overrightarrow{OP}$. If $\overrightarrow{OP}$ is the position vector of the point P with coordinates (x, y), and if $\mathbf{i}$ and $\mathbf{j}$ are unit vectors along Ox and Oy

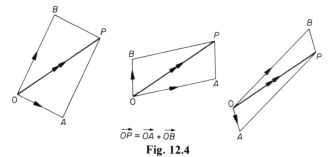

$$\overrightarrow{OP} = \overrightarrow{OA} + \overrightarrow{OB}$$
Fig. 12.4

respectively, then $\overrightarrow{OP} = x\mathbf{i} + y\mathbf{j}$. $x\mathbf{i}$ and $y\mathbf{j}$ are the components of $\overrightarrow{OP}$ along Ox and Oy respectively. A matrix is often used to display the components of a vector.

e.g.
$$\overrightarrow{OP} = x\mathbf{i} + y\mathbf{j} = \begin{pmatrix} x \\ y \end{pmatrix}$$

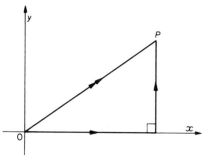

Fig. 12.5

We can see (Fig. 12.6) that the parallelogram law of addition is the same as if we define vector addition by adding the corresponding entries in the matrix that describes each vector. This equivalent definition of vector addition is used later when manipulating vectors with more than two components.

Magnitude of a vector

Using the above notation, the magnitude of the vector $\begin{pmatrix} x \\ y \end{pmatrix} \equiv x\mathbf{i} + y\mathbf{j}$ is $\sqrt{x^2 + y^2}$, since the length of the line OP is $\sqrt{x^2 + y^2}$, by Pythagoras' theorem.

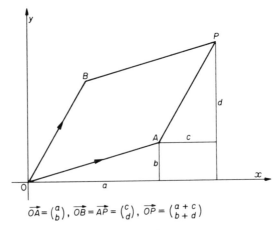

$$\overrightarrow{OA} = \begin{pmatrix} a \\ b \end{pmatrix}, \ \overrightarrow{OB} = \overrightarrow{AP} = \begin{pmatrix} c \\ d \end{pmatrix}, \ \overrightarrow{OP} = \begin{pmatrix} a+c \\ b+d \end{pmatrix}$$

Fig. 12.6

Position vectors

Fig. 12.7 shows points A and B with coordinates $(3,1)$ and $(1,-2)$ respectively. Their position vectors relative to the origin O are $3\mathbf{i} + \mathbf{j}$ and $\mathbf{i} - 2\mathbf{j}$ respectively. To find the vector that describes the displacement $\overrightarrow{AB}$, since $\overrightarrow{OB} = \overrightarrow{OA} + \overrightarrow{AB}$, we use

$$\overrightarrow{AB} = \overrightarrow{OB} - \overrightarrow{OA}$$
$$= \begin{pmatrix} 1 \\ -2 \end{pmatrix} - \begin{pmatrix} 3 \\ 1 \end{pmatrix} = \begin{pmatrix} -2 \\ -3 \end{pmatrix}$$

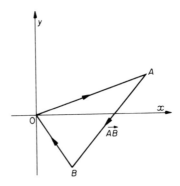

Fig. 12.7

Example 1. *Find the position vector of M, the midpoint of OP, where $\overrightarrow{OP} = 2\mathbf{i} + 6\mathbf{j}$.*

Since M is the mid-point of OP,

$$\overrightarrow{OM} = \tfrac{1}{2}\overrightarrow{OP} = \tfrac{1}{2}(2\mathbf{i} + 6\mathbf{j})$$
$$= \mathbf{i} + 3\mathbf{j}.$$

Example 2. *If the position vectors of A and B are $2\mathbf{i} + 5\mathbf{j}$ and $-4\mathbf{i} + \mathbf{j}$ respectively, find the position vector $\overrightarrow{OM}$, where M is the mid-point of AB.*

$$\overrightarrow{OM} = \overrightarrow{OA} + \overrightarrow{AM}, \text{ where } \overrightarrow{AM} = \tfrac{1}{2}\overrightarrow{AB}$$

The displacement vector of B from A is $-6\mathbf{i} - 4\mathbf{j}$, so half of this is $-3\mathbf{i} - 2\mathbf{j}$. Thus

$$\overrightarrow{OM} = 2\mathbf{i} + 5\mathbf{j} + (-3\mathbf{i} - 2\mathbf{j})$$
$$= -\mathbf{i} + 3\mathbf{j}$$

Example 3. *Show that the points A, position vector $3\mathbf{i} + 5\mathbf{j}$, B, position vector $4\mathbf{i} + 7\mathbf{j}$, and C, position vector $6\mathbf{i} + 11\mathbf{j}$, are collinear.*

Since $AB = \mathbf{i} + 2\mathbf{j}$ and $BC = 2\mathbf{i} + 4\mathbf{j}$, AB and BC are in the same direction. But both vectors are localized through the point B, so that ABC must be a straight line.

Example 4. *Show that the points A, B, C, and D, position vectors $-\mathbf{j}, \mathbf{i} + \mathbf{j}, 4\mathbf{i} + 2\mathbf{j}$, and $3\mathbf{i}$, are the vertices of a parallelogram.*

To prove that ABCD is a parallelogram, we can prove that AB is equal and parallel to DC.

$$\overrightarrow{AB} = \overrightarrow{OB} - \overrightarrow{OA} = \mathbf{i} + \mathbf{j} - (-\mathbf{j}) = \mathbf{i} + 2\mathbf{j}$$
$$\overrightarrow{DC} = \overrightarrow{OC} - \overrightarrow{OD} = 4\mathbf{i} + 2\mathbf{j} - 3\mathbf{i} = \mathbf{i} + 2\mathbf{j}$$

Since these two vectors are equal, AB and DC must be both equal and parallel, so ABCD is a parallelogram.

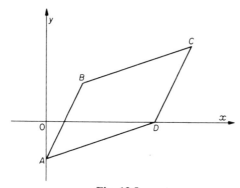

Fig. 12.8

Exercise 12a

1. Points A, B, C, and D have the following coordinates: A, $(1,0)$; B, $(3,1)$; C, $(4,-1)$; D, $(-2,3)$. Plot these points, and write down in matrix form the position vectors relative to the origin of A, B, C, and D.

   (i) Write down the position vectors relative to the point A of points B, C, and D.

   (ii) Write down the position vectors relative to the point B of points A, C, and D.

2. Plot the points A $(1,0)$, B $(2,2)$, C $(3,6)$ and D $(1,2)$. Write down in **i**, **j** form the position vector of each point relative to the origin O.

   (i) Write down the position vectors of B, C, and D relative to A.

   (ii) Write down the position vectors of A, B, and D relative to the point C.

   (iii) What relation exists between $\overrightarrow{AB}$ and $\overrightarrow{BA}$?

   (iv) What relation exists between $\overrightarrow{AB}$ and $\overrightarrow{DC}$?

3. If the position vector of the point A relative to the origin O is $\begin{pmatrix} 3 \\ 2 \end{pmatrix}$ write down the position vector of (i) the midpoint M of OA; (ii) the point N such that A is the midpoint of ON.

4. Find the magnitude of each of the following vectors

   (i) $\begin{pmatrix} 3 \\ -4 \end{pmatrix}$ (ii) $\begin{pmatrix} -2 \\ -3 \end{pmatrix}$ (iii) $\begin{pmatrix} 1 \\ -4 \end{pmatrix}$ (iv) $\begin{pmatrix} -1 \\ 1 \end{pmatrix}$

5. The position vectors of A, B, C, and D are $\begin{pmatrix} 1 \\ 0 \end{pmatrix}$, $\begin{pmatrix} 3 \\ 2 \end{pmatrix}$, $\begin{pmatrix} 4 \\ 4 \end{pmatrix}$, and $\begin{pmatrix} 2 \\ 2 \end{pmatrix}$ respectively. Show that $ABCD$ is a parallelogram.

6. The position vectors of points A, B, C, and D are $\begin{pmatrix} 1 \\ 0 \end{pmatrix}$, $\begin{pmatrix} 3 \\ 2 \end{pmatrix}$, $\begin{pmatrix} 5 \\ 5 \end{pmatrix}$, and $\begin{pmatrix} 2 \\ 2 \end{pmatrix}$ respectively. Show that $ABCD$ is a trapezium.

7. The position vectors of points A, B, C, and D are $\begin{pmatrix} 1 \\ 0 \end{pmatrix}$, $\begin{pmatrix} 4 \\ 1 \end{pmatrix}$, $\begin{pmatrix} 3 \\ 4 \end{pmatrix}$, and $\begin{pmatrix} 0 \\ 3 \end{pmatrix}$ respectively. Prove that $ABCD$ is a square.

8. The position vectors of points A, B, C, and D are $\begin{pmatrix} 1 \\ 0 \end{pmatrix}$, $\begin{pmatrix} 4 \\ 1 \end{pmatrix}$, $\begin{pmatrix} 5 \\ 4 \end{pmatrix}$ and $\begin{pmatrix} 2 \\ 3 \end{pmatrix}$ respectively. Prove that $ABCD$ is a rhombus.

9. The position vectors of points A and B relative to an origin O are $3\mathbf{i} + 6\mathbf{j}$ and $-6\mathbf{i} + 9\mathbf{j}$ respectively. Find the position vectors of (i) the point M in OA such that $OM:MA = 1:2$; (ii) the point N in OB such that $ON:NB = 1:2$. Find the vectors AB and MN. Show that they are parallel, and find the ratio of the lengths $AB:MN$.

10. If $\mathbf{a} = \begin{pmatrix} 3 \\ 4 \end{pmatrix}$ and $\mathbf{b} = \begin{pmatrix} -1 \\ 2 \end{pmatrix}$ show that $\mathbf{a} - 5\mathbf{b}$ is perpendicular to $\mathbf{a}$, using

(i) matrix multiplication, and (ii) trigonometry.

$\left(\textit{Hint: a rotation through } 90° \text{ in a certain sense is described by the matrix} \right.$

$\left. \begin{pmatrix} 0 & -1 \\ 1 & 0 \end{pmatrix}. \right)$

Addition of vectors, given their magnitudes and directions

In the previous questions we were able to add or subtract easily vectors given in terms of their components along two perpendicular axes. If we are given vectors defined by their magnitude and directions then we have either to use scale drawing, or trigonometry, or geometry.

Example 1. *Two unit vectors* $\mathbf{u}$ *and* $\mathbf{v}$ *are inclined at 60°. Find the magnitude of* (i) $\mathbf{u} + \mathbf{v}$, (ii) $\mathbf{u} - \mathbf{v}$, *and the angle each of these makes with the vector* $\mathbf{u}$.

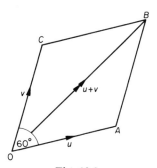

Fig. 12.9

(i) If $\overrightarrow{OA}$ represents $\mathbf{u}$ and $\overrightarrow{OC}$ represents $\mathbf{v}$, the resultant, $\mathbf{u} + \mathbf{v}$, will be represented by the diagonal OB of the parallelogram $OABC$. Since triangle OAC is isosceles we can draw the line through A perpendicular to OB. Thus

$$OB = 2 \cos 30°$$
$$= \sqrt{3}, \simeq 1.73.$$

The angle between $\mathbf{u} + \mathbf{v}$ and $\mathbf{u}$ will clearly be 30°.

(ii) If $\overrightarrow{OB'}$ represents $-\mathbf{v}$, then $\overrightarrow{OC'}$ represents $\mathbf{u} - \mathbf{v}$. Drawing the perpendicular as before, or spotting that OAC' is an equilateral triangle, we have $OC' = 1$ unit, i.e. the magnitude of $\mathbf{u} - \mathbf{v} = 1$. We can write this as $|\mathbf{u} - \mathbf{v}| = 1$.

The angle between OC' and OA we see is 60°.

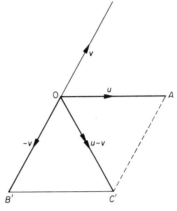

Fig. 12.10

Example 2. *If* **u** *and* **v** *are unit vectors inclined at 50°, find the magnitude of* 2**u** + 3**v**, *and the angle between* 2**u** + 3**v** *and* **u**.

From Fig. 12.11, using the cosine formula,

$$OC^2 = 2^2 + 3^2 - 2 \times 2 \times 3 \cos 130°$$
$$= 13 + 12 \cos 50°$$
$$OC \simeq 4.55$$

Check: This is reasonable, as we expect a length a little less than 5.

To find the angle θ, use the sine formula:

$$\sin \theta = \frac{2 \sin 50°}{4.551}$$

$$\simeq 20°$$

This problem can also be solved by scale drawing.

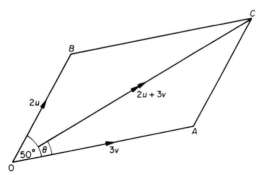

Fig. 12.11

Exercise 12b

1. If **u** and **v** are unit vectors inclined at 45°, find (a) by calculation, and (b) by scale drawing, the magnitude of each of the following, and the angle it makes with the vector **u**.

   (i) **u** + **v**   (ii) **u** − **v**   (iii) **u** + 2**v**   (iv) 2**u** − **v**

2. If **u** and **v** are unit vectors inclined at 140°, find (a) by calculation, and (b) by scale drawing, the magnitude of each of the following, and the angle it makes with the vector **u**.

   (i) **u** + **v**   (ii) 2**u** − **v**   (iii) 2**u** + 2**v**   (iv) **u** + 2**v**

3. If **u** and **v** are unit vectors inclined at 30°, find by drawing or by calculation the magnitude of the vector 3**u** − **v**. Deduce the magnitude of the following vectors:

   (i) 3**v** − **u**   (ii) 6**u** − 2**v**   (iii) 4**v** − 12**u**

4. Two vectors **a** and **b** are such that $|\mathbf{a}| = 7$, $|\mathbf{b}| = 3$, and $|\mathbf{a} + \mathbf{b}| = 5$. Find the angle between **a** and **b**.

5. Vectors **a** and **b** are such that **a** + **b** is perpendicular to **b**. If $|\mathbf{a}| = 13$ and $|\mathbf{a} + \mathbf{b}| = 12$, find $|\mathbf{b}|$.

6. Two vectors **a** and **b** are such that $|\mathbf{a}| = 4$, $|\mathbf{b}| = 3$, and **a** + **b** is inclined at 40° to **a**. Find the magnitude of **a** + **b**. (Two possible answers.)

7. Two vectors **a** and **b** are such that **a** + 2**b** is perpendicular to **a** − **b** and $|\mathbf{a} + 2\mathbf{b}| : |\mathbf{a} - \mathbf{b}| = 3:2$. Find by drawing the ratio $|\mathbf{a}| : |\mathbf{b}|$.

8. A helicopter flies at 150 km h⁻¹ in still air. On a certain day the wind is blowing at 30 km h⁻¹ from the north.

   (i) How long would the helicopter take to fly a distance of 100 km due east?

   (ii) How long would the helicopter take to fly 100 km due west, and what course should the pilot set in order to fly due west?

   (iii) What course should the pilot set to fly south-west, and what will be his groundspeed then?

9. An explorer is on an ice-pack. He walks at 4 km h⁻¹ on a bearing of 005°, but the ice is drifting south-east at 1 km h⁻¹. On what course is the explorer travelling, and what is his speed?

10. A private aircraft leaves an aerodrome X to fly to Y, which is 300 km away on a bearing of 050°. The pilot flies at a constant speed of 250 km h⁻¹ on a course of 050°, not allowing for a constant wind of 40 km h⁻¹ from the north.

    (i) Find by drawing or calculation where he will be in relation to X after flying for 1 hour.

    (ii) He eventually arrives at Y. Find the course he must set if he is to fly back to X in a straight line.

11. Show geometrically that $|\mathbf{a}| + |\mathbf{b}| \geqslant |\mathbf{a} + \mathbf{b}|$. When does equality hold?

12. Show geometrically that $|\mathbf{a}| + |\mathbf{b}| + |\mathbf{c}| \geqslant |\mathbf{a} + \mathbf{b} + \mathbf{c}|$. When does equality hold?

13 Geometrical Applications of Vectors

There are many geometrical results that can be proved easily using vectors. Some require that we first prove the Section Theorem.

Section theorem

We require to find the position vector of the point dividing a line AB in a given ratio $\lambda:\mu$, $\mathbf{a}$ and $\mathbf{b}$ being the position vectors of A and B respectively.

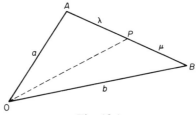

Fig. 13.1

From Fig. 13.1, if P is the required point,

$$\overrightarrow{OP} = \overrightarrow{OA} + \overrightarrow{AP}$$

$$= \overrightarrow{OA} + \frac{\lambda}{\lambda + \mu} \overrightarrow{AB}$$

$$= \mathbf{a} + \frac{\lambda}{\lambda + \mu} (\mathbf{b} - \mathbf{a})$$

$$= \frac{\mu\mathbf{a} + \lambda\mathbf{b}}{\lambda + \mu}$$

Deductions

(a) The position vector of the midpoint of AB is $\frac{1}{2}(\mathbf{a} + \mathbf{b})$.

(b) If $\mu = k$ and $\lambda = 1 - k$, then any point in AB has position vector $k\mathbf{a} + (1 - k)\mathbf{b}$, so that the equation of the straight line through points with position vectors $\mathbf{a}$ and $\mathbf{b}$ is $\mathbf{r} = k\mathbf{a} + (1 - k)\mathbf{b}$, where k is a parameter.

(c) Points with position vectors $\mathbf{a}$, $\mathbf{b}$, and $p\mathbf{a} + q\mathbf{b}$ are collinear if and only if $p + q = 1$.

The diagonals of a parallelogram bisect each other

From the law of vector addition, if $OABC$ is a parallelogram and $\mathbf{a}$ and $\mathbf{c}$ are the position vectors relative to O of A and C, then the position vector of B is $\mathbf{a} + \mathbf{c}$, and the position vector of the midpoint

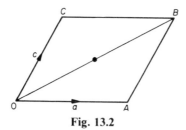

Fig. 13.2

of OB is $\frac{1}{2}(\mathbf{a} + \mathbf{c})$. The position vector of the midpoint of AC, is $\frac{1}{2}(\mathbf{a} + \mathbf{c})$, by the first corollary of the Section Theorem. Therefore OB and AC have the same midpoint, so they must bisect each other.

To find the condition that four points, P, Q, R, S, with position vectors $\mathbf{p}$, $\mathbf{q}$, $\mathbf{r}$, $\mathbf{s}$, are the vertices of a parallelogram.

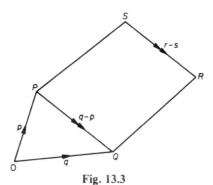

Fig. 13.3

$\overrightarrow{PQ} = \mathbf{q} - \mathbf{p}$ and $\overrightarrow{SR} = \mathbf{r} - \mathbf{s}$. If $PQRS$ is a parallelogram, then $\mathbf{q} - \mathbf{p} = \mathbf{r} - \mathbf{s}$. This is the condition that P, Q, R, and S are the vertices of a parallelogram. Notice that we could have used $\overrightarrow{PS} = \overrightarrow{QR}$ and obtained $\mathbf{s} - \mathbf{p} = \mathbf{r} - \mathbf{q}$.

Midpoint theorem

The straight line joining the midpoints of two sides of a triangle is parallel to the third side and equal to half of it.

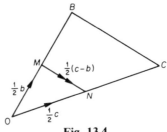

Fig. 13.4

Let the position vectors of B and C relative to O be **b** and **c** respectively. If the midpoints of OB and OC are M and N (Fig. 13.4) then the position vectors of M and N relative to O are $\frac{1}{2}$**b** and $\frac{1}{2}$**c**. Thus $\overrightarrow{BC} = $ **c** $-$ **b** and $\overrightarrow{MN} = \frac{1}{2}($**c** $-$ **b**$)$, so that MN is parallel to BC and equal to half of BC.

To prove that the medians of a triangle are concurrent

If the position vectors relative to an origin O of points A, B, and C are **a**, **b**, and **c** respectively, then the position vector of A', the midpoint of

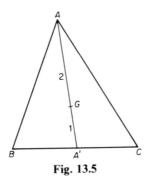

Fig. 13.5

BC, is $\frac{1}{2}($**b** $+$ **c**$)$. The point G which divides AA' in the ratio $2:1$ has position vector

$$\frac{1 \times \mathbf{a} + 2 \times \frac{1}{2}(\mathbf{b} + \mathbf{c})}{1 + 2}$$

i.e.
$$\frac{1}{3}(\mathbf{a} + \mathbf{b} + \mathbf{c})$$

Since this expression is symmetrical in **a**, **b**, and **c**, G must also divide BB' in the ratio $2:1$ and divide CC' in the ratio $2:1$.

$\therefore$ The medians AA', BB', and CC' concur at G.

This can easily be generalized to show the concurrence of similar lines in a tetrahedron.

Example 1. *ABCDEF are the vertices of a regular hexagon, whose diagonals intersect at the point O. The position vectors relative to O of A and B are **a** and **b** respectively. Find the position vectors of each of the other vertices.*

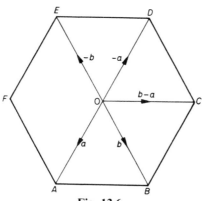

Fig. 13.6

Since OC is equal and parallel to AB, $\overrightarrow{OC} = \overrightarrow{AB}$. But $\overrightarrow{AB} = \mathbf{b} - \mathbf{a}$

$$\therefore \qquad \overrightarrow{OC} = \mathbf{b} - \mathbf{a}$$

Since $OD = OA$, but is in the opposite direction,

$$\therefore \qquad \overrightarrow{OD} = -\mathbf{a}$$

Similarly, $\qquad \overrightarrow{OE} = -\mathbf{b}$

and $\qquad \overrightarrow{OF} = \mathbf{a} - \mathbf{b}$

Example 2. *OPQR is a parallelogram. Points X and Y are such that $RX:XO$ $= RY:YQ = 1:3$. If $\overrightarrow{OP} = \mathbf{p}$, $\overrightarrow{OR} = \mathbf{r}$, find the vector representing $\overrightarrow{XY}$, and show that it is parallel to $\overrightarrow{OQ}$.*

The position vector $\overrightarrow{OY} = \mathbf{r} + \frac{1}{4}\mathbf{p}$, the position vector $\overrightarrow{OX} = \frac{3}{4}\mathbf{r}$,

$$\therefore \qquad \overrightarrow{XY} = \overrightarrow{OY} - \overrightarrow{OX} = \mathbf{r} + \frac{1}{4}\mathbf{p} - \frac{3}{4}\mathbf{r}$$
$$= \frac{1}{4}(\mathbf{p} + \mathbf{r})$$

Now $\qquad \overrightarrow{OQ} = \overrightarrow{OR} + \overrightarrow{RQ} = \mathbf{r} + \mathbf{p}$
$$= 4\,\overrightarrow{XY},$$

$\therefore \overrightarrow{XY}$ is parallel to $\overrightarrow{OQ}$.

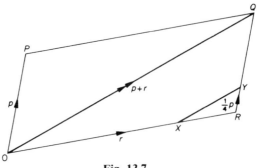

Fig. 13.7

Example 3. *L and M are the midpoints of the sides AB and BC of a square OABC. OL and AM meet at P; BP meets OA at N. Prove that ON = $\frac{2}{3}$ OA.*

Let the position vectors of A be $2\mathbf{i}$; of C be $2\mathbf{j}$. Then the position vector of L is $2\mathbf{i} + \mathbf{j}$, and the position vector of any point on the line OL is $\lambda(2\mathbf{i} + \mathbf{j})$. The vector $\overrightarrow{AM}$ is $-\mathbf{i} + 2\mathbf{j}$, and the position vector of any point on AM is $\overrightarrow{OA} + \mu\overrightarrow{AM}$, i.e. $2\mathbf{i} + \mu(-\mathbf{i} + 2\mathbf{j})$. Thus the position vector of the point of intersection of OL and AM is given by

$$\lambda(2\mathbf{i} + \mathbf{j}) = 2\mathbf{i} + \mu(-\mathbf{i} + 2\mathbf{j})$$

Since $\mathbf{i}$ and $\mathbf{j}$ are independent, equating the coefficients of each,

$$2\lambda = 2 - \mu$$
and $$\lambda = 2\mu$$
$$\therefore \qquad \lambda = \tfrac{4}{5}, \mu = \tfrac{2}{5}$$

and the position vector $\overrightarrow{OP}$ is $\frac{8}{5}\mathbf{i} + \frac{4}{5}\mathbf{j}$.

Since $\overrightarrow{BP} = -\frac{2}{5}\mathbf{i} - \frac{6}{5}\mathbf{j}$ and $\overrightarrow{ON} = \overrightarrow{OP} + \overrightarrow{PN}$

$$\overrightarrow{ON} = \tfrac{8}{5}\mathbf{i} + \tfrac{4}{5}\mathbf{j} + v(-\tfrac{2}{5}\mathbf{i} - \tfrac{6}{5}\mathbf{j})$$

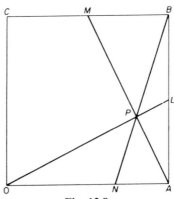

Fig. 13.8

But N lies on OA, so its position vector must not have a $\mathbf{j}$ component, i.e.
$\frac{4}{5} - \frac{6}{5}v = 0, v = \frac{2}{3}$.

$$\therefore \overrightarrow{ON} = \tfrac{8}{5}\mathbf{i} - \tfrac{4}{15}\mathbf{i} = \tfrac{4}{3}\mathbf{i} = \tfrac{2}{3}\,\overrightarrow{OA}$$

i.e. $$ON = \tfrac{2}{3}\,OA$$

Exercise 13: Miscellaneous

1. Find the position vector of the midpoint of AB if the position vectors of A and B are
   (i) $\mathbf{i} + 3\mathbf{j}, 3\mathbf{i} + \mathbf{j}$ (ii) $2\mathbf{i}, 4\mathbf{j}$ (iii) $\mathbf{i} - 5\mathbf{j}, \mathbf{i} + 3\mathbf{j}$

2. Find the position vector of P, the point of trisection of AB nearer to A, when the position vectors of A and B are
   (i) $\mathbf{i} + 3\mathbf{j}, 4\mathbf{i} + 6\mathbf{j}$ (ii) $\mathbf{i} + 3\mathbf{j}, -2\mathbf{i}$ (iii) $3\mathbf{i} + \mathbf{j}, -2\mathbf{j}$

3. If the position vectors of points A and B are $15\mathbf{i} + 2\mathbf{j}$ and $17\mathbf{j}$ respectively, find the position vector of the point that divides AB in the ratio (i) $1:2$; (ii) $2:3$; (iii) $2:1$.

4. Find the vector equation of the straight line through the points whose position vectors are
   (i) $\mathbf{i} + \mathbf{j}, 2\mathbf{i} + 4\mathbf{j}$ (ii) $\mathbf{i} - \mathbf{j}, 2\mathbf{i} - 4\mathbf{j}$ (iii) $\mathbf{i}, \mathbf{j}$

5. Find whether P, Q, and R are collinear if the position vectors of P, Q, and R are
   (i) $\mathbf{i} + \mathbf{j}, 2\mathbf{i} + 3\mathbf{j}, 3\mathbf{i} + 5\mathbf{j}$ (ii) $\mathbf{i}. \tfrac{1}{3}(\mathbf{i} + \mathbf{j}), \mathbf{i} + \mathbf{j}$ (iii) $\mathbf{i}, \tfrac{1}{2}\mathbf{i}, \mathbf{j}$.

6. Show that points with position vectors $2\mathbf{i} + 3\mathbf{j}, 3\mathbf{i} + \mathbf{j}, 5\mathbf{i} + 2\mathbf{j}$, and $4\mathbf{i} + 4\mathbf{j}$ lie at the vertices of a parallelogram.

7. The position vectors of points A, B, and C are $\mathbf{i}, 2\mathbf{i} - \mathbf{j}$, and $3\mathbf{i} + 2\mathbf{j}$ respectively. Find the fourth vertex D of (i) the parallelogram $ABCD$, (ii) the parallelogram $ABDC$.

8. Show that the points with position vectors $\mathbf{i}, 3\mathbf{i} + 4\mathbf{j}$, and $3\mathbf{i} - 4\mathbf{j}$ are vertices of an isosceles triangle.

9. Show that the points with position vectors $2\mathbf{i} + \mathbf{j}, 3\mathbf{i} + 4\mathbf{j}$, and $5\mathbf{i}$ are the vertices of an isosceles triangle.

10. Points A and B have position vectors $\begin{pmatrix} 2 \\ 0 \end{pmatrix}$ and $\begin{pmatrix} -4 \\ 0 \end{pmatrix}$ respectively. Find the position vector of C, the point such that triangle ABC is equilateral. (Two possible answers.)

11. Find the value of λ if the points with position vectors $\begin{pmatrix} -1 \\ 2 \end{pmatrix}$, $\begin{pmatrix} 2 \\ 6 \end{pmatrix}$, and $\begin{pmatrix} 5 \\ \lambda \end{pmatrix}$ are collinear.

12. Find the value of μ if the points with position vectors are $\begin{pmatrix} -2 \\ 0 \end{pmatrix}$, $\begin{pmatrix} 0 \\ 3 \end{pmatrix}$, and $\begin{pmatrix} \frac{1}{2} \\ \mu \end{pmatrix}$ are collinear.

13. Show that the points with position vectors $\begin{pmatrix} -1 \\ -1 \end{pmatrix}$, $\begin{pmatrix} 2 \\ -2 \end{pmatrix}$, and $\begin{pmatrix} 3 \\ -3 \end{pmatrix}$ are not collinear.

14. Show that the points with position vectors $\begin{pmatrix} 1 \\ 0 \end{pmatrix}$, $\begin{pmatrix} 4 \\ 1 \end{pmatrix}$, $\begin{pmatrix} 2 \\ -3 \end{pmatrix}$, and $\begin{pmatrix} 5 \\ -2 \end{pmatrix}$ are the vertices of a square.

In the following questions, points A, B, C, and D have position vectors **a, b, c,** *and* **d** *respectively.*

15. Find the condition that C is the midpoint of AB.

16. Find the condition that B and C are points of trisection of the straight line AD, B being nearer to A than to D.

17. Find the condition that $ABCD$ is a trapezium with AB parallel to CD.

18. If E is the midpoint of CD, find the condition that AC is parallel to BE.

19. If F is the midpoint of BD, find the condition that AF is parallel to CD.

20. The line BX is drawn equal and parallel to AC. Find the position vector of X.

21. The line CY is drawn parallel to and twice the length of AD. Find the position vector of Y.

22. Points P and Q divide OA, OB respectively in the ratio $m:n$. Prove using vectors that PQ is parallel to AB, and find the ratio $PQ:AB$.

23. In the parallelogram $OABC$, X is the midpoint of OA and Y the midpoint of CB. Show that CX is equal and parallel to YA.

24. In the parallelogram $OABC$, X is the point of trisection of OA nearer O, and P and Q are the points of trisection of CB, P being nearer to C than to B. Find whether CX is parallel to PA or to QA.

25. In the parallelogram $OABC$, X is the midpoint of AB. Find in terms of **a** and **c** the position vector of Z, the point of intersection of OX and AC.

26. In the parallelogram $OABC$, X and Y are the midpoints of AB and BC respectively. Find, in terms of **a** and **b**, the position vectors of the points of intersection of

(i) OX and AC; (ii) OY and AC; (iii) AY and OX; (iv) CX and OY.

(*Hint:* in (ii) and (iv), use symmetry and **c** = **b** − **a**.)

27. In the parallelogram $OABC$, L and M are points in AB and CB respectively such that $AL:LB = CM:MB = 1:2$. Find, in terms of **a** and **b**, the position vectors of the points of intersection of

(i) OL and AC; (ii) OM and AC.

28. Points D and E divide OA and OB respectively in the ratio $2:3$. BD and AE meet at P. Find, in terms of **a** and **b**, the position vector of P.

29. Given that X, Y and Z are the midpoints of the sides of a triangle ABC, prove that $\mathbf{a} + \mathbf{b} + \mathbf{c} = \mathbf{x} + \mathbf{y} + \mathbf{z}$, where **x**, **y**, and **z** are the position vectors of X, Y, and Z respectively.

30. Using the data of question 29, deduce from $\mathbf{a} + \mathbf{b} + \mathbf{c} = \mathbf{x} + \mathbf{y} + \mathbf{z}$ that
$$\overrightarrow{AX} + \overrightarrow{BY} + \overrightarrow{CZ} = 0.$$

14 Vectors in Three Dimensions

The great advantage of working with vectors given in terms of their components is that we can solve problems in three dimensions as easily as those in two dimensions, and the world in which we live is largely three dimensional. It is hard to describe direction in three dimensions, and so we invariably define a vector in terms of its components, and write it in terms of $\mathbf{i}$, $\mathbf{j}$, and $\mathbf{k}$, or in matrix form as a 'column-vector' $\begin{pmatrix} x \\ y \\ z \end{pmatrix}$ or occasionally as a 'row vector' $(x \quad y \quad z)$.

Magnitude

The magnitude of the vector $x\mathbf{i} + y\mathbf{j} + z\mathbf{k}$, $\begin{pmatrix} x \\ y \\ z \end{pmatrix}$, or $(x \quad y \quad z)$ is defined as $\sqrt{x^2 + y^2 + z^2}$. This is clearly equal to the length of the line joining the point with coordinates (x, y, z) to the origin.

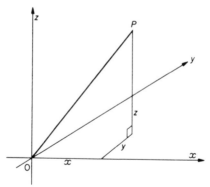

Fig. 14.1

Direction

Try to describe, in three dimensions, the direction of a straight line, say the path of an aircraft while climbing. We may say '40° above the horizontal, in a northerly direction'. It is only because we have a fixed plane ('horizontal') and some fixed direction that we can describe it at all.

Direction cosines

We usually describe direction in terms of 'direction cosines', the cosines of the angles a vector makes with each coordinate axis. If α is the angle between the vector $\overrightarrow{OP}$ and the x-axis, then from Fig. 14.2,

$$\cos \alpha = \frac{x}{\sqrt{x^2 + y^2 + z^2}}$$

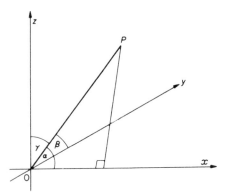

Fig. 14.2

Similarly if β is the angle between $\overrightarrow{OP}$ and the y-axis

$$\cos \beta = \frac{y}{\sqrt{x^2 + y^2 + z^2}}$$

and

$$\cos \gamma = \frac{z}{\sqrt{x^2 + y^2 + z^2}}$$

where γ is the angle between $\overrightarrow{OP}$ and the z-axis. Notice that

$$\cos^2\alpha + \cos^2 \beta + \cos^2 \gamma = \frac{x^2}{x^2 + y^2 + z^2} + \frac{y^2}{x^2 + y^2 + z^2} + \frac{z^2}{x^2 + y^2 + z^2}$$

$$= 1.$$

Example 1. *The position vectors of points A, B, and C relative to an origin O are*

$\begin{pmatrix} 2 \\ 1 \\ 0 \end{pmatrix}$, $\begin{pmatrix} 3 \\ 2 \\ 1 \end{pmatrix}$, $\begin{pmatrix} -1 \\ 1 \\ -1 \end{pmatrix}$ *respectively. Find the position vectors of A and B*

relative to C, and find the length of the lines AC and BC.

The rules for combining three-dimensional vectors are exactly the same as for combining those in two dimensions. So $\overrightarrow{CA} = \overrightarrow{OA} - \overrightarrow{OC}$

$\therefore$
$$\overrightarrow{CA} = \begin{pmatrix} 2 \\ 1 \\ 0 \end{pmatrix} - \begin{pmatrix} -1 \\ 1 \\ -1 \end{pmatrix} = \begin{pmatrix} 3 \\ 0 \\ 1 \end{pmatrix}$$

and
$$\overrightarrow{CB} = \begin{pmatrix} 3 \\ 2 \\ 1 \end{pmatrix} - \begin{pmatrix} -1 \\ 1 \\ -1 \end{pmatrix} = \begin{pmatrix} 4 \\ 1 \\ 2 \end{pmatrix}$$

The magnitude of $\overrightarrow{CA} = \sqrt{(3^2 + 0^2 + 1^2)} = \sqrt{10}$, and the magnitude of $\overrightarrow{CB} = \sqrt{(4^2 + 1^2 + 2^2)} = \sqrt{21}$.

Example 2. *Find the equation of the straight line through the point position*

vector $\begin{pmatrix} 2 \\ 3 \\ 1 \end{pmatrix}$, *parallel to the vector* $\begin{pmatrix} 1 \\ -1 \\ -1 \end{pmatrix}$. *Does the point P, whose position*

vector is $\begin{pmatrix} 4 \\ 0 \\ -1 \end{pmatrix}$, *lie on this straight line?*

From the Section Theorem (page 102) we saw that the equation of the straight line through points with position vectors **a**, **b**, was $\mathbf{r} = k\mathbf{a} + (1 - k)\mathbf{b}$. This can be written $\mathbf{r} = \mathbf{b} + k(\mathbf{a} - \mathbf{b})$, the equation of the straight line through a point position vector **b**, in the direction of a vector $\mathbf{a} - \mathbf{b}$. Thus the equation of the straight line through a point with position vector **c**, in the direction of a given vector **d**, is $\mathbf{r} = \mathbf{c} + k\mathbf{d}$, where k is a variable parameter.

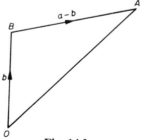

Fig. 14.3

Therefore the equation of the required straight line is $\mathbf{r} = \begin{pmatrix} 2 \\ 3 \\ 1 \end{pmatrix} + k\begin{pmatrix} 1 \\ -1 \\ -1 \end{pmatrix}$.

The point P lies on this line if we can find a single value k such that

$$\begin{pmatrix} 4 \\ 0 \\ -1 \end{pmatrix} = \begin{pmatrix} 2 \\ 3 \\ 1 \end{pmatrix} + k\begin{pmatrix} 1 \\ -1 \\ -1 \end{pmatrix}$$

From the first elements, we have $4 = 2 + k \times 1$, i.e. $k = 2$. But this value of k makes the second element in the right hand matrix equal to 1 instead of 0, so that P does not lie on the straight line. Notice that the third entries are satisfied by $k = 2$, since $1 + 2 \times (-1) = -1$, as required. The value of k has to be such that *all three* entries are correct.

Example 3. *Show that the points*

$$A, \begin{pmatrix} 2 \\ 1 \\ 0 \end{pmatrix}; B, \begin{pmatrix} 1 \\ 0 \\ -1 \end{pmatrix}; C, \begin{pmatrix} 3 \\ 2 \\ 1 \end{pmatrix}; D, \begin{pmatrix} 4 \\ 3 \\ 2 \end{pmatrix}; A', \begin{pmatrix} 3 \\ 2 \\ 2 \end{pmatrix}; B', \begin{pmatrix} 2 \\ 1 \\ 1 \end{pmatrix}; C', \begin{pmatrix} 4 \\ 3 \\ 3 \end{pmatrix};$$

and $D', \begin{pmatrix} 5 \\ 4 \\ 4 \end{pmatrix}$ *are the vertices of a parallelepiped.*

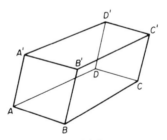

Fig. 14.4

A parallelepiped can be regarded as a three dimensional parallelogram, in which each face is parallel to the opposite face. A solid is a parallelepiped if four edges, e.g. AB, DC, $A'B'$, and $D'C'$ are equal and parallel. Subtracting the position vectors, using $\overrightarrow{AB} = \overrightarrow{OB} - \overrightarrow{OA}$, we have $\overrightarrow{AB} = \begin{pmatrix} -1 \\ -1 \\ -1 \end{pmatrix}$;

$\overrightarrow{DC} = \begin{pmatrix} -1 \\ -1 \\ -1 \end{pmatrix}$; $\overrightarrow{A'B'} = \begin{pmatrix} -1 \\ -1 \\ -1 \end{pmatrix}$ and $\overrightarrow{D'C'} = \begin{pmatrix} -1 \\ -1 \\ -1 \end{pmatrix}$, so that $AB = DC = A'B' = D'C'$, and the four edges are equal and parallel, making the solid a parallelepiped.

Exercise 14: Miscellaneous

1. The position vectors relative to an origin O of points A, B, C, and D are

$$\begin{pmatrix} 2 \\ 1 \\ -1 \end{pmatrix}, \quad \begin{pmatrix} 4 \\ 3 \\ -2 \end{pmatrix}, \quad \begin{pmatrix} 0 \\ -1 \\ 0 \end{pmatrix}, \quad \text{and} \quad \begin{pmatrix} -3 \\ -2 \\ 3 \end{pmatrix} \quad \text{respectively. Find the}$$

vectors (i) $\overrightarrow{AB}$, (ii) $\overrightarrow{AC}$, (iii) $\overrightarrow{AD}$, (iv) $\overrightarrow{BC}$. Find also the magnitude of each of these vectors.

2. Find the magnitude of each of the following vectors, and the cosine of the angle each makes with the coordinate axes:

(i) $\begin{pmatrix} 1 \\ 2 \\ 2 \end{pmatrix}$ (ii) $\begin{pmatrix} 2 \\ -1 \\ -2 \end{pmatrix}$ (iii) $\begin{pmatrix} -1 \\ 1 \\ 1 \end{pmatrix}$ (iv) $\begin{pmatrix} 0 \\ 3 \\ -4 \end{pmatrix}$ (v) $\begin{pmatrix} -1 \\ 2 \\ -2 \end{pmatrix}$

3. If $\mathbf{a} = \begin{pmatrix} 1 \\ 2 \\ 1 \end{pmatrix}$, $\mathbf{b} = \begin{pmatrix} 3 \\ 1 \\ 4 \end{pmatrix}$, and $\mathbf{c} = \begin{pmatrix} -5 \\ 0 \\ 2 \end{pmatrix}$, find the following vectors:

(i) $\mathbf{a} + \mathbf{b}$, (ii) $\mathbf{a} + 2\mathbf{b}$, (iii) $\mathbf{a} + 2\mathbf{b} - \mathbf{c}$, (iv) $3\mathbf{a} + 2\mathbf{b} + \mathbf{c}$.

4. If the position vectors of points A and B are $\begin{pmatrix} 2 \\ -3 \\ 1 \end{pmatrix}$ and $\begin{pmatrix} 5 \\ 3 \\ 7 \end{pmatrix}$ respectively, find the position vectors of the points that divide AB in the ratio (i) $1:2$, (ii) $2:1$.

5. Find the position vector of the midpoint of the lines joining the points $\begin{pmatrix} 4 \\ 1 \\ 0 \end{pmatrix}$ and $\begin{pmatrix} -2 \\ -1 \\ 2 \end{pmatrix}$.

6. Find the position vector of the point which divides AB in the ratio $2:1$ externally, if the position vectors of A and B are $\begin{pmatrix} 2 \\ 1 \\ 0 \end{pmatrix}$ and $\begin{pmatrix} 1 \\ 0 \\ \frac{1}{2} \end{pmatrix}$ respectively.

7. The position vectors of four points A, B, C, and D are $\begin{pmatrix} 4 \\ 2 \\ 0 \end{pmatrix}$, $\begin{pmatrix} 0 \\ 0 \\ 0 \end{pmatrix}$, $\begin{pmatrix} 6 \\ 4 \\ -2 \end{pmatrix}$, and $\begin{pmatrix} 4 \\ 4 \\ -4 \end{pmatrix}$ respectively. Show that the midpoints of AB, BC, and CD lie on a straight line.

Draw a diagram to illustrate this result, and see whether this could have been shown without using vectors.

8. Using **a**, **b**, and **c** from question 3, find constants λ and μ such that

$$\mathbf{a} + \lambda\mathbf{b} + \mu\mathbf{c} = \begin{pmatrix} 0 \\ 5 \\ 17 \end{pmatrix}$$

9. Using **a**, **b**, and **c** from question 3, find constants λ and μ such that

$$\lambda\mathbf{a} + \mathbf{b} + \mu\mathbf{c} = \begin{pmatrix} 12 \\ 9 \\ 6 \end{pmatrix}$$

10. Using the data from question 3, find constants λ, μ, and ν such that

$$\lambda\mathbf{a} + \mu\mathbf{b} + \nu\mathbf{c} = \begin{pmatrix} 4 \\ 3 \\ -4 \end{pmatrix}.$$

11. If $\mathbf{x} = \begin{pmatrix} 1 \\ 2 \\ -1 \end{pmatrix}$, $\mathbf{y} = \begin{pmatrix} 3 \\ -1 \\ 2 \end{pmatrix}$, and $\mathbf{z} = \begin{pmatrix} 5 \\ 3 \\ 0 \end{pmatrix}$ show that it is not possible to

find constants λ, μ, and ν such that $\lambda\mathbf{x} + \mu\mathbf{y} + \nu\mathbf{z} = \begin{pmatrix} 1 \\ 0 \\ 1 \end{pmatrix}$.

12. If $\mathbf{x} = \begin{pmatrix} 1 \\ 0 \\ -1 \end{pmatrix}$, $\mathbf{y} = \begin{pmatrix} 2 \\ 0 \\ 0 \end{pmatrix}$, and $\mathbf{z} = \begin{pmatrix} -2 \\ 0 \\ 3 \end{pmatrix}$ show that it is not possible

to find constants λ, μ, and ν such that $\lambda\mathbf{x} + \mu\mathbf{y} + \nu\mathbf{z} = \begin{pmatrix} 3 \\ 1 \\ 0 \end{pmatrix}$.

Illustrate this result geometrically.

13. Show that if three vectors **x**, **y**, and **z** lie in a plane, then it is always possible to find scalars λ, μ, and ν so that $\lambda\mathbf{x} + \mu\mathbf{y} + \nu\mathbf{z} = 0$.
Show also that if **x**, **y**, and **z** do not lie in a plane, then it is never possible to find non-zero scalars λ, μ, and ν such that $\lambda\mathbf{x} + \mu\mathbf{y} + \nu\mathbf{z} = 0$.

14. Find the equation of the straight line through the point with position

vector $\begin{pmatrix} 3 \\ 0 \\ -2 \end{pmatrix}$ in the direction of the vector $\begin{pmatrix} 2 \\ 1 \\ -1 \end{pmatrix}$. Find also (i)

whether the point position vector $\begin{pmatrix} -1 \\ -2 \\ -4 \end{pmatrix}$ lies on this line; (ii) the position

vector of the point in which the straight line meets the plane $z = 0$.

15. Find the vector equation of the straight line through the points with position vectors $\begin{pmatrix} 0 \\ -5 \\ -3 \end{pmatrix}$ and $\begin{pmatrix} 2 \\ -1 \\ -1 \end{pmatrix}$, also the equation of the straight line through the points with position vectors $\begin{pmatrix} 2 \\ 1 \\ -1 \end{pmatrix}$ and $\begin{pmatrix} 1 \\ 1 \\ -2 \end{pmatrix}$.

Find the position vector of the point in which these lines intersect.

16. $OABCXYZW$ is a cubical box, with edge-length 4 units. Let the position vector of A relative to O be $\begin{pmatrix} 4 \\ 0 \\ 0 \end{pmatrix}$, and set up similar position vectors for the points B, C, X, Y, Z, and W. Let L be the midpoint of AY, M the midpoint of XW, and N the point in WZ such that $WN:NZ = 3:1$. Find the lengths of the lines LM, MN, and NL, and, using the cosine formula, find the cosine of the angle NLM.

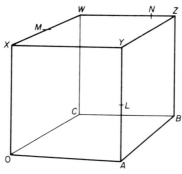

Fig. 14.5

15 The Scalar Product

Definition

If **a** and **b** are two vectors and θ is the angle between them, then we define the scalar product of **a** and **b** as $ab \cos \theta$. This is usually written $\mathbf{a} \cdot \mathbf{b} = ab \cos \theta$, and is sometimes called the 'dot-product'.

Since $\cos \theta = \cos (360° - \theta)$, it does not matter whether we rotate from **a** to **b** or from **b** to **a**, so that $\mathbf{a} \cdot \mathbf{b} = \mathbf{b} \cdot \mathbf{a}$, i.e. the scalar product is commutative.

Distributive law

Since $\mathbf{a} \cdot \mathbf{b}$ is a scalar, it is meaningless to write $\mathbf{a} \cdot (\mathbf{b} \cdot \mathbf{c})$, so that the associative law cannot apply, but we can show that the scalar product is distributive over addition, i.e.

$$\mathbf{c} \cdot (\mathbf{a} + \mathbf{b}) = \mathbf{c} \cdot \mathbf{a} + \mathbf{c} \cdot \mathbf{b}$$

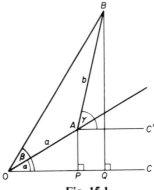

Fig. 15.1

In Fig. 15.1, let $\overrightarrow{OA} = \mathbf{a}$, $\overrightarrow{AB} = \mathbf{b}$, and $\overrightarrow{OC} = \mathbf{c}$, so that $\overrightarrow{OB} = \mathbf{a} + \mathbf{b}$. If $A\hat{O}C = \alpha$, $B\hat{O}C = \beta$, and $B\hat{A}C' = \gamma$,

$$
\begin{aligned}
\mathbf{c} \cdot \mathbf{a} + \mathbf{c} \cdot \mathbf{b} &= OC \cdot OA \cos \alpha + OC \cdot AB \cos \gamma \\
&= OC \cdot OP + OC \cdot PQ \\
&= OC \cdot (OP + PQ) \\
&= OC \cdot OQ \\
&= OC \cdot OB \cos \beta \\
&= \mathbf{c} \cdot (\mathbf{a} + \mathbf{b}), \text{ as required.}
\end{aligned}
$$

Unit vectors

With the usual notation, $i.i = j.j = 1$ and $i.j = j.i = 0$. This enables us to find easily the scalar products of vectors given in component form.

Example *If* $a = 3i + j$ *and* $b = 4i - 2j$, *find* (i) $a.b$, (ii) a^2.

(i)
$$\begin{aligned} a.b &= (3i + j).(4i - 2j) \\ &= 12i.i + 4j.i - 6i.j - 2j.j \\ &= 12 - 2 \\ &= 10 \end{aligned}$$

(ii)
$$a.a = a^2 \cos 0° = a^2,$$

So
$$\begin{aligned} a^2 &= (3i + j).(3i + j) \\ &= 9i.i + 3i.j + 3j.i + j.j \\ &= 9 + 1 \\ &= 10 \end{aligned}$$

Parallel and perpendicular vectors

If a and b are two parallel vectors, $a.b = ab$, since the angle between them is $0°$, and $\cos 0° = 1$. In particular, $a.a = a^2$, for all a.

If a and b are two perpendicular vectors, $a.b = ab \cos 90° = 0$.

Angle between two vectors

To find the angle between two vectors, let us first find the angle between the vector $3i + 5j$ and the unit vector i. Then using the scalar product,

$$(3i + 5j).i = |3i + 5j|.1 \cos \theta$$

i.e.
$$3 = \sqrt{34} \cos \theta$$
$$\theta = \text{arc } \cos 3/\sqrt{34}.$$

Looking at Fig. 15.2, we see, using trigonometry, that $\cos \theta = 3/\sqrt{34}$.

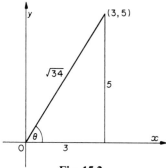

Fig. 15.2

To find the angle between two vectors, neither of which is along a coordinate axis, there is considerable advantage in using the scalar product.

Example *Find the angle between the vectors* $3\mathbf{i} + \mathbf{j}$ *and* $\mathbf{i} + 5\mathbf{j}$.

$$(3\mathbf{i} + \mathbf{j}) \cdot (\mathbf{i} + 5\mathbf{j}) = |3\mathbf{i} + \mathbf{j}| \cdot |\mathbf{i} + 5\mathbf{j}| \cos \theta$$
$$3 + 5 = \sqrt{10} \cdot \sqrt{26} \cos \theta$$
$$\theta = \text{arc cos } 8/\sqrt{260}$$
$$\simeq \text{about } 60°$$

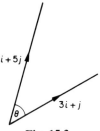

Fig. 15.3

Proof of trigonometric formulae

Scalar products give us the easiest way of proving the formula for $\cos (\theta - \varphi)$*. For if $\mathbf{a} = a(\cos \varphi \mathbf{i} + \sin \varphi \mathbf{j})$, and $\mathbf{b} = b(\cos \theta \mathbf{i} + \sin \theta \mathbf{j})$, then

$$\mathbf{a} \cdot \mathbf{b} = ab(\cos \varphi \mathbf{i} + \sin \varphi \mathbf{j}) \cdot (\cos \theta \mathbf{i} + \sin \theta \mathbf{j})$$
$$= ab(\cos \varphi \cos \theta + \sin \varphi \sin \theta)$$

But $\mathbf{a} \cdot \mathbf{b} = ab \cos (\theta - \varphi)$, so from the definition of a scalar product

$$\cos (\theta - \varphi) = \cos \theta \cos \varphi + \sin \theta \sin \varphi$$

Replacing φ by $-\varphi$, we have

$$\cos (\theta + \varphi) = \cos \theta \cos \varphi - \sin \theta \sin \varphi.$$

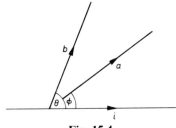

Fig. 15.4

* pages 91 and 298.

Replacing θ by $\dfrac{\pi}{2} - \theta$

$$\cos\left(\frac{\pi}{2} - \theta - \varphi\right) = \cos\left(\frac{\pi}{2} - \theta\right)\cos\varphi + \sin\left(\frac{\pi}{2} - \theta\right)\sin\varphi$$

i.e. $\sin(\theta + \varphi) = \sin\theta\cos\varphi + \cos\theta\sin\varphi.$

Equation of a straight line

If we require the equation of the straight line through a point A with position vector $\mathbf{a}$, perpendicular to a given vector $\mathbf{b}$, then, if $\mathbf{r}$ is the

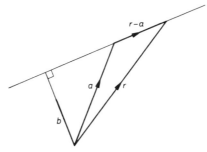

Fig. 15.5

position vector of any point P in that line, $\mathbf{r} - \mathbf{a}$ must be perpendicular to $\mathbf{b}$, i.e. $(\mathbf{r} - \mathbf{a}).\mathbf{b} = 0$. The equation of the line can be written in that form, or as $\mathbf{r}.\mathbf{b} = \mathbf{a}.\mathbf{b}$.

Example *Find the equation of the straight line through the point position vector* $2\mathbf{i} + 3\mathbf{j}$ *perpendicular to the vector* $\mathbf{i} - 2\mathbf{j}$.

The equation of the straight line is

$$\mathbf{r}.(\mathbf{i} - 2\mathbf{j}) = (2\mathbf{i} + 3\mathbf{j}).(\mathbf{i} - 2\mathbf{j})$$

i.e. $\mathbf{r}.(\mathbf{i} - 2\mathbf{j}) = -4$

If we write $\mathbf{r} = x\mathbf{i} + y\mathbf{j}$, then the cartesian equation of the line is found to be

$$x - 2y + 4 = 0$$

This can be checked by coordinate geometry.

Geometrical application

Now that we have a means of describing the angle between two vectors, we can prove many geometrical results easily.

Cosine formula

If **a** and **b** are two vectors inclined at an angle θ to each other, and **c** is a third vector such that $\mathbf{a} - \mathbf{b} = \mathbf{c}$, then

$$
\begin{aligned}
c^2 &= (\mathbf{a} - \mathbf{b}) \cdot (\mathbf{a} - \mathbf{b}) \\
&= a^2 + b^2 - 2\mathbf{a} \cdot \mathbf{b} \\
&= a^2 + b^2 - 2ab \cos \theta
\end{aligned}
$$

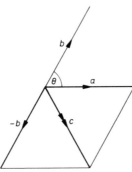

Fig. 15.6

The altitudes of a triangle are concurrent

Given a triangle ABC, let the line through a vertex A perpendicular to BC and the line through B perpendicular to AC meet at H. Let the position vectors of A, B, and C relative to H be **a**, **b**, and **c** respectively.

Since AH is perpendicular to BC,

$$\mathbf{a} \cdot (\mathbf{b} - \mathbf{c}) = 0$$

Since BH is perpendicular to AC,

$$\mathbf{b} \cdot (\mathbf{c} - \mathbf{a}) = 0$$
$$\mathbf{c} \cdot (\mathbf{b} - \mathbf{a}) = 0$$

CH is perpendicular to AB, so that H is the point of concurrence of the three altitudes.

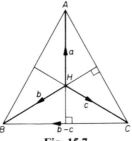

Fig. 15.7

Exercise 15a

1. If $\mathbf{a} = 2\mathbf{i} + \mathbf{j}$, $\mathbf{b} = 2\mathbf{i} - 3\mathbf{j}$, and $\mathbf{c} = -2\mathbf{j}$, find
   (i) a^2 (ii) $\mathbf{a} \cdot \mathbf{b}$ (iii) $\mathbf{a} \cdot \mathbf{c}$
   (iv) b^2 (v) $\mathbf{b} \cdot \mathbf{c}$ (vi) $\mathbf{c} \cdot \mathbf{b}$

2. If $\mathbf{a} = 3\mathbf{i} + 2\mathbf{j}$, $\mathbf{b} = 4\mathbf{i} - 5\mathbf{j}$, find the length of the projection of $\mathbf{a}$ onto $\mathbf{b}$, illustrating your result by a diagram.

3. Find the cosine of the angle between each of the following pairs of vectors

   (i) $\begin{pmatrix} 3 \\ 1 \end{pmatrix}, \begin{pmatrix} 2 \\ 4 \end{pmatrix}$ (ii) $\begin{pmatrix} -1 \\ 1 \end{pmatrix}, \begin{pmatrix} -2 \\ 3 \end{pmatrix}$ (iii) $\begin{pmatrix} 3 \\ 3 \end{pmatrix}, \begin{pmatrix} 1 \\ 2 \end{pmatrix}$

4. Find which pairs of the following vectors are perpendicular:
   $$\mathbf{a} = \mathbf{i} - 2\mathbf{j}; \ \mathbf{b} = 2\mathbf{i} - \mathbf{j}; \ \mathbf{c} = 2\mathbf{i} + \mathbf{j}; \ \mathbf{d} = \mathbf{i} + 2\mathbf{j}$$

5. Find the equation, in vector and in cartesian form, of each of the following straight lines:
   (i) through $\mathbf{i} - \mathbf{j}$, perpendicular to $\mathbf{i} + \mathbf{j}$;
   (ii) through $2\mathbf{i} + 5\mathbf{j}$, perpendicular to $2\mathbf{i} + 5\mathbf{j}$;
   (iii) through $2\mathbf{i} + 5\mathbf{j}$, perpendicular to $5\mathbf{i} - 2\mathbf{j}$.

6. Find the cosine of the angle between the straight lines $\mathbf{r} \cdot (\mathbf{i} - 2\mathbf{j}) = 1$ and $\mathbf{r} \cdot (3\mathbf{i} + \mathbf{j}) = 3$.

7. Find the cosine of the angle between $3x - 2y + 1 = 0$ and $x + 4y = 0$.

8. Find in vector form the equation of the circle on points A and B as diameter, where A and B have position vectors $\mathbf{a}$ and $\mathbf{b}$ respectively.

9. Find a unit vector perpendicular to the vector $3\mathbf{i} - 4\mathbf{j}$ (i) using scalar products, (ii) graphically.

10. Find a unit vector inclined at $60°$ to the vector $\mathbf{i} - 2\mathbf{j}$.

11. Use the scalar product to prove that the diagonals of a rhombus are perpendicular.

12. Using the scalar products, prove Ptolemy's Theorem, that if A, B, C, and D are vertices of a cyclic quadrilateral,
    $$AC \cdot BD = AB \cdot CD + BC \cdot AD$$

Scalar products in three dimensions

In Chapter 14 we displayed the components of a vector in a matrix with three rows and one column, a 3×1 matrix, because it was easier to relate corresponding elements in matrices than to relate multiples of $\mathbf{i}$, $\mathbf{j}$, and $\mathbf{k}$. However, when using scalar products of vectors it is important to use results such as $\mathbf{i} \cdot \mathbf{j} = 0$, so that the $\mathbf{i}, \mathbf{j}, \mathbf{k}$ form usually proves more helpful.

Scalar product

From the definition of the scalar product, we saw that $\mathbf{i} \cdot \mathbf{j} = 0$ and $\mathbf{i} \cdot \mathbf{i} = 1$. If $\mathbf{k}$ is a unit vector perpendicular to both $\mathbf{i}$ and $\mathbf{j}$, then
$$\mathbf{i} \cdot \mathbf{i} = \mathbf{j} \cdot \mathbf{j} = \mathbf{k} \cdot \mathbf{k} = 1$$

and
$$i.j = j.k = k.i = i.k = k.j = j.i = 0.$$
Thus the scalar product of $x_1i + y_1j + z_1k$ and $x_2i + y_2j + z_2k$ is
$$x_1x_2 + y_1y_2 + z_1z_2$$

Example 1. *To find the scalar product of $2i + 3j - k$ and $4i - 5j - 3k$.*
Using $i.i = j.j = k.k = 1$, and $i.j$ etc. $= 0$,
$$(2i + 3j - k).(4i - 5j - 3k) = 8 - 15 + 3$$
$$= -4$$

To find the angle between two vectors

From the definition $a.b = ab \cos \theta$ we can find the angle between two vectors in three dimensions.

Example 2. *Find the angle between the two vectors $i + j + 2k$ and $2i - j + k$.*

$|i + j + 2k| = \sqrt{6}$ and $|2i - j + k| = \sqrt{6}$ so that if θ is the angle between these two vectors,
$$(i + j + 2k).(2i - j + k) = \sqrt{6} \sqrt{6} \cos \theta$$
But
$$(i + j + 2k).(2i - j + k) = 2 - 1 + 2 = 3,$$
$$3 = 6 \cos \theta$$
$$\cos \theta = \tfrac{1}{2}$$
$$\theta = 60°$$

To find the equation of the plane through a given point, perpendicular to a given vector

We saw (page 119) how to find the equation of a straight line through a given point in a given direction. If we require the equation of the plane through the point position vector **a**, perpendicular to the vector **n**, then, if **r** is the position vector of any point in the plane, (**r** − **a**) is perpendicular to **n**,

i.e. $$(r - a).n = 0$$
i.e. $$r.n = a.n$$

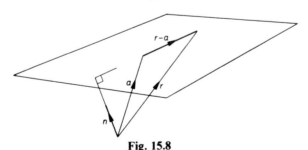

Fig. 15.8

Example 3. *Find the vector and cartesian equations of the plane through the point with position vector* $3\mathbf{i} - \mathbf{j} + 4\mathbf{k}$, *perpendicular to the vector* $\mathbf{i} + 2\mathbf{j} - \mathbf{k}$.

Using the above result,

$$\mathbf{r}.(\mathbf{i} + 2\mathbf{j} - \mathbf{k}) = (3\mathbf{i} - \mathbf{j} + 4\mathbf{k}).(\mathbf{i} + 2\mathbf{j} - \mathbf{k})$$
$$= 3 - 2 - 4$$
$$= -3$$

the vector equation is $\mathbf{r}.(\mathbf{i} + 2\mathbf{j} - \mathbf{k}) = -3$.

Writing $\mathbf{r} = x\mathbf{i} + y\mathbf{j} + z\mathbf{k}$, the cartesian equation is

$$(x\mathbf{i} + y\mathbf{j} + z\mathbf{k}).(\mathbf{i} + 2\mathbf{j} - \mathbf{k}) = -3$$

i.e.
$$x + 2y - z = -3$$

To find the angle between two planes

The angle between two planes is equal to the angle between two straight lines, each perpendicular to one of the given planes.

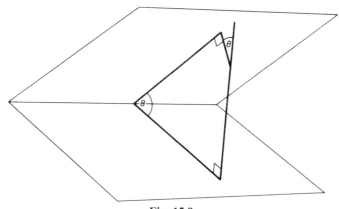

Fig. 15.9

Example 4. *Find the angle between the planes* $2x + y - 2z = 0$ *and* $3x + y + z = 2$.

Write the planes in vector form,

$$\mathbf{r}.(2\mathbf{i} + \mathbf{j} - 2\mathbf{k}) = 0$$

and
$$\mathbf{r}.(3\mathbf{i} + \mathbf{j} + \mathbf{k}) = 2$$

Then the vector $2\mathbf{i} + \mathbf{j} - 2\mathbf{k}$ is perpendicular to the first plane and the vector $3\mathbf{i} + \mathbf{j} + \mathbf{k}$ is perpendicular to the second plane. If θ is the angle between these vectors,

$$(2\mathbf{i} + \mathbf{j} - 2\mathbf{k}).(3\mathbf{i} + \mathbf{j} + \mathbf{k}) = |2\mathbf{i} + \mathbf{j} - 2\mathbf{k}|.|3\mathbf{i} + \mathbf{j} + \mathbf{k}| \cos \theta$$
$$6 + 1 - 2 = 3\sqrt{11} \cos \theta,$$

$$\cos \theta = \frac{5}{3\sqrt{11}}$$

$$\theta \simeq 60°$$

This example illustrates that the procedure to find the angle between two vectors in three dimensions is the same as in two dimensions. Without using vectors it is very much harder.

Exercise 15b

1. If $\mathbf{a} = 2\mathbf{i} + 2\mathbf{j} - \mathbf{k}$, $\mathbf{b} = 2\mathbf{i} - 10\mathbf{j} + 11\mathbf{k}$, and $\mathbf{c} = 3\mathbf{i} - 4\mathbf{k}$, find
   (i) $\mathbf{a} \cdot \mathbf{b}$ (ii) $\mathbf{a} \cdot \mathbf{c}$ (iii) $\mathbf{b} \cdot \mathbf{c}$
   (iv) a^2 (v) b^2 (vi) c^2
   Find also the cosine of the angle between
   (vii) $\mathbf{a}$ and $\mathbf{b}$ (viii) $\mathbf{b}$ and $\mathbf{c}$ (ix) $\mathbf{c}$ and $\mathbf{a}$.

2. If $\mathbf{a} = 2\mathbf{i} + 3\mathbf{j} + \mathbf{k}$, $\mathbf{b} = \mathbf{i} + 4\mathbf{j} - \mathbf{k}$, and $\mathbf{c} = \mathbf{i} - \mathbf{j} + 2\mathbf{k}$, find
   (i) $\mathbf{a} \cdot \mathbf{b}$ (ii) $\mathbf{b} \cdot \mathbf{c}$ (iii) $\mathbf{c} \cdot \mathbf{a}$
   (iv) a^2 (v) b^2 (vi) c^2
   Find also the cosine of the angle between
   (vii) $\mathbf{a}$ and $\mathbf{b}$ (viii) $\mathbf{b}$ and $\mathbf{c}$ (ix) $\mathbf{c}$ and $\mathbf{a}$.

3. Find the cosine of the angle between the vectors $2\mathbf{i} + \mathbf{j} + \mathbf{k}$ and $3\mathbf{i} - 2\mathbf{j} + 2\mathbf{k}$.

4. The position vectors of points A, B, and C are $\begin{pmatrix} 2 \\ 3 \\ -1 \end{pmatrix}$, $\begin{pmatrix} 4 \\ 1 \\ 0 \end{pmatrix}$, and $\begin{pmatrix} 6 \\ -1 \\ 1 \end{pmatrix}$ respectively. Find the cosine of the angle ABC.

5. Show that the vectors $2\mathbf{i} + 4\mathbf{j} - \mathbf{k}$ and $\mathbf{i} - \mathbf{j} - 2\mathbf{k}$ are perpendicular.

6. Find a unit vector perpendicular to both the vectors $2\mathbf{i} + \mathbf{j} + 2\mathbf{k}$ and $-2\mathbf{i} + 3\mathbf{j} + 2\mathbf{k}$.

7. Find a unit vector perpendicular to both the vectors $2\mathbf{i} - \mathbf{j} + \mathbf{k}$ and $3\mathbf{i} - 4\mathbf{j} - \mathbf{k}$.

8. Find a vector, magnitude 3 units, perpendicular to both $2\mathbf{i} + \mathbf{j} + \mathbf{k}$ and $\mathbf{i} - \mathbf{j} - \mathbf{k}$.

9. Using vectors, prove that the cosine of the angle between the diagonals of a cube is $1/3$.

10. Find the vector equation of the following planes:
    (i) Through the point position vector $\mathbf{i} - \mathbf{j} - \mathbf{k}$, perpendicular to $3\mathbf{i} + 2\mathbf{j} + \mathbf{k}$.
    (ii) Through the point position vector $2\mathbf{i} + 3\mathbf{j} + 4\mathbf{k}$ perpendicular to $\mathbf{i} + \mathbf{j} + \mathbf{k}$.
    (iii) Through the point position vector $\mathbf{i} - \mathbf{j} - \mathbf{k}$ perpendicular to $\mathbf{i}$.

11. Find the cartesian equation of the plane through the point with position vector $3\mathbf{i} - 2\mathbf{j} + \mathbf{k}$, perpendicular to the vector $\mathbf{i} + 2\mathbf{j} - 4\mathbf{k}$.

12. Use the scalar product to find the cosines of the angles in triangle LMN in question 16, page 115.

13. Find the angle between the planes $x + 2y + 2z = 1$ and $3x - 4y - 5z = 2$.

14. Fig. 15.10 shows a cube $OABCXYZW$, edge 6 units. Points M, N, and P are such that M is the midpoint of AY, N is the point in XY such that

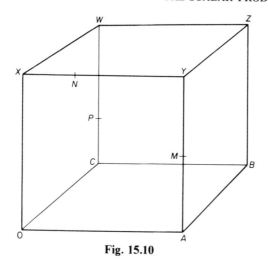

Fig. 15.10

$XN:NY = 1:2$ and P is the point in CW such that $CP:PW = 1:2$. Choosing a suitable system of vectors, find the cosines of the following angles:

(i) *MON*   (ii) *XPA*   (iii) *NPM*.

15. Find the vector equation of the sphere on points with position vectors **a**, **b** as diameter.

16. Find the cartesian equation of the sphere on points with position vectors $\mathbf{i} + \mathbf{k}$, $2\mathbf{i} - \mathbf{j}$ as diameter.

17. In the tetrahedron *OABC*, *OA* is perpendicular to *BC* and *OB* is perpendicular to *AC*. Prove that *OC* is perpendicular to *AB*.

18. Write down the equation of the plane through the point position vector **a** perpendicular to the unit vector **u**. Show that the perpendicular distance of the point position vector **b** from the plane is $\mathbf{u}.(\mathbf{b} - \mathbf{a})$.

19. Prove that the perpendicular bisectors of the sides of a triangle meet at a point (the circumcentre of the triangle).

20. If *OABC* is a parallelogram, show that $AC^2 + OB^2 = 2(OA^2 + OC^2)$.

CALCULUS

16 Gradient

Gradient of a line

The gradient of a line referred to perpendicular axes of x and y is the ratio of the increase of y between any two points of the line to the increase of x between the same two points. If y decreases as x increases, the gradient is negative. Written as a fraction, the gradient equals $\dfrac{\text{increase of } y}{\text{increase of } x}$.

Consider the line $y = 3x + 4$.

When $x = 0$, $y = 4$. So the point where $x = 0$ and $y = 4$ lies on the line. The coordinates of a point are written as an ordered pair with the x-coordinate first. So we say that the lines passes through the point (0, 4).

When $x = 1$, $y = 7$, and so the line also passes through the point (1, 7). The line is shown in Fig. 16.1.

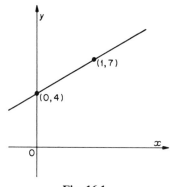

Fig. 16.1

The increase of y between these two points is $7 - 4$ or 3.
The increase of x between these two points is $1 - 0$ or 1.
The gradient of the line is $\frac{3}{1}$ or 3.

The reader should prove, using similar triangles, that the gradient is independent of the position of the two points on the line.

Now consider the line $3y + x = 4$.
When $x = 1$, $y = 1$. Therefore the point (1, 1) lies on the line.
When $y = 0$, $x = 4$. Therefore the point (4, 0) lies on the line.
The line is shown in Fig. 16.2.

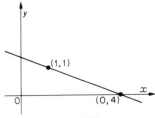

Fig. 16.2

The increase of x between these two points is $4 - 1$ or 3.

The *decrease* of y between these two points is $1 - 0$ or 1.

The increase of y is therefore -1 and the gradient is $-\frac{1}{3}$.

In general, a line which makes an acute angle with the positive x direction has a positive gradient; a line which makes an obtuse angle with the positive x direction has a negative gradient.

The gradient is also equal to the tangent of the angle which the line makes with the positive direction of the x-axis. Remember, however, that the scales chosen alter this angle and measuring the angle and finding its tangent will give the correct gradient only when the scales chosen for x and y are equal.

In general, the gradient of the line joining the points (x_1, y_1) and (x_2, y_2) is $\dfrac{y_2 - y_1}{x_2 - x_1}$.

The order in which the points are taken is immaterial provided the same order is chosen for both numerator and denominator, i.e. the gradient is also equal to $\dfrac{y_1 - y_2}{x_1 - x_2}$.

Gradient of a curve

The gradient of a curve at a point is equal to the gradient of the tangent at that point.

Consider the point $(1, 0)$ on the curve $y = x^2 - x$, whose shape is shown in Fig. 16.3.

Draw the curve accurately and draw the tangent at the point $(1, 0)$. Check that the tangent passes through the point $(3, 2)$.

The gradient of the tangent is therefore $\dfrac{2 - 0}{3 - 1}$ or 1. This is also the gradient of the curve at the point $(1, 0)$.

This drawing method of finding the gradient of a curve is not very satisfactory because the tangent has to be drawn by guess-work and is seldom accurate. We shall now consider how the gradient may be calculated.

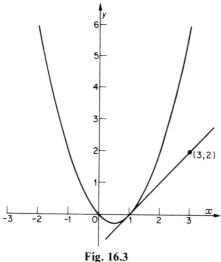

Fig. 16.3

The distance–time graph

Suppose that s metres is the distance travelled in t seconds by a particle moving in a straight line where s is measured from a fixed point of the line. If s is plotted vertically against t horizontally, the resulting curve is called a distance–time graph.

Figure 16.4 represents such a graph in which $s = 0$ when $t = 0$. Suppose we consider two points, P and Q, on the graph, and let the line through P parallel to the x-axis meet the line through Q parallel to the y-axis at N. Then the gradient of the line $PQ = \dfrac{QN}{PN}$ where QN is the distance represented in metres and PN the time in seconds.

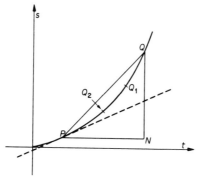

Fig. 16.4

If QN represents 8 m and PN represents 2 seconds, the gradient of the line PQ is $\dfrac{8 \text{ m}}{2 \text{ s}}$ or 4 m s^{-1}. The line QN represents the distance travelled between P and Q and the line PN the time taken to travel this distance. The gradient therefore represents the average velocity of the particle in travelling the distance between P and Q.

Now suppose that we take points Q_1, Q_2 ... on the graph between Q and P and that these points get progressively closer to P. The gradient of PQ_1 represents the average velocity between P and Q_1, the gradient of PQ_2 the average velocity between P and Q_2 and so on. When Q gets so close to P that the distance PQ is negligible, what will the line joining P and Q become? It will, in the limit, become the tangent at P and the gradient of the tangent at P therefore represents the velocity of the particle at the point P. This velocity may be found by accurate drawing and calculation of the gradient of the tangent at P from the graph. In examples where there is no known algebraic law connecting distance and time, the velocity at a point must be found by graphical methods. If, however, s is known as an algebraic function of t, the velocity can be found by calculation and we shall now investigate how this may be done.

Velocity at a point

Suppose the connection between s and t is the simple one $s = t^2$. The distance–time graph is shown in Fig. 16.5.

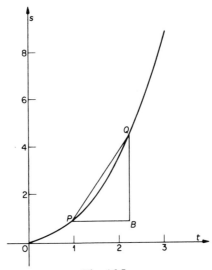

Fig. 16.5

When $t = 1$, $s = 1$ and let P be the point $(1, 1)$ on the curve. Let Q be any other point on the curve and let Q get nearer and nearer to the point P. The gradient of the line PQ becomes gradually nearer to the gradient of the tangent at P and, by taking numerical values, we should be able to find a good approximation for the gradient of the tangent.

Suppose we start with the point on the curve where $t = 2$. Since $s = t^2$, $s = 4$ at this point. Between P and Q the particle has travelled 3 m in 1 second and so the average velocity over the distance PQ is 3 m s^{-1}. This is not a good approximation for the velocity at P because Q is too far from P. We now tabulate the results obtained from several positions of Q getting nearer and nearer P.

Value of t	1.5	1.4	1.3	1.2	1.1	1.05	1.01	1.001
Value of s	2.25	1.96	1.69	1.44	1.21	1.1025	1.0201	1.002001
QN	1.25	0.96	0.69	0.44	0.21	0.1025	0.0201	0.002001
PN	0.5	0.4	0.3	0.2	0.1	0.05	0.01	0.001
Average velocity $\left(\dfrac{QN}{PN}\right)$ (m s^{-1})	2.5	2.4	2.3	2.2	2.1	2.05	2.01	2.001

The average velocity becomes very nearly equal to 2 m s^{-1} but will never become exactly 2 m s^{-1} as Q cannot be made to coincide with P. Both quantities PN and QN would then become zero and the fraction $\frac{0}{0}$ cannot be evaluated.

Suppose now we have a further addition to the table and consider the position of the particle when $t = 1 + h$.

The distance s is given by $(1 + h)^2$ or $1 + 2h + h^2$. Then
$$QN = 2h + h^2 \quad \text{and} \quad PN = h.$$

The average velocity $= \dfrac{QN}{PN} = \dfrac{2h + h^2}{h} \text{ m s}^{-1}.$

The point Q cannot be made to coincide with P and therefore h is not zero. The fraction $\dfrac{2h + h^2}{h}$ equals $(2 + h)$ unless $h = 0$. The average velocity between P and Q is therefore $(2 + h) \text{ m s}^{-1}$. This numerically is one greater than the value of t, namely $(1 + h)$.

Notice that the values in the bottom row of the table are always one greater than the corresponding values in the top row. The average velocity becomes very nearly equal to 2 m s^{-1} when h becomes very small. We say that the velocity tends to 2 m s^{-1} as h tends to 0 or that the velocity $\rightarrow 2 \text{ m s}^{-1}$ as $h \rightarrow 0$.

We may also say that the limit of the gradient of the line PQ as h

tends to 0 is 2 m s^{-1}. So the gradient of the curve $s = t^2$ at the point $(1, 1)$ is 2.

In this example, the gradient has been linked to a distance–time graph, but obviously the same argument would apply to a graph between x and y. So the gradient of the curve $y = x^2$ at the point $(1, 1)$ is 2.

The gradient of $y = x^2$ at any point

In Fig. 16.4, suppose that P is the point at time t and Q the point at time $(t + h)$.

The corresponding distances at these times are t^2 and $(t + h)^2$.

$$\therefore \ QN = (t + h)^2 - t^2 = 2th + h^2,$$

and
$$PN = (t + h) - t = h.$$

The gradient of $PQ = \dfrac{2th + h^2}{h} \text{ m s}^{-1}$

$$= (2t + h) \text{ m s}^{-1} \text{ (if } h \text{ is not zero).}$$

As $h \to 0$, the gradient tends to $2t$.

The gradient of the tangent at t is $2t$ or the gradient of the curve $s = t^2$ at the point (t, s) is $2t$.

Similarly the gradient of the curve $y = x^2$ at the point (x, y) is $2x$. It follows that the gradient of this curve at the point $(2, 4)$ is 4; its gradient at the point $(3, 9)$ is 6 and so on.

The gradient of the line $y = k$

Any two points on this line are (a, k) and (b, k). The gradient of the line is therefore $\dfrac{k - k}{b - a}$ or 0.

This line is parallel to the axis of x and the gradient of such a line is always zero.

The gradient of the line $y = mx$

Any two points on this line are (a, ma) and (b, mb). The gradient of the line is therefore

$$\frac{mb - ma}{b - a} = \frac{m(b - a)}{b - a}.$$

Such a line passes through the origin and its gradient is given by finding the ratio of y to x.

The gradient of the line $y = mx + c$

Any two points on this line are $(a, ma + c)$ and $(b, mb + c)$. The gradient of this line is therefore

$$\frac{(mb + c) - (ma + c)}{b - a} = \frac{m(b - a)}{b - a} = m$$

This suggests that the gradient of the line $y = mx + c$ could be found by adding the gradient of the line $y = mx$ which is m and the gradient of the line $y = c$ which is zero. Let us investigate whether this rule would apply to a more complicated example.

Example. *Find the gradient of the curve $y = 3x^2 - 2x + 5$ at the point (x, y).*

Suppose that (x, y) and $(x + h, y + k)$ are two points on the curve.

Then
$$y + k = 3(x + h)^2 - 2(x + h) + 5 \qquad \text{(i)}$$
and
$$y = 3x^2 - 2x + 5 \qquad \text{(ii)}$$

Subtracting (ii) from (i), $k = 6xh + 3h^2 - 2h.$

Therefore, $\dfrac{k}{h} = 6x + 3h - 2$ (if h is not zero).

$\dfrac{k}{h}$ is the gradient of the chord joining the two points and tends to $(6x - 2)$ as h tends to zero. The gradient of the curve at the point (x, y) is therefore $(6x - 2)$.

For example, $(2, 13)$ is a point on the curve. The gradient at this point is found by putting $x = 2$ in the formula for the gradient $(6x - 2)$. The gradient at this point is 10.

The same result would have been obtained by breaking up the equation $y = 3x^2 - 2x + 5$ into the three components $3x^2$, $-2x$ and 5.

The gradient of the curve $y = 3x^2$ is $3(2x)$; the gradient of the line $y = -2x$ is -2 and the gradient of the line $y = 5$ is 0. The sum of these three gradients is $6x - 2$ and agrees with the result found from first principles.

Exercise 16a

Find the gradients of the following curves at the points specified:

1. $y = 2x^2$ at $(1, 2)$.
2. $y = x^2 + 4x$ at $(1, 5)$.
3. $y = 3x^2 - 2x$ at $(2, 8)$.
4. $y = 5x^2 + 1$ at $(2, 21)$.
5. $y = 4x^2$ at $(3, 36)$.
6. $y = 7x^2 - 8x$ at $(2, 12)$.
7. $y = x^2 + x + 1$ at $(1, 3)$.
8. $y = x^2 - x - 1$ at $(2, 1)$.
9. $y = 3x^2 + 4x + 5$ at (x', y').
10. $y = ax^2 + bx + c$ at (x', y').
11. If $s = t^2$, find the velocity when $t = 3$.
12. If $s = 2t^2 + 3$, find the velocity when $t = 2$.
13. If $s = 4t^2 + 3t + 2$, find the velocity after 2 seconds.
14. If $s = 5t^2 - 4t$, find the velocity after 3 seconds.
15. If $s = 4t^2 - 6t$, find the average velocity over the first 4 seconds.

16. If $s = 4t^2 - 3t$, find the average velocity between the times $t = 1$ and $t = 3$.

17. If the velocity, v m s^{-1}, and the time t seconds of a particle moving in a straight line are connected by the equation $v = 4t + 3$, find the acceleration of the particle.

18. If $v = t^2 + 4t$, find the initial acceleration of the particle.

19. If $v = 2t^2 + 3t$, find the acceleration of the particle when $t = 2$.

20. If $s = 3t^2 + 6t$, find the velocity and the acceleration of the particle.

The line $y = mx + c$

We have already seen that the gradient of the line $y = mx + c$ is m. What does c represent? When $x = 0$, $y = c$ and therefore c is the intercept made by the line on the axis of y.

$y = mx + c$ is the line of gradient m which makes an intercept of c on the y-axis.

Example. The line $3x + 4y = 5$ may be put in the form $y = -\frac{3}{4}x + \frac{5}{4}$. The gradient of the line is $-\frac{3}{4}$ and the intercept made on the axis of y is $\frac{5}{4}$.

Equation of a line of given gradient through a given point

Suppose we wish to find the equation of the line of gradient 2 which passes through the point $(2, 3)$. Since the line has gradient 2, its equation must be of the form $y = 2x + c$.

Since the line passes through the point $(2, 3)$, the coordinates of the point satisfy the line.

$$\therefore 3 = 2(2) + c.$$
$$\therefore c = -1.$$

The equation of the line is $y = 2x - 1$.

Let us consider the general case of the line of gradient m which passes through the point (x', y').

The equation of the line is $y = mx + c$, where c must be chosen so that the point (x', y') lies on the line.

$$\therefore y' = mx' + c$$

or
$$c = y' - mx'.$$

The equation of the line is

$$y = mx + y' - mx'$$

or
$$y - y' = m(x - x').$$

Example. *Find the line of gradient 3 through the point $(1, -1)$.*

Substituting in the formula, the equation of the line is

$$y - (-1) = 3(x - 1)$$

or
$$y + 1 = 3x - 3$$

or
$$y = 3x - 4.$$

Equation of tangent at a given point to a curve

Earlier in the chapter, it was stated that the tangent at the point $(1, 0)$ to the curve $y = x^2 - x$ passes through the point $(3, 2)$. We shall now find the equation of the tangent at the given point to this curve and verify that it does pass through the point $(3, 2)$. We have already proved that the gradient of the tangent to the curve $y = x^2 - x$ at the point (x', y') is given by $2x' - 1$.

The gradient of the tangent at the point $(1, 0)$ is therefore $2 - 1$ or 1. The equation of the line of gradient 1 through the point $(1, 0)$ is

$$y - 0 = 1(x - 1)$$
or
$$y = x - 1.$$

By substitution, this line obviously passes through the point $(3, 2)$.

This is an example of the general method of finding the equation of the tangent at a given point to a curve. First the general formula for the gradient is found; the value of the gradient at the particular point is then calculated and the equation of the tangent found by substituting in the formula $y - y' = m(x - x')$.

Example. *Find the equation of the tangent at the point $(3, 2)$ to the curve* $y = x^2 - 4x + 5.$

The gradient of the tangent at the point (x', y') to the curve $y = x^2 - 4x + 5$ is $2x' - 4$.

At the point $(3, 2)$ the gradient of the tangent is therefore $2(3) - 4$ or 2. The equation of the tangent is

$$y - 2 = 2(x - 3)$$
or
$$y = 2x - 4.$$

Exercise 16b: Miscellaneous

1. Find the gradient of the tangent at the point $(2, -2)$ to the curve $y = x^2 - 3x$.

2. Find the gradient of the tangent at the point $(1, 2)$ to the curve $y = 3x^2 - 1$.

3. Find the gradient of the tangent at the point $(2, 18)$ to the curve $y = 4x^2 + x$.

4. Find the equation of the line of gradient 3 through the point $(4, 2)$.

5. Find the equation of the line of gradient 4 which makes an intercept of 3 on the y-axis.

6. Find the equation of the line of gradient 2 which makes an intercept of -2 on the y-axis.

7. Find the equation of the line of gradient 1 which makes an intercept of 3 on the x-axis.

8. Find the equation of the line of gradient -2 which makes an intercept of -2 on the x-axis.

9. Find the equation of the tangent at the point $(2, 0)$ to the curve $y = x^2 - 2x$.

10. Find the point in which the tangent to the curve $y = 3x^2 - x$ at the point $(1, 2)$ cuts the axis of x.

11. Find the point in which the tangent to the curve $y = x^2 - 4x + 1$ at the point $(2, -3)$ cuts the axis of y.

12. Find the coordinates of the point on $y = x^2 - x$ at which the tangent is parallel to the axis of x.

13. Find the coordinates of the point on $y = 2x^2 - 3x$ at which the tangent has gradient 1.

14. If distance s in metres and time t in seconds are connected by the equation $s = 4t^2 - 2t$, show that the particle has a constant acceleration and find it.

15. Given that $s = 2t - 3t^2$, show that the particle has constant retardation and find it.

16. Find the point of intersection of the tangents at the points $(1, 1)$ and $(2, 0)$ to the curve $y = 2x - x^2$.

17. Find the point of intersection of the tangents at the points $(1, 1)$ and $(2, 4)$ to the curve $y = x^2$.

18. A point moves in a straight line so that the distance from a fixed point O on the line is $(16t - 2t^2)$ metres after t s. Show that it moves away from O for 4 seconds.

19. The distance s metres moved by a point on a straight line in t s is given by $s = t^2 + 3t$. Find its velocity after 2 seconds.

20. The velocity v metres per second of a point moving in a straight line after t s is given by $v = 3t^2 - t$. Find the acceleration after 2 s.

21. Find the equation of the line through the point $(1, 2)$ parallel to the line $y = 2x + 3$.

22. Find the equation of the tangent at the point (t, t^2) to the curve $y = x^2$. Find the equations of the tangents to this curve which pass through the point $(1, -3)$.

23. If $s = 6 + 2t - t^2$, find the velocity and acceleration when $t = 2$.

24. The graph of $y = Ax^2 + Bx + C$ passes through the origin and its gradient there is 3. The graph also passes through the point $(1, 1)$. Find A, B and C.

25. The graph of $y = ax^2 + bx + c$ passes through the origin. The gradient of the tangent when $x = 1$ is 4 and when $x = 2$ is 5. Find a, b and c.

17 The Derived Function

The derived function

The gradient of the tangent to a curve at the point (x, y) on it is called its derived function. We have already seen that the derived function of x^2 is $2x$ and that the derived function of $4x^2 - 2x + 3$ may be found by adding the derived functions of $4x^2$, $-2x$ and 3. The derived function of $4x^2 - 2x + 3$ is therefore $4(2x) - 2$ or $8x - 2$.

The derived function of x^2 is written $D(x^2)$ and the derived function of $4x^2 - 2x + 3$ is written $D(4x^2 - 2x + 3)$.

$$\therefore\ D(x^2) = 2x$$

and
$$D(4x^2 - 2x + 3) = 8x - 2.$$

The derived function of a constant

In the last chapter, we saw that a line such as $y = k$ is a line parallel to the x-axis and has zero gradient. Its derived function is therefore 0.

$$\therefore\ D(\text{any constant}) = 0.$$

The derived function of $mx + c$

We know that $y = mx + c$ where m and c are constants is a line of gradient m. The derived function of $mx + c$ is therefore m.

$$\therefore\ D(mx + c) = m.$$

The derived function of $ax^2 + bx + c$

Suppose that (x, y) and $(x + h, y + k)$ are two points on the curve $y = ax^2 + bx + c$, where a, b and c are constants. Then

$$y + k = a(x + h)^2 + b(x + h) + c.$$
$$= ax^2 + 2axh + ah^2 + bx + bh + c \qquad \text{(i)}$$
and
$$y = ax^2 + bx + c \qquad \text{(ii)}$$

Subtracting (ii) from (i),

$$k = 2axh + ah^2 + bh.$$

$$\therefore\ \frac{k}{h} = 2ax + ah + b \quad \text{(if h is not zero).}$$

$\frac{h}{k}$ is the gradient of the chord joining the two points and the limit of the gradient as h tends to 0 is $(2ax + b)$.

$$\therefore\ D(ax^2 + bx + c) = 2ax + b.$$

This gives a formal proof that

$$D(ax^2 + bx + c) = D(ax^2) + D(bx) + D(c)$$
$$= aD(x^2) + bD(x) + D(c)$$
$$= a(2x) + b(1) + 0$$
$$= 2ax + b.$$

Example 1. *Find the derived function of* $7x^2 - 8x + 15$.

$$D(7x^2 - 8x + 15) = 7D(x^2) - 8D(x) + D(15)$$
$$= 7(2x) - 8(1) + 0$$
$$= 14x - 8.$$

Example 2. *Find the derived function of* $(4x - 1)(3x + 2)$.

At this stage, the product must be evaluated before the derived function can be found

$$(4x - 1)(3x + 2) = 12x^2 + 5x - 2.$$
$$D(12x^2 + 5x - 2) = 12D(x^2) + 5D(x) - D(2)$$
$$= 12(2x) + 5(1) - 0$$
$$= 24x + 5.$$

Exercise 17a

Find the derived functions of the following:

1. $x(x + 1)$.   2. $4x^2 - 6x + 5$.   3. $(2x + 1)(x - 1)$.

4. $2x(x - 1)$.   5. $x + \frac{1}{2}x^2$.   6. $x\left(x + \frac{2}{x}\right)$.

7. $x(x + 2) - x$.   8. $(2x + 3)^2$.   9. $(3x - 4)^2$.

10. $(x + 2)(1 - x)$.   11. $(3x - 1)(4x + 6)$.   12. $x(x + 3)$.

13. $5x^2 + 14x - 1$.   14. $2x\left(x + \frac{3}{x}\right)$.   15. $x(5x - 7)$.

The derived function of x^3

Suppose that (x, y) and $(x + h, y + k)$ are two points on the curve $y = x^3$. Then

$$y + k = (x + h)^3 = x^3 + 3x^2h + 3xh^2 + h^3 \qquad \text{(i)}$$

and
$$y = x^3 \qquad \text{(ii)}$$

Subtracting (ii) from (i):

$$k = 3x^2h + 3xh^2 + h^3.$$

$$\therefore \frac{k}{h} = 3x^2 + 3xh + h^2 \quad \text{(if } h \text{ is not zero)}.$$

But $\dfrac{k}{h}$ is the gradient of the chord joining the two points and the limit of the gradient as h tends to 0 is $3x^2$.

$$\therefore D(x^3) = 3x^2.$$

The derived function of 1/x

Suppose that (x, y) and $(x + h, y + k)$ are two points on the curve $y = \dfrac{1}{x}$. Then

$$y + k = \frac{1}{x + h} \tag{i}$$

and

$$y = \frac{1}{x} \tag{ii}$$

Subtracting (ii) from (i):

$$k = \frac{1}{x + h} - \frac{1}{x} = \frac{x - (x + h)}{x(x + h)}$$

$$= -\frac{h}{x(x + h)}.$$

$$\therefore \frac{k}{h} = -\frac{1}{x(x + h)} \quad \text{(if } h \text{ is not zero).}$$

But $\dfrac{k}{h}$ is the gradient of the chord joining the two points and the

limit of the gradient as h tends to 0 is $-\dfrac{1}{x^2}$.

$$\therefore D\left(\frac{1}{x}\right) = -\frac{1}{x^2}.$$

The derived function of 1/x²

Suppose that (x, y) and $(x + h, y + k)$ are two points on the curve $y = \dfrac{1}{x^2}$. Then

$$y + k = \frac{1}{(x + h)^2} \tag{i}$$

and

$$y = \frac{1}{x^2} \tag{ii}$$

Subtracting (ii) from (i):

$$k = \frac{1}{(x + h)^2} - \frac{1}{x^2} = \frac{x^2 - (x + h)^2}{x^2(x + h)^2}$$

$$= \frac{-2xh - h^2}{x^2(x + h)^2}.$$

$$\therefore \frac{k}{h} = \frac{-2x - h}{x^2(x + h)^2} \quad \text{(if } h \text{ is not zero).}$$

$\dfrac{k}{h}$ is the gradient of the line joining the two points, and as h tends to

zero, the gradient tends to $-\dfrac{2x}{x^4}$ or $-\dfrac{2}{x^3}$.

$$\therefore D\left(\frac{1}{x^2}\right) = -\frac{2}{x^3}.$$

The general rule

Tabulating the results already proved and remembering that $x^1 = x$ and $x^0 = 1$:

$$D(x^3) = 3x^2.$$
$$D(x^2) = 2x.$$
$$D(x^1) = 1.$$
$$D(x^0) = 0.$$
$$D(x^{-1}) = -1x^{-2}.$$
$$D(x^{-2}) = -2x^{-3}.$$

It will be seen that all these results obey the general rule

$$D(x^n) = nx^{n-1}.$$

Examples. $D(x^4) = 4x^3$; $D\left(\dfrac{1}{x^3}\right) = D(x^{-3}) = -3x^{-4} = -\dfrac{3}{x^4}$;

$$D(\sqrt{x}) = D(x^{\frac{1}{2}}) = \tfrac{1}{2}x^{-\frac{1}{2}} = \frac{1}{2\sqrt{x}}.$$

The proof of the general rule is now given; this may be omitted at this stage if the binomial theorem has not yet been studied.

The derived function of x^n

Suppose that (x, y) and $(x + h, y + k)$ are two points on the curve $y = x^n$. Then

$$y + k = (x + h)^n \qquad\qquad\text{(i)}$$

and $$y = x^n \qquad\qquad\text{(ii)}$$

Subtracting (ii) from (i):

$$k = (x + h)^n - x^n$$

$$= x^n\left(1 + \frac{h}{x}\right)^n - x^n$$

$$= x^n\left(1 + n\frac{h}{x} + \text{higher powers of } h\right) - x^n.$$

This is true for all values of n provided that $\dfrac{h}{x}$ is small.

$$\therefore k = nhx^{n-1} + \text{higher powers of } h$$

and

$$\frac{k}{h} = nx^{n-1} + \text{positive powers of } h.$$

$\dfrac{k}{h}$ is the gradient of the chord joining the two points and as h tends to zero, the gradient tends to nx^{n-1}.

$$\therefore D(x^n) = nx^{n-1} \quad \text{for all values of } n.$$

Notation

If $f(x)$ denotes any function of x, that is any expression containing x, we have called its derived function $D\{f(x)\}$. The derived function of $f(x)$ may also be called $f'(x)$.

If $y = f(x)$, the most common way of expressing its derived function is $\dfrac{dy}{dx}$. This notation will be used in future. The reader must realize that $\dfrac{dy}{dx}$ is not an ordinary fraction which may be cancelled; that the d of dy is not a multiple (compare $\sin y$) and that the dy cannot be separated from its denominator, dx. The expression $\dfrac{dy}{dx}$ actually compares the rate of growth of y with that of x.

Finding the derived function of an expression is called differentiating the expression and the result may be called its differential co-efficient. The advantage of the notation $\dfrac{dy}{dx}$ is that it tells us what quantities are being compared; in fact, that we have differentiated y with respect to x. For example if $z = t^3 - t^2$, then $\dfrac{dz}{dt} = 3t^2 - 2t$, or the result may be written in the form $\dfrac{d}{dt}(t^3 - t^2) = 3t^2 - 2t$.

The increment notation

The conventional notation is that δx stands for a small increase in x and that δy stands for the corresponding small increment in y (which may be positive or negative). If P is the point (x, y) on a given curve, another point Q close to it is said to have coordinates $(x + \delta x, y + dy)$.

The gradient of the chord $PQ = \dfrac{\text{increase of } y}{\text{increase of } x} = \dfrac{\delta y}{\delta x}$

where δx and δy are both small quantities and connected in such a way that both P and Q are points of the given curve. As the point Q gets nearer to P on the curve, δx and δy will both become smaller but the result of dividing δy by δx will not necessarily be small. This ratio $\dfrac{\delta y}{\delta x}$ will get nearer and nearer to a limit which is the gradient of the limiting chord PQ as Q approaches P. This limiting chord is the tangent at P and so the limiting value of $\dfrac{\delta y}{\delta x}$ as δx and δy both become small is the gradient of the tangent to the curve at the point (x, y).

So far no new ideas have been introduced except that δx and δy are small quantities. The limiting value of $\dfrac{\delta y}{\delta x}$ is called $\dfrac{dy}{dx}$ and here it is important to realize that dy and dx are no longer quantities as were δx and δy but that $\dfrac{dy}{dx}$ represents the value which the fraction $\dfrac{\delta y}{\delta x}$ approaches as δx and δy become very small.

The quantities dx and dy for example cannot be cancelled. Perhaps it is more instructive to write $\dfrac{d}{dx}(y)$ which shows that we have a function of y just as $\sin y$, $\log y$ and $\sqrt{y}$ are functions of y, and that $\dfrac{d}{dx}$ operates in some way upon y just as $\sin$ or $\log$ does.

The differential coefficient

The quantity $\dfrac{dy}{dx}$ or $\dfrac{d}{dx}(y)$ or Dy is called the **differential coefficient** (in future, written as D.C.) of y with respect to x, or derivative of y with respect to x. It can exist only if y is a function of x and tells us the rate of increase of y compared with that of x at the particular point of the curve, or the gradient of the tangent to the curve at that point. If $\dfrac{dy}{dx}$ is negative, it means that y must decrease as x increases.

The result of differentiating $f(x)$ with respect to x may be written as $f'(x)$ or $f_1(x)$.

Worked examples

Example 1. *Find the derived function of* $4x^3 - 3x^2 + 7x - 8$.
$$D(4x^3 - 3x^2 + 7x - 8) = 4D(x^3) - 3D(x^2) + 7D(x) - D(8)$$
$$= 4(3x^2) - 3(2x) + 7(1) - 0$$
$$= 12x^2 - 6x + 7.$$

Alternative method

Let
$$y = 4x^3 - 3x^2 + 7x - 8.$$

Then
$$\frac{dy}{dx} = 4(3x^2) - 3(2x) + 7(1) - 0$$

$$= 12x^2 - 6x + 7$$

or simply
$$\frac{d}{dx}(4x^3 - 3x^2 + 7x - 8) = 12x^2 - 6x + 7.$$

Example 2. *Find the derived function of* $\dfrac{3}{x^3} - \dfrac{1}{\sqrt{x}}$.

$$D\left(\frac{3}{x^3} - \frac{1}{\sqrt{x}}\right) = D(3x^{-3} - x^{-\frac{1}{2}}) = 3(-3x^{-4}) - (-\tfrac{1}{2})(x^{-\frac{3}{2}})$$

$$= -9x^{-4} + \tfrac{1}{2}x^{-\frac{3}{2}} = -\frac{9}{x^4} + \frac{1}{2\sqrt{x^3}}.$$

Alternatively, if $y = \dfrac{3}{x^3} - \dfrac{1}{\sqrt{x}}$,

$$\frac{dy}{dx} = -\frac{9}{x^4} + \frac{1}{2\sqrt{x^3}}$$

or
$$\frac{d}{dx}\left(\frac{3}{x^3} - \frac{1}{\sqrt{x}}\right) = -\frac{9}{x^4} + \frac{1}{2\sqrt{x^3}}.$$

Example 3. *Find the derived function of* $\dfrac{(1 + x)(1 + x^2)}{x}$.

Let $y = \dfrac{(1 + x)(1 + x^2)}{x}$.

Then $y = \dfrac{1 + x + x^2 + x^3}{x} = \dfrac{1}{x} + 1 + x + x^2 = x^{-1} + 1 + x + x^2$.

$$\therefore \frac{dy}{dx} = -1x^{-2} + 1 + 2x = -\frac{1}{x^2} + 1 + 2x.$$

Example 4. *Find the equation of the tangent to the curve $y = x^3 - x^2$ at the point $(1, 0)$.*

Since
$$y = x^3 - x^2,$$

$$\frac{dy}{dx} = 3x^2 - 2x.$$

The gradient of the tangent at $(1, 0)$ is found by putting $x = 1$ and is $3(1)^2 - 2(1)$ or 1.

The tangent has gradient 1 and passes through the point $(1, 0)$.

Its equation is therefore

$$y - 0 = 1(x - 1) \quad \text{or} \quad y = x - 1.$$

Exercise 17b: Miscellaneous

Find the derived functions of:

1. $3x^3 - 2x^2 - 4x - 1$.

2. $5x^4 - 6x^2 - 1$.

3. $3x^5 - 4x^3 - 7x$.

4. $1 + \dfrac{2}{x^2}$.

5. $\left(2 + \dfrac{1}{x}\right)^2$.

6. $(x^2 - 1)^2$.

7. $\sqrt{x} + \dfrac{1}{\sqrt{x}}$.

8. $\dfrac{x + 2}{\sqrt{x}}$.

9. $\dfrac{(x + 1)(\sqrt{x} + 1)}{\sqrt{x}}$.

10. $(x^2 + 1)(2x - 1)$.

Find the following:

11. $\dfrac{d}{dt}(3t^3 - 7t - 2)$.

12. $\dfrac{d}{dy}(1 - 2y - 5y^3)$.

13. $\dfrac{d}{dz}\left(\dfrac{1}{z} - \dfrac{1}{z^2}\right)$.

14. $\dfrac{d}{dx}\left(1 + x + \dfrac{1}{\sqrt{x}}\right)$.

15. $\dfrac{d}{du}\{(u + 1)(u + 2)(u + 3)\}$.

16. $\dfrac{d}{dz}(z^3 - 3z + 1)$.

17. $\dfrac{d}{dv}\left\{\dfrac{v + 1}{v}\right\}$.

18. $\dfrac{d}{dt}(t^5 - t)$.

19. $\dfrac{d}{dy}\left(\dfrac{1}{y^2} - \dfrac{1}{y^3}\right)$.

20. $\dfrac{d}{dx}\left\{\dfrac{(x + 2)(x + 3)}{x}\right\}$.

Differentiate with respect to t:

21. $1 + 5t - t^2 + t^3$. **22.** $\dfrac{1}{t^2} + \dfrac{1}{t} + 1$. **23.** $(t + 2)(t - 1)$.

24. $\dfrac{t^2 + t + 1}{t}$.   **25.** $(t + 1)(2t^2 - 1)$.

Differentiate with respect to y:

26. $\dfrac{1}{y^6}$.   **27.** $y^{15} - y^{13}$.   **28.** $y^{7/2}$

29. $\dfrac{1}{y^5} - \dfrac{1}{y^4}$.   **30.** $(y + 1)^3$.

31. Find the gradient of the tangent to the curve $y = x^3 - 3x^2$ at the point $(1, -2)$.

32. Find the gradient of the curve $y = x^3 + x^2$ at the point $(2, 12)$.

33. Find the coordinates of the points on the curve $y = x^3 - 3x$ at which the tangent is parallel to the axis of x.

34. Find the coordinates of the points on the curve $y = x^3$ at which the tangent has gradient 3.

35. Find the equation of the tangent to the curve $y = x^3 - 5x$ at the point where $x = 2$.

36. Find the equations of the tangents to the curve $y = x^3 - 12x$ which are parallel to the axis of x.

37. Given that $xy = 4$, find the value of $\dfrac{dy}{dx}$ when $x = 1$.

38. Find the equation of the tangent to the curve $xy = 1$ at the point $(1, 1)$.

39. Find the equation of the tangent to the curve $yx^2 = 1$ at the point $(1, 1)$.

40. Given that $u + v = 8$, find $\dfrac{du}{dv}$.

41. If $y^2 = x$, find $\dfrac{dy}{dx}$ (i) in terms of x, (ii) in terms of y.

42. Given that $pv = 50$, find the value of $\dfrac{dp}{dv}$ when $v = 5$.

43. Find the gradient of the tangent to the curve
$$y = x^3 - 4x^2 + 3x + 1$$
at the point where $x = -1$.

44. Given that $f(t) = 4t^3 - 6t^2$, find $f'(t)$.

45. Find the points on the curve $y = x^3 - x^2$ at which the tangents are parallel to the axis of x.

46. If $\dfrac{dy}{dx}$ is differentiated with respect to x, the result is called the second differential coefficient of y with respect to x and is written $\dfrac{d^2y}{dx^2}$. Given that $y = x^3 - 3x^2 + 4x$, find $\dfrac{d^2y}{dx^2}$.

47. If $s = t^3 - 3t^2$, find $\dfrac{d^2s}{dt^2}$.

48. The differential coefficient of $f'(x)$ may be written $f''(x)$. Given that $f(x) = 4x^4 - 3x^2$, find $f''(x)$.

49. If $xy = 4$, find $\dfrac{d^2y}{dx^2}$.

50. If $x^2y = 12$, find the value of $\dfrac{d^2y}{dx^2}$ when $x = 2$.

18 Applications of the Derived Function

Maxima and minima

At the point P to the curve shown in Fig. 18.1, the tangent is parallel to the axis of x. The gradient of the tangent is zero and therefore the value of the derived function at this point is zero. Such a point is called a **maximum point.** The value of y at such a point is called a maximum value of y. The value of y at a maximum point is larger than at points on either side of it; it is not necessarily the greatest value of all. For example, in the curve shown, the value of y at P is less than that at R.

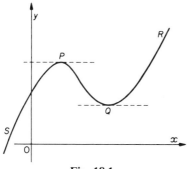

Fig. 18.1

The gradient at the point Q is also zero. At this point, the value of y is less than at points immediately on either side of it. It is not the least of all possible values of y, because, for example, the value of y at S is smaller still. Such a point is called a **minimum point** and the value of y at such a point is called a minimum value of y.

At a point near P and between S and P, the gradient of the curve is positive; at a point between P and Q, the gradient of the curve is negative.

So, at a maximum point, the gradient changes from positive to negative through the zero value.

At a point near Q between P and Q, the gradient is negative; at Q it is zero and between Q and R it is positive.

So, at a minimum point, the gradient changes from negative to positive through the zero value.

Maximum and minimum points are called **turning points**; the values of y at these points are called turning values.

At a maximum point, $\frac{dy}{dx}$ is zero; the value of $\frac{dy}{dx}$ changes from positive to negative as x increases through the point.

At a minimum point, $\frac{dy}{dx}$ is also zero; but the value of $\frac{dy}{dx}$ changes from negative to positive as x increases through the point.

There is one other type of point at which $\frac{dy}{dx} = 0$. This is shown in Fig. 18.2. It is not a turning point because the gradient does not change sign (or turn).

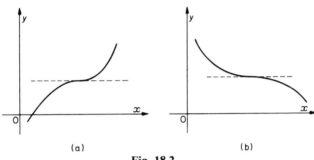

(a) (b)

Fig. 18.2

In Fig. 18.2(a), $\frac{dy}{dx}$ is positive both before and after its zero value.

In Fig. 18.2(b), $\frac{dy}{dx}$ is negative both before and after its zero value.

A point where $\frac{dy}{dx}$ is zero but does not change sign is called a **point of inflexion**.

Example 1. *Find the maximum or minimum value of $x^2 - 2x + 7$.*

If
$$y = x^2 - 2x + 7$$

$$\frac{dy}{dx} = 2x - 2.$$

$$\therefore \frac{dy}{dx} = 0 \quad \text{when } x = 1.$$

When x is slightly less than 1, $\frac{dy}{dx}$ is negative.

When $x = 1$, $\dfrac{dy}{dx}$ is zero.

When x is slightly greater than 1, $\dfrac{dy}{dx}$ is positive.

Therefore the point where $x = 1$ is a minimum point.
The value of y at this point is $1 - 2 + 7$ or 6.
Therefore the minimum value of $(x^2 - 2x + 7)$ is 6 and occurs when $x = 1$.

Alternative method

It is worth noting that this example may be solved by completing the square.

$$x^2 - 2x + 7 = (x - 1)^2 + 6.$$

$(x - 1)^2$ is a perfect square and therefore cannot be negative. Its smallest value, zero, occurs when $x = 1$ and the value of the expression is then 6.

Example 2. *Find the maximum and minimum values of $x^2(x - 2)$.*

If
$$y = x^2(x - 2) = x^3 - 2x^2$$

$$\frac{dy}{dx} = 3x^2 - 4x \quad \text{or} \quad x(3x - 4).$$

$$\frac{dy}{dx} = 0 \quad \text{when } x = 0 \text{ or when } x = \tfrac{4}{3}.$$

(i) When x is slightly less than 0, x is negative and $3x - 4$ is also negative. $\therefore \dfrac{dy}{dx}$ is positive.

When $x = 0$, $\dfrac{dy}{dx}$ is zero.

When x is slightly greater than 0, x is positive and $3x - 4$ is negative.

$\therefore \dfrac{dy}{dx}$ is negative.

So at $x = 0$, $\dfrac{dy}{dx}$ changes from positive to negative. This is a maximum point and the corresponding value of y is 0.

(ii) When x is slightly less than $\tfrac{4}{3}$, x is positive and $3x - 4$ is negative.

$\therefore \dfrac{dy}{dx}$ is negative.

When $x = \tfrac{4}{3}$, $\dfrac{dy}{dx} = 0$.

When x is slightly greater than $\tfrac{4}{3}$, x is positive and $3x - 4$ is also positive.

$\therefore \dfrac{dy}{dx}$ is positive.

So at $x = \frac{4}{3}$, $\dfrac{dy}{dx}$ changes from negative to positive.

This is a minimum point and the corresponding value of y is $(\frac{4}{3})^2(\frac{4}{3} - 2)$ or $\frac{16}{9}(-\frac{2}{3})$ or $-\frac{32}{27}$.

Therefore the maximum value of $x^2(x - 2)$ is 0 and the minimum value is $-\frac{32}{27}$.

From this information and the fact that the curve cuts the axis of x at the points where $x = 0$ and $x = 2$, it is possible to draw a rough sketch of the curve. This is shown in Fig. 18.3.

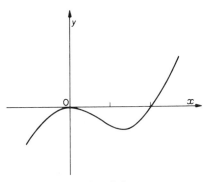

Fig. 18.3

Example 3. *Find the turning values of* $x^3(x - 2)$.

If
$$y = x^3(x - 2) = x^4 - 2x^3$$

$$\frac{dy}{dx} = 4x^3 - 6x^2 = 2x^2(2x - 3).$$

$$\therefore \frac{dy}{dx} = 0 \quad \text{when } x = 0 \text{ or } 1\tfrac{1}{2}.$$

(i) When x is slightly less than 0, x^2 is positive and $2x - 3$ is negative.

$\therefore \dfrac{dy}{dx}$ is negative.

When $x = 0$, $\dfrac{dy}{dx} = 0$.

When x is slightly greater than 0, x^2 is positive and $2x - 3$ is negative.

$\therefore \dfrac{dy}{dx}$ is negative.

Since $\dfrac{dy}{dx}$ does not change sign, $x = 0$ is a point of inflexion.

(ii) When x is slightly less than $1\frac{1}{2}$, x^2 is positive and $2x - 3$ is negative.

$\therefore \dfrac{dy}{dx}$ is negative.

When $x = 1\frac{1}{2}$, $\dfrac{dy}{dx} = 0$.

When x is slightly greater than $1\frac{1}{2}$, x^2 is positive and $2x - 3$ is positive.

$\therefore \dfrac{dy}{dx}$ is positive.

$\therefore \dfrac{dy}{dx}$ changes sign from negative to positive.

So $x = 1\frac{1}{2}$ gives a minimum point and the minimum value of y is $(1\frac{1}{2})^3(-\frac{1}{2})$ or $-\frac{27}{16}$.

The curve has no maximum point.
A sketch of the curve is given in Fig. 18.4.

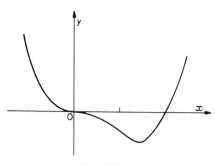

Fig. 18.4

The next two examples are problems concerning maxima and minima.

Example 4. *A rectangular box with a square base is to be made from sheet metal to have a volume of 216 cm³. Find the least area of sheet metal needed.*

Suppose the base of the box is of edge x cm and the height of the box is h cm. Then

$$x^2h = 216.$$

The area of sheet metal, A cm², is given by

$$A = 2x^2 + 4xh$$

Express A in terms of x only by substituting $h = \dfrac{216}{x^2}$. Then

$$A = 2x^2 + 4x\left(\frac{216}{x^2}\right)$$

$$= 2x^2 + \frac{864}{x}.$$

$$\therefore \frac{dA}{dx} = 4x - \frac{864}{x^2} = \frac{4}{x^2}(x^3 - 216).$$

$$\therefore \frac{dA}{dx} = 0 \quad \text{when } x^3 = 216 \text{ or } x = 6.$$

When x is just less than 6, $\dfrac{dA}{dx}$ is negative.

When x is just greater than 6, $\dfrac{dA}{dx}$ is positive.

$$\therefore x = 6 \text{ gives a minimum.}$$

When $x = 6$, $A = 2(36) + \dfrac{864}{6} = 72 + 144 = 216$.

The least area of sheet metal is 216 cm^2.

Example 5. *A right circular cylinder is to be made so that the sum of its diameter and height is 3 metres. Prove that, when the cylinder has maximum volume, the height is equal to the radius.*

Let r metres be the radius and h metres the height. Then
$$2r + h = 3 \quad \text{or} \quad h = 3 - 2r.$$
Let V be the volume in m^3. Then
$$V = \pi r^2 h$$
$$= \pi r^2 (3 - 2r)$$
$$= \pi (3r^2 - 2r^3)$$
$$\therefore \frac{dV}{dr} = \pi(6r - 6r^2) = 6\pi r(1 - r)$$

For a maximum, $r = 0$ or $r = 1$.
$r = 0$ gives $V = 0$ which is obviously not a maximum.

When r is just less than 1, $\dfrac{dV}{dr}$ is positive.

When r is just greater than 1, $\dfrac{dV}{dr}$ is negative.

$$\therefore r = 1 \text{ gives a maximum.}$$

When the radius is 1 metre, the height is 1 metre. Therefore the height is equal to the radius.

Exercise 18a

Find the maximum and minimum values of the following:

1. $x^2 - 4x$.   2. $1 - 6x - x^2$.

3. $x^2 - 8x + 5$.   4. $3x^2 - x^3$.

5. $9x + \dfrac{1}{x}$.   6. $x + \dfrac{2}{\sqrt{x}}$.

7. $2x^3 - 9x^2 + 12x + 4$.   8. $x^3 - 27x$.

9. $48x - x^3$.   10. $2x^3 - 3x^2$.

11. The sum of two numbers is 36. Find their maximum product.

12. A rectangle has a perimeter of 80 cm. Find its maximum area.

13. The volume of a right circular cylinder open at one end is to be 27π cm^3. Find its minimum surface area.

14. The volume of a right circular cylinder closed at both ends is to be 54π cm^3. Find its minimum surface area.

15. A farmer encloses sheep in a rectangular pen using hurdles for three sides and a long wall for the other side. If he has 80 metres of hurdles, find the greatest area he can enclose.

16. A rectangular box is to have a volume of 72 cm^3. If its length is twice its breadth, find its least possible surface area.

17. Find the equation of the line through the point $(1, 2)$ of gradient m. Find also the area of the smallest triangle that this line and the axes can form in the first quadrant.

18. The sum of the base radius and the height of a right circular cone is to be 6 metres. Find the greatest possible volume of the cone.

19. A square sheet of metal has sides of length 10 cm. Equal square pieces are removed from each corner and the remaining piece is bent into the form of an open box. Find the maximum volume of the box.

20. Find the equation of the tangent at the point $(-at^2, 2at)$ to the curve $y^2 = -4ax$. If this tangent is such that the sum of the intercepts made by it on the axes is least, find the value of t.

Velocity and acceleration

We have already seen that, when distance is plotted against time, the gradient is equal to the velocity. Velocity is therefore equal to the rate of change of the distance, or

$$v = \frac{ds}{dt}.$$

Similarly, in a velocity–time graph, the gradient gives the rate of change of velocity or the acceleration.

$$\therefore a = \frac{dv}{dt} = \frac{d^2s}{dt^2}.$$

$\dfrac{d^2 s}{dt^2}$ is the notation used for the second differential coefficient of s with respect to t and is found by differentiating $\dfrac{ds}{dt}$ with respect to t.

If the velocity of a particle increases from 3 m s^{-1} to 7 m s^{-1} in 2 seconds, the increase of velocity is 4 m s^{-1} in 2 seconds and the average acceleration is said to be 2 m s^{-2}. If a body is slowing down, it has a deceleration or a **retardation**.

Example. *The distance s metres travelled by a body moving in a straight line in t s is given by* $s = 3t^3 - 4t^2$. *Find the velocity after 2 seconds and the initial acceleration.*

$$s = 3t^3 - 4t^2.$$

$$v = \frac{ds}{dt} = 9t^2 - 8t.$$

$$a = \frac{dv}{dt} = 18t - 8.$$

The velocity after 2 s $= 9(2^2) - 8(2)$ or 20 m s^{-1}.
The initial acceleration is the acceleration when $t = 0$ and is -8 m s^{-2}.

Rate of change

$\dfrac{dy}{dx}$ compares the rate of change of y with respect to that of x. If $\dfrac{dy}{dx} = 6$, y is increasing 6 times as fast as x, numerically. If x and y are distances and x is increasing at 4 m s^{-1}, then y is increasing at 24 m s^{-1}.

If $\dfrac{dy}{dx} = -6$, then y decreases 6 times as fast as x increases.

Example 1. *The radius of a circle is increasing at the rate of 0.1 cm s^{-1}. Find the rate of increase of the area of the circle when its radius is 5 cm.*

$$A = \pi r^2.$$

$$\therefore \frac{dA}{dr} = 2\pi r.$$

When $r = 5$,
$$\frac{dA}{dr} = 10\pi.$$

The area increases 10π times as fast as the radius.
The radius increases at 0.1 cm s^{-1}.
Therefore the area increases at $10\pi \times 0.1$ cm^2 s^{-1} or at π cm^2 s^{-1}.

Example 2. *When a metal cube is heated, the length of each side increases at the rate of 0.05 cm s⁻¹. Find the rate of increase of the volume of the cube if the length of its edge is 12 cm.*

If x cm is the length of edge and V cm³ the volume,

$$V = x^3.$$

$$\therefore \frac{dV}{dx} = 3x^2.$$

When $x = 12$,
$$\frac{dV}{dx} = 3 \times 144 = 432.$$

Therefore, numerically, the volume increases 432 times as fast as the length of edge.

The edge increases at 0.05 cm s⁻¹.

Therefore the volume increases at 432×0.05 cm s⁻¹.

The volume increases at 21.6 cm³ s⁻¹.

Exercise 18b

1. The side of a square is increasing at the rate of 0.1 cm s⁻¹. Find the rate of increase of the perimeter of the square.

2. The radius of a circle is increasing at the rate of 0.2 cm s⁻¹. Find the rate of increase of the circumference.

3. The radius of a circle is increasing at the rate of 0.4 cm s⁻¹. Find the rate of increase of the area when the radius is 5 cm.

4. The radius of a sphere is increasing at the rate of 0.5 cm s⁻¹. Find the rate of increase of the surface area of the sphere when its radius is 4 cm.

5. The side of a square is 4 cm long and is decreasing at the rate of 0.1 cm s⁻¹. Find the rate of decrease of the area.

6. The radius of a soap bubble is increasing at 0.3 cm s⁻¹. Find the rate of increase of its volume when the radius is 5 cm.

7. The line $y + tx = t^2$ cuts the x-axis at A and the y-axis at B. If OA is 4 cm long and is increasing at the rate of 0.2 cm s⁻¹, find the rate of increase of OB.

8. The distance s m and the time t s are connected by the equation $s = t^3 - 2t$. Find the acceleration after 2 seconds.

9. Given that $s = ut + \frac{1}{2}at^2$ where u and a are constants, show that the acceleration is constant and equal to a.

10. The velocity of a particle is proportional to the square of the time. If the velocity is 4 m s⁻¹ when the time is 2 s, find the acceleration at that time.

Exercise 18c: Miscellaneous

1. Find the gradient of the curve $y = x^3 - 3x^2 + 2x$ at the points where it cuts the x-axis and hence sketch the curve.

2. Find the maximum and minimum values of $2x^3 - 3x^2 - 12x + 6$.

3. The velocity, v m s^{-1}, of a particle moving in a straight line t s after the beginning of the motion is given by $v = 2 + 4t + t^3$. Find the initial acceleration.

4. The graph of $y = Ax + Bx^2$ passes through the point $(2, 0)$ and its gradient at that point is -2. Calculate the maximum value of y.

5. The distance s metres moved by a particle travelling in a straight line in t s is given by $s = 2t + t^3$. Calculate.

   (i) the average velocity over the first 3 seconds;
   (ii) the velocity after 3 seconds.

6. Calculate the gradient of the tangent to the curve $y = 4x + \dfrac{1}{x}$ at the point $(\frac{1}{2}, 4)$. Calculate also the coordinates of the other point on the curve at which the tangent is parallel to that at $(\frac{1}{2}, 4)$.

7. Show that the differential coefficient of $x^3 - 6x^2 + 12x + 5$ is never negative. Hence show that the expression is never less than 5 when x is positive.

8. If a ship consumes fuel at the rate of $(2000 + 20v^2)$ kg h^{-1} where v is the speed in km h^{-1}, find the least amount of fuel needed for a journey of 1000 km.

9. Find the maximum and minimum values of $(x + 1)^2(2x - 7)$.

10. The perimeter of a sector of a circle of radius r and angle x radians is $(2r + rx)$ and its area is $\frac{1}{2}r^2x$. Show that, if the perimeter is fixed, the area will be a maximum when $x = 2$.

11. The least value of $x^2 + ax + b$ is 4 when $x = 2$. Find the values of a and b.

12. The greatest value of $p + qx - x^2$ is 5 and occurs when $x = 3$. Find p and q.

13. If $y^2 = 16x$, find $\dfrac{dy}{dx}$ when $x = 4$.

14. Find the points on the graph of $y = 5x - x^2$ at which the tangent makes equal angles with the axes.

15. If $\dfrac{d}{dx}(x^3 - 1) = \dfrac{d}{dx}(x^2 - 1)$, find x.

16. Show that the gradient of $y = 1 + 2x + 2x^2$ at $x = 1$ is three times the gradient at the point where $x = 0$.

17. Two concentric circles are such that the radius of one is always three times that of the other. If the radius of the smaller is increasing at 0.2 cm s^{-1}, find the rate at which the area between the circles is increasing when the radius of the larger is 3 cm.

18. Water is poured into an inverted cone of semi-vertical angle $45°$ at a constant rate of 4 cm^3 s^{-1}. At what rate is the depth of water increasing when it is 1 cm deep?

19. The volume of a cap of a sphere of height h is given by $V = \dfrac{\pi h^2}{3}(3r - h)$, where r is the radius. Show that for a given value of r, the volume is a maximum when $h = 2r$. Is this what you would expect?

20. A rectangular box whose length is three times its breadth is to have a volume of 288 cm^3. Find its least possible surface area.

19 Integration

The inverse of differentiation

We know that the derived function of x is 1. Therefore, if we are asked to find an expression whose derived function is 1, one answer is x. But it is not the only answer, because $x + 1$, $x + 2$, and $x + 19\frac{3}{4}$ are all expressions whose derived functions equal 1. In fact, $(x + c)$ is the general solution, where c is called an **arbitrary constant** and may be equal to any number. Its precise value may only be found if we are given further information.

Figure 19.1 shows a number of straight lines each of gradient 1. Any line parallel to these lines will also have gradient 1. The equation of any such line is $y = x + c$. Therefore, if $\dfrac{dy}{dx} = 1$, $y = x + c$.

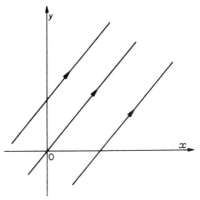

Fig. 19.1

If we are also told that $y = 4$ when $x = 1$, by substituting these values in $y = x + c$, we see that $4 = 1 + c$ or that $c = 3$. The solution in this case is therefore $y = x + 3$.

Again the derived function of x^2 is $2x$. The general expression whose derived function equals $2x$ is therefore $x^2 + c$, where c is an arbitrary constant. The curves $y = x^2 + c$ are 'parallel' curves and are shown in Fig. 19.2. The curves are all exactly the same shape with the vertex moved along the axis of y. All the curves have the same gradient for any given value of x.

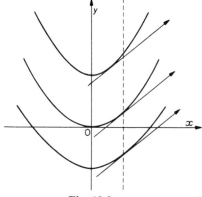

Fig. 19.2

The process of obtaining the function whose derived function is given is called integrating. Integration is the inverse of differentiation.

If $\dfrac{dy}{dx} = 1,$ $\qquad\qquad\qquad y = x + c.$

If $\dfrac{dy}{dx} = x,$ $\qquad\qquad\qquad y = \tfrac{1}{2}x^2 + c.$

If $\dfrac{dy}{dx} = x^n,$ $\qquad\qquad\qquad y = \dfrac{x^{n+1}}{n+1} + c.$

This applies for all values of n except -1.

Notation

If y is a function of x, $\displaystyle\int y\,dx$ stands for the integral of y with respect to

x. The integral sign $\displaystyle\int$ cannot be divorced from dx if the integral is with respect to x.

Examples.
$$\int x\,dx = \frac{x^2}{2} + c;$$

$$\int t^2\,dt = \frac{t^3}{3} + c;$$

$$\int (v^2 + v + 1)\,dv = \frac{v^3}{3} + \frac{v^2}{2} + v + c.$$

Example 1. *Integrate $5x^2 - 7x + 8$.*

Integrate each term separately.

The integral of x^2 is $\dfrac{x^3}{3}$.

The integral of x is $\frac{1}{2}x^2$.

The integral of 8 is $8x$.

The integral of $(5x^2 - 7x + 8)$ is $\frac{5}{3}x^3 - \frac{7}{2}x^2 + 8x + c$.

(Note that one arbitrary constant only is sufficient.)

Example 2. *Integrate $2\sqrt{x} - \dfrac{1}{x^2}$.*

The integral of $\sqrt{x}$ or $x^{\frac{1}{2}}$ is $\dfrac{x^{\frac{3}{2}}}{\frac{3}{2}}$.

The integral of $\dfrac{1}{x^2}$ or x^{-2} is $\dfrac{x^{-1}}{-1}$.

The integral of $2\sqrt{x} - \dfrac{1}{x^2}$ is $2\left(\dfrac{x^{\frac{3}{2}}}{\frac{3}{2}}\right) - \left(\dfrac{x^{-1}}{-1}\right) + c$

$$\text{or} \quad \tfrac{4}{3}\sqrt{x^3} + \dfrac{1}{x} + c.$$

Exercise 19a

Integrate with respect to x:

1. x^5. **2.** $x^3 - 2x$. **3.** $x^3 + 6$.

4. $\dfrac{1}{x^3}$. **5.** $ax^2 + bx + c$. **6.** $(x + 1)(x + 2)$.

7. $\dfrac{1}{\sqrt{x}}$. **8.** $\dfrac{1 + x}{x^4}$. **9.** $\dfrac{x^4 + x}{x^3}$. **10.** $\dfrac{1 + x}{\sqrt{x}}$.

Integrate with respect to y:

11. y^6. **12.** $y^3 - 3y^2 + 1$. **13.** $y^4 - 2y^2 + 6$.

14. $\dfrac{1}{y^4}$. **15.** $y(y - 1)$. **16.** $\dfrac{y^2 + 1}{y^2}$. **17.** $\dfrac{y^2 + 1}{\sqrt{y}}$.

18. $\dfrac{1}{y^2} + \dfrac{1}{y^3}$. **19.** $5y - 1$. **20.** $y\left(1 + \dfrac{1}{y}\right)$.

21. The gradient of a line which passes through the point $(3, -1)$ is 2. Find the equation of the line.

22. The gradient of a curve which passes through the point $(1, 1)$ is given by $(2 + x)$. Find the equation of the curve.

23. If $\dfrac{dy}{dx} = 4$ and $y = 3$ when $x = -1$, find y in terms of x.

24. If $\dfrac{dy}{dx} = 3x - 2$ and $y = 1$ when $x = 1$, find y in terms of x.

25. The gradient of a curve at the origin is zero and at any point of the curve $\dfrac{dy}{dx} = 2x^2 - x$. Find the equation of the curve.

Velocity and acceleration

If we are given the distance moved along a straight line by a particle in time t, the velocity is found by differentiating the distance with respect to the time. Inversely, if we are given the velocity in terms of the time, the distance may be found by integration. Similarly, if we are given the acceleration in terms of the time, integration gives the velocity.

Example. *The acceleration of a particle moving in a straight line is given by $a = 6t + 4$. Find formulae for the velocity and distance, given that $s = 0$ and $v = 6$ when $t = 0$.*

$$a = 6t + 4.$$

$$\therefore \frac{dv}{dt} = 6t + 4.$$

Integrating, $v = 3t^2 + 4t + c.$
When $t = 0$, $v = 6$.

$$\therefore c = 6.$$
Since $v = 3t^2 + 4t + 6,$

$$\frac{ds}{dt} = 3t^2 + 4t + 6.$$

Integrating, $s = t^3 + 2t^2 + 6t + k.$
Since $s = 0$ when $t = 0$,

$$k = 0.$$
$$\therefore v = 3t^2 + 4t + 6 \quad \text{and} \quad s = t^3 + 2t^2 + 6t.$$

Exercise 19b

In this exercise, s, v, a, t represent the distance, velocity, acceleration and time in SI units of a particle moving in a straight line.

1. If $v = 2t + 5$, find s given that $s = 0$ when $t = 0$.

2. If $v = 3t^2 + 1$, find s given that $s = 0$ when $t = 0$.

3. If $a = 6t^2$, find v given that $v = 4$ when $t = 0$.

4. If $a = t^2 + t$, find s given that $s = 0$ and $v = 5$ when $t = 0$.

5. If $a = t^3 + t^2$, find s given that $s = 0$ and $v = 8$ when $t = 0$.

6. Find $a = t + 3$, and $v = 8$ when $t = 0$, find the distance travelled in the first 4 s.

7. If $a = 3t^2 + 1$ and $v = 10$ when $t = 0$, find the velocity when $t = 2$.

8. If $v = 3t^2 + 2t$, find the distance travelled in the first 3 seconds.

9. If the velocity of a particle is proportional to the time and the velocity is 4 m s^{-1} after 2 seconds, find the distance travelled in the first 2 seconds.

10. If the acceleration of a particle is proportional to the time and the velocity is 4 m s^{-1} after 1 s and 7 m s^{-1} after 2 s, find the velocity after 3 s.

11. A stone falling under its own weight has a constant acceleration of 10 ms^{-2}. If it starts from rest, find its velocity after 5 s.

12. Find the velocity of a stone falling under gravity for 3 s if its initial velocity is 10 m s^{-1} down.

13. Find the distance fallen in 6 s by a stone starting from rest.

14. Find the distance fallen by a stone in 2 s if its initial velocity is 8 m s^{-1} down.

15. Find the distance fallen by a stone in 2 s if its initial velocity is 8 m s^{-1} up.

16. A stone is thrown vertically upwards with a velocity of 20 m s^{-1}. Find how high it rises.

17. A stone is thrown vertically upwards with a velocity of 40 m s^{-1}. Find how long it takes to reach its highest point.

18. A stone is thrown upwards. Its velocities at two instants are 42 m s^{-1} and 10 m s^{-1}. Find the time which has elapsed between these two instants.

19. If $v = 3(1 + t^2)$, find the time taken to cover the first 36 m. (*Hint:* the answer is a whole number of seconds.)

20. If $v = 3t^2 + 2t$, find the distance travelled in the fourth second.

Area

Consider the area under the curve $y = x^2$ and enclosed by the lines $x = 1$ and $x = 2$. Let HN, KM be the ordinates at $x = 1$ and $x = 2$ and let PQ be an ordinate somewhere between HN and KM. Suppose the distance of PQ from the axis of y is x, and let the area $NHPQ$ be A. This area A is obviously a function of x and is zero when $x = 1$. When $x = 2$, A is equal to the area required.

Suppose in Fig. 19.3 that PQ moves to another position $P'Q'$ whose distance from the axis of y is $(x + h)$.

Since P and P' are on the curve $y = x^2$,

$$PQ = x^2 \quad \text{and} \quad P'Q' = (x + h)^2.$$

The increase in A due to the movement from PQ to $P'Q'$ lies between the areas of two rectangles, one of length x^2, the other of length $(x + h)^2$ and each of breadth h as can be seen from the figure.

$$\therefore x^2 < \frac{\text{increase in } A}{h} < (x + h)^2.$$

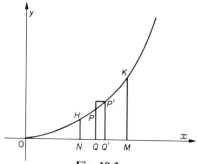

Fig. 19.3

Now suppose that h tends to zero, so that $(x + h)^2$ tends to x^2. It follows that $\dfrac{\text{increase in } A}{h}$ also tends to x^2.

But $\dfrac{\text{increase in } A}{h}$, as h tends to 0, tends to the derived function of A.

$$\therefore \frac{dA}{dx} = x^2,$$

and
$$A = \tfrac{1}{3}x^3 + c.$$

When $x = 1$, $A = 0$.

Substituting
$$0 = \tfrac{1}{3} + c \quad \text{and} \quad c = -\tfrac{1}{3}.$$
$$\therefore A = \tfrac{1}{3}x^3 - \tfrac{1}{3}.$$

The area required is the value of A when $x = 2$, which is $(\tfrac{8}{3} - \tfrac{1}{3})$ or $\tfrac{7}{3}$. Therefore the area under the graph is $2\tfrac{1}{3}$ sq. units.

In general the problem of finding the area under a curve $y = f(x)$ between the two given ordinates $x = a$ and $x = b$ is equivalent to the following problem.

'Find the value of A when $x = b$, given that $\dfrac{dA}{dx} = f(x)$ and that $A = 0$ when $x = a$.'

The definite integral

Notice that the area under the curve $y = x^2$ between the ordinates $x = 1$ and $x = 2$ is equal to $\dfrac{2^3}{3} - \dfrac{1^3}{3}$. This is equal to the value of the integral of x^2 when $x = 1$ subtracted from the value of the integral when $x = 2$. This area may be written in the form $\displaystyle\int_1^2 x^2\, dx$, which means the integral between the limits 1 and 2 for the variable x.

Such an integral is called a **definite integral.** To evaluate a definite integral, integrate the expression; from the value of the integral at the upper limit subtract the value at the lower limit. In evaluating a definite integral, there is no need for an arbitrary constant. If the arbitrary constant is included, it will disappear in the subtraction.

An integral which does not have the limits at the ends of the integral sign is called an **indefinite integral** and the arbitrary constant should always be inserted in such an integration.

Example. *Evaluate* $\int_2^3 (3x^2 + x)\, dx$.

$$\int_2^3 (3x^2 + x)\, dx = [x^3 + \tfrac{1}{2}x^2]_2^3$$

$$= (3^3 + \tfrac{1}{2}.3^2) - (2^3 + \tfrac{1}{2}.2^2)$$
$$= 31\tfrac{1}{2} - 10$$
$$= 21\tfrac{1}{2}.$$

The area under the curve $y = f(x)$ between the ordinates $x = a$ and $x = b$ is given by

$$A = \int_a^b y\, dx \quad \text{or} \quad \int_a^b f(x)\, dx.$$

Example. *Find the area under the curve* $y = 3x^2 + x$ *between the ordinates* $x = 2$ *and* $x = 3$.

$$\frac{dA}{dx} = 3x^2 + x.$$

$$\therefore A = \int_2^3 (3x^2 + x)\, dx$$

$$= 21\tfrac{1}{2} \quad \text{from the previous example.}$$

Area under a velocity–time graph

The area under a velocity–time graph represents the integral of the velocity with respect to the time. We have already seen that this is equal to the distance travelled. If we are given a number of corresponding readings of velocity and time, the area under the graph may be found by the addition of the small squares under the curve. A small square is included in the addition if more than half of it is in the required area, otherwise it is excluded. If we know the relation between velocity and time, the distance can be calculated exactly by integration.

Area under an acceleration–time graph

Similarly the area under an acceleration–time graph is equal to the integral of the acceleration with respect to the time. Therefore this area gives the velocity. The area between the two ordinates equals the change of velocity between the corresponding times. This may be determined either by adding squares or by integration.

Example 1. *Two particles moving in a straight line start from the same origin. The velocity, v m s^{-1}, of one particle is connected with the time taken, t seconds, by the equation $v = t^2$; for the other particle, $v = t$. Find the distance between the particles after 1 second.*

If $v = t^2$,
$$\frac{ds}{dt} = t^2.$$
$$\therefore s = \tfrac{1}{3}t^3 + c,$$
and since $s = 0$ when $t = 0$,
$$s = \tfrac{1}{3}t^3.$$
The distance travelled by this particle in 1 s is $\tfrac{1}{3}$ metre.

For the other particle, $v = t$ and therefore $\dfrac{ds}{dt} = t$.
$$\therefore s = \tfrac{1}{2}t^2 + k,$$
and since $s = 0$ when $t = 0$, the constant of integration is zero.
$$\therefore s = \tfrac{1}{2}t^2$$
and the distance travelled by this particle in 1 s is $\tfrac{1}{2}$ metre.
The distance between the particles after 1 s is $(\tfrac{1}{2} - \tfrac{1}{3})$ m or $\tfrac{1}{6}$ m.
This problem is analogous to finding the area between the curve $y = x^2$ and the line $y = x$.
Figure 19.4 shows a sketch of the curve and the line. They meet where $x = x^2$, i.e. at the points $(0, 0)$ and $(1, 1)$.

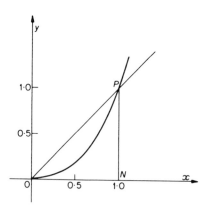

Fig. 19.4

The area under the curve $y = x^2$ between $x = 0$ and $x = 1$ is

$$\int_0^1 x^2 \, dx = [\tfrac{1}{3}x^3]_0^1 = \tfrac{1}{3}.$$

The area under the line $y = x$ between $x = 0$ and $x = 1$ is

$$\int_0^1 x \, dx = [\tfrac{1}{2}x^2]_0^1 = \tfrac{1}{2}.$$

(This may also be calculated by finding the area of the triangle OPN.) The difference between these two areas is $(\tfrac{1}{2} - \tfrac{1}{3})$ or $\tfrac{1}{6}$. This answer may be verified by counting squares. Be careful to draw as large a figure as possible on the graph paper.

Example 2. *A body moving in a straight line starts from rest and accelerates uniformly at $2 \, m \, s^{-1}$ for 4 s. It then travels at constant velocity for 3 s and finally decelerates uniformly and comes to rest after decelerating for 6 s. Find the total distance travelled.*

The gradient of the velocity–time graph is the acceleration. When the acceleration is constant, the graph is a straight line; when the acceleration is zero, the graph is a straight line parallel to the time axis.

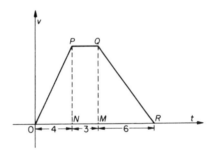

Fig. 19.5

The graph is shown in Fig. 19.5 and consists of three lines, OP of gradient 2, PQ parallel to the time axis and QR. The area under the graph represents the distance travelled.

$$\frac{PN}{ON} = \text{the gradient of } OP = 2. \quad \therefore PN = 8.$$

The area under the graph $= ONP + NMQP + MRQ$
$$= \tfrac{1}{2}(4)(8) + 3(8) + \tfrac{1}{2}(6)(8)$$
$$= 16 + 24 + 24 = 64.$$

The distance travelled is 64 metres.

Sign of the area

In the integral $\int_a^b y\,dx$ it is conventional to assume that the increment in x is taken to be positive. If then the limit b is larger than the limit a and y is positive, the value of the integral is positive; if y is negative, the value of the integral is negative. This means that an area above the x-axis will have a positive sign and that an area below the x-axis will have a negative sign. If the area required is partly above and partly below the axis of x, the areas of the two parts must be considered separately.

Example. *Find the area included between the curve* $y = x^3 - 3x^2 + 2x$ *and the axis of* x.

The graph is shown in Fig. 19.6. It cuts the axis of x at A $(1, 0)$ and B $(2, 0)$.

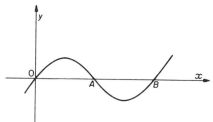

Fig. 19.6

The area above $OA = \int_0^1 (x^3 - 3x^2 + 2x)\,dx$

$$= \left[\frac{x^4}{4} - x^3 + x^2\right]_0^1 = (\tfrac{1}{4} - 1 + 1) - 0$$

$$= \tfrac{1}{4}.$$

The area below $AB = \int_1^2 (x^3 - 3x^2 + 2x)\,dx$

$$= \left[\frac{x^4}{4} - x^3 + x^2\right]_1^2$$

$$= (4 - 8 + 4) - (\tfrac{1}{4} - 1 + 1).$$

$$= -\tfrac{1}{4} \quad \text{(note the negative sign)}.$$

The area required is the sum of these two areas, both regarded as positive, and is $\tfrac{1}{2}$ square unit.

Note that $\int_0^2 (x^3 - 3x^2 + 2x)\,dx = \left[\dfrac{x^4}{4} - x^3 + x^2\right]_0^2$

$$= 4 - 8 + 4 = 0,$$

i.e. the *algebraic* sum of the two areas.

Exercise 19c

Evaluate:

1. $\int_{1}^{2} \frac{1}{x^2} \, dx.$

2. $\int_{1}^{2} (x + 1)(x + 2) \, dx.$

3. $\int_{0}^{4} \sqrt{x} \, dx.$

4. $\int_{1}^{3} \frac{1 + x^2}{x^2} \, dx.$

5. $\int_{1}^{2} (x^2 - x) \, dx.$

6. $\int_{2}^{3} (x^3 + x) \, dx.$

7. $\int_{-1}^{1} (1 - t^2) \, dt.$

8. $\int_{0}^{2} (1 + u)^2 \, du.$

9. $\int_{1}^{4} v\sqrt{v} \, dv.$

10. $\int_{-1}^{1} (1 + 2t + 3t^2) \, dt.$

11. Find the area under the curve $y = x + 3x^2$ between $x = 1$ and $x = 2$.

12. Find the area enclosed by the curve $y = 2x - x^2$ and the x-axis.

13. Find the area enclosed by the curve $y = x(x - 4)$ and the axis of x.

14. Find the area between the curve $y = \sqrt{x}$, the x-axis and the ordinates $x = 1$ and $x = 4$.

15. Find the area under the curve $yx^2 = 1$ between $x = 1$ and $x = 2$.

16. Find the area enclosed by the line $y = 4x$ and the curve $y = x^2$.

17. Find the area in the first quadrant enclosed by the line $y = x$ and the curve $y = x^3$.

18. Find the area enclosed by the curve $y^2 = x$, the axis of y and the line $y = 4$.

19. Find the area between the curve $x^2 = y$ and the line $y = 2$.

20. Find the area in the positive quadrant enclosed by the curves $y = x^2$ and $y = x^3$.

Solid of revolution

If the region under a curve and included between two ordinates is rotated about the axis of x, the resulting solid is called a solid of revolution. A section of this solid by a plane perpendicular to the axis of x is a circle.

The region under a line parallel to the axis of x when rotated gives a right circular cylinder.

The region under a line through the origin when rotated gives a right circular cone.

We shall now find the volume of a solid of revolution, given the equation of the curve and the bounding ordinates.

As an example, consider the curve $y = x^2$ between the ordinates

$x = 1$ and $x = 2$. Suppose the bounding ordinates are HN and KM and that PQ is any ordinate between them, as shown in Fig. 19.7.

If we rotate PQ, we shall get a circle. If this circle moves from the position in which HN is a radius to the position when KM is a radius, it will trace out the volume required. Let V be equal to the volume between HN and the position when the distance of PQ from the y-axis is equal to x. Then V is a function of x which is zero when $x = 1$; we

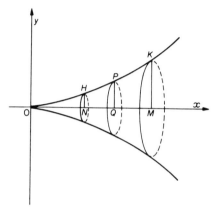

Fig. 19.7

are asked to find the value of V when $x = 2$. If PQ moves to a position $P'Q'$ whose distance from the axis of y is $(x + h)$, V is increased by a volume whose plane ends are circles of radii x^2 and $(x + h)^2$ respectively. This increase of volume lies between that of a cylinder of radius x^2 and length h and a cylinder of radius $(x + h)^2$ and length h.

$$\therefore \pi h x^4 < \text{increase in } V < \pi h(x + h)^4.$$

$$\therefore \pi x^4 < \frac{\text{increase in } V}{h} < \pi(x + h)^4.$$

$\therefore \dfrac{\text{increase in } V}{h}$ tends to πx^4 as h tends to 0.

But $\dfrac{\text{increase in } V}{h}$ tends to the derived function of V as h tends to 0.

$$\therefore \frac{dV}{dx} = \pi x^4.$$

Integrating $\qquad V = \tfrac{1}{5}\pi x^5 + c.$

But $V = 0$ when $x = 1$,

$$\therefore 0 = \tfrac{1}{5}\pi + c$$

and
$$V = \pi\left(\frac{x^5 - 1}{5}\right).$$

When $x = 2$
$$V = \pi\left(\frac{32 - 1}{5}\right) = \frac{31\pi}{5}.$$

The problem of finding the volume obtained by rotating about the axis of x the region under the curve $y = f(x)$ between $x = a$ and $x = b$ is equivalent to the following problem:

'Find V when $x = b$, given that $\dfrac{dV}{dx} = \pi\{f(x)\}^2$ and that $V = 0$ when $x = a$.'

The volume may also be expressed as a definite integral. The volume of rotation about the x-axis of the area under the curve $y = f(x)$ between $x = a$ and $x = b$ is given by

$$V = \pi \int_a^b y^2 \, dx \quad \text{or} \quad \pi \int_a^b \{f(x)\}^2 \, dx.$$

Example 1. *Find the volume of a cone of height h and base radius r.*

The cone is formed by the revolution about the axis of x of the region under a line which passes through the origin. The gradient of the line is $\dfrac{r}{h}$ and its equation is $y = \dfrac{r}{h}x$ (see Fig. 19.8).

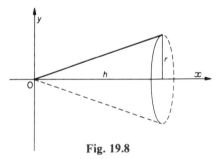

Fig. 19.8

If V is the volume of the cone,

$$V = \pi \int_0^h y^2 \, dx$$

$$= \pi \int_0^h \frac{r^2}{h^2}x^2 \, dx = \pi\left[\frac{r^2}{h^2} \cdot \frac{x^3}{3}\right]_0^h$$

$$= \tfrac{1}{3}\pi r^2 h.$$

The volume of a cone is therefore $\tfrac{1}{3}\pi r^2 h$.

Example 2. *The equation of a circle is $x^2 + y^2 = r^2$. Deduce from this equation a formulae for the volume of a sphere.*

By rotating a quadrant of a circle as shown in Fig. 19.9 a hemisphere is obtained.

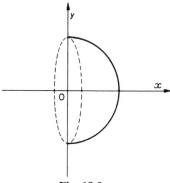

Fig. 19.9

The volume of the hemisphere $= \pi \int_0^r y^2 \, dx$

$$= \pi \int_0^r (r^2 - x^2) \, dx$$

$$= \pi [r^2 x - \tfrac{1}{3} x^3]_0^r$$
$$= \pi (r^3 - \tfrac{1}{3} r^3) = \tfrac{2}{3} \pi r^3.$$

The formula for the volume of a sphere is therefore $\tfrac{4}{3} \pi r^3$.

Example 3. *The region between the line $y = x$ and the curve $y = x^2$ is rotated about the axis of x. Find the volume formed.*

The line and the curve are shown in Fig. 19.10.

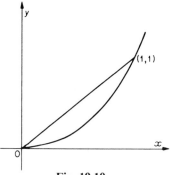

(1,1)

Fig. 19.10

The line and the curve meet at $(0, 0)$ and $(1, 1)$.
The volume obtained by rotating the region under the curve

$$= \pi \int_0^1 y^2 \, dx = \pi \int_0^1 x^4 \, dx = \pi \left[\frac{x^5}{5} \right]_0^1 = \frac{\pi}{5}.$$

The volume obtained by rotating the region under the line

$$= \pi \int_0^1 y^2 \, dx = \pi \int_0^1 x^2 \, dx = \pi \left[\frac{x^3}{3} \right]_0^1 = \frac{\pi}{3}.$$

The volume required $= \dfrac{\pi}{3} - \dfrac{\pi}{5} = \dfrac{2\pi}{15}.$

Centre of gravity

To find the centre of gravity of a region or volume, the principle of moments is used. The region or volume is divided up into a number of elements and the sum of the moments of these elements about either axis is equal to the moment of the whole about that axis. Suppose we wish to find the centre of gravity of the region between the curve $y = f(x)$, the x-axis and the ordinates at $x = a$ and $x = b$.

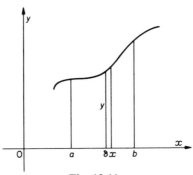

Fig. 19.11

The region is divided up into elements parallel to the y-axis. The area of the element shown in Fig. 19.11 lies between $y \, \delta x$ and $(y + \delta y) \, \delta x$, and its mass between $my \, \delta x$ and $m(y + \delta y) \, \delta x$ where m is the mass per unit area. If A is the area under the curve, taking moments about the y-axis,

$$\sum mxy \, \delta x < mA\bar{x} < \sum mx(y + \delta y) \, \delta x.$$

In the limit, therefore,

$$A\bar{x} = \int_a^b xy \, dx.$$

The distance of the centre of gravity of the element from the x-axis is approximately $\frac{1}{2}y$ and so

$$A\bar{y} = \int_a^b y\,dx(\tfrac{1}{2}y) \quad \text{or} \quad A\bar{y} = \tfrac{1}{2}\int_a^b y^2\,dx.$$

Example 1. *Find the coordinates of the centre of gravity of the region included in the first quadrant by the parabola $y^2 = 4x$, the x-axis and the ordinate $x = 1$.*

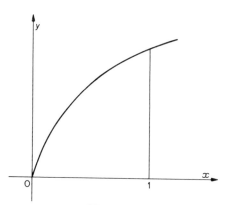

Fig. 19.12

The region defined is given by

$$A = \int_0^1 y\,dx$$

$$= \int_0^1 2x^{\frac{1}{2}}\,dx = [\tfrac{4}{3}x^{\frac{3}{2}}]_0^1 = \tfrac{4}{3}$$

$$\therefore A\bar{x} = \int_0^1 xy\,dx = \int_0^1 2x^{\frac{3}{2}}\,dx = [\tfrac{4}{5}x^{\frac{5}{2}}]_0^1 = \tfrac{4}{5}.$$

$$\therefore \bar{x} = \tfrac{3}{5}.$$

Also

$$A\bar{y} = \tfrac{1}{2}\int_0^1 y^2\,dx = \int_0^1 2x\,dx = [x^2]_0^1 = 1.$$

$$\therefore \bar{y} = \tfrac{3}{4}.$$

The centre of gravity is $(\tfrac{3}{5}, \tfrac{3}{4})$.

Example 2. *Find the coordinates of the centre of gravity of the region included between the x-axis and the curve $y = x^2 - 3x + 2$.*

$$x^2 - 3x + 2 = (x - 1)(x - 2).$$

The curve therefore cuts the x-axis where $x = 1$ and $x = 2$.

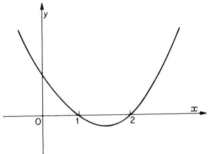

Fig. 19.13

The region defined is given by

$$A = \int_1^2 y\, dx = \int_1^2 (x^2 - 3x + 2)\, dx = \left[\frac{x^3}{3} - \frac{3x^2}{2} + 2x\right]_1^2$$
$$= \tfrac{7}{3} - 4\tfrac{1}{2} + 2 = 4\tfrac{1}{3} - 4\tfrac{1}{2} = -\tfrac{1}{6}.$$

Since $x = 1\tfrac{1}{2}$ is the axis of symmetry, $\bar{x} = 1\tfrac{1}{2}$.
Alternatively,

$$A\bar{x} = \int_1^2 xy\, dx = \int_1^2 (x^3 - 3x^2 + 2x)\, dx = \left[\frac{x^4}{4} - x^3 + x^2\right]_1^2$$
$$= 3\tfrac{3}{4} - 7 + 3 = -\tfrac{1}{4}.$$
$$\therefore \bar{x} = 1\tfrac{1}{2}.$$

Also
$$A\bar{y} = \tfrac{1}{2}\int_1^2 y^2\, dx = \tfrac{1}{2}\int_1^2 (x^4 - 6x^3 + 13x^2 - 12x + 4)\, dx$$

$$= \tfrac{1}{2}\left[\frac{x^5}{5} - \tfrac{3}{2}x^4 + \tfrac{13}{3}x^3 - 6x^2 + 4x\right]_1^2$$

$$= \tfrac{1}{2}(6\tfrac{1}{5} - 22\tfrac{1}{2} + 30\tfrac{1}{3} - 18 + 4)$$
$$= \tfrac{1}{2}(40\tfrac{8}{15} - 40\tfrac{1}{2}) = \tfrac{1}{60}.$$
$$\therefore \bar{y} = -\tfrac{1}{10}.$$

The centre of gravity is $(1\tfrac{1}{2}, -\tfrac{1}{10})$.

Exercise 19d: Miscellaneous

1. Evaluate $\displaystyle\int_{-1}^1 (t - t^3)\, dt$.

2. Find the volume obtained by rotating the region under the curve $y^2 = 1 + x$ between $x = 1$ and $x = 2$ about the axis of x.

3. Find the volume obtained by rotating the region under the curve $y = 1 + x^2$ between $x = 0$ and $x = 3$ about the axis of x.

4. The curve $y^2 = 4ax$ passes through the point P in the first quadrant. PN and PM are perpendiculars to the axes of x and y respectively. Prove that the area under the curve between O, the origin, and P is two thirds of the area of the rectangle $ONPM$.

5. Show that the volume obtained by rotating the region under $y^2 = 4ax$ about the axis of x is equal to one-half the volume of the cylinder of base radius PN and height ON (see question 4).

6. Find the volume obtained by rotating the region included between the curve $y = x - x^2$ and the x-axis about the axis of x.

7. P is the point $(4, 4)$ on the curve $y^2 = 4x$. The perpendiculars to the axes of x and y are PN, PM respectively. Find the volume obtained by rotating the region OPM about the axis of y.

8. With the data of question 7, find the volume obtained by rotating the region OPN about the axis of y.

9. Find the area included between the x-axis, the curve $y = x^3$ and the line $x + y = 2$ which meets the curve at the point $(1, 1)$.

10. Find the volume obtained by rotating the region defined in question 9 about the axis of x.

11. Find the volume obtained by rotating the region under the line $y = 1 + 2x$ between $x = 1$ and $x = 3$ about the axis of x.

12. Find the equation of the line joining the points $(0, r)$ and (h, R). Hence find a formula for the volume of a frustum of a cone of height h and base radii R and r.

13. Evaluate $\displaystyle\int_{-1}^{3} (1 + u)^2 \, du$.

14. The points $(1, 1)$ and $(2, 8)$ on the curve $y = x^3$ are joined. Find the region between this line and the curve.

15. Find the volume obtained by rotating the region defined in question 14 about the axis of x.

16. If $\dfrac{dy}{dx} = x + 1$ and $y = 4$ when $x = 1$, find y in terms of x.

17. If $\dfrac{dy}{dx} = 2x$ and $y = 3$ when $x = 0$, find y when $x = 2$.

18. A curve has a gradient of $x^2 - x$ at the point (x, y). If the curve passes through the point $(1, 1)$, find its equation.

19. The area under the curve $y = Ax + Bx^2$ between $x = 0$ and $x = 1$ is 2 square units. If the curve passes through the point $(1, 5)$, find the values of A and B.

20. A point moves in a straight line so that its acceleration in m s^{-2} is given by $a = t^2 + 2t$, where t is the time in seconds. If the initial velocity is 4 m s^{-1}, find the velocity after 3 seconds.

21. Find the volume formed by the revolution between $x = 0$ and $x = 3$ of the region under the ellipse $y^2 = 4\left(1 - \dfrac{x^2}{9}\right)$.

22. By rotating a quadrant of the ellipse $\dfrac{x^2}{a^2} + \dfrac{y^2}{b^2} = 1$ about the axis of x, find a formula for the volume of an ellipsoid of semi-axes b, b and a.

23. Integrate $(4x - 3)^2$.

24. Find the volume of rotation of the region under the curve $y = \dfrac{1}{x}$ between $x = 1$ and $x = 2$ about the axis of x.

25. Find the volume of rotation of the region under the curve $y = \dfrac{1}{x^2}$ between $x = \frac{1}{2}$ and $x = 1$ about the axis of x.

26. Find the centre of gravity of the region included between the curve $y = x^2$, the x-axis and the ordinate $x = 1$.

27. Find the centre of gravity of the region included between the curve $y = x^3$, the x-axis and the ordinates $x = 1$ and $x = 2$.

28. Find the centre of gravity of the region between the curve $y = x^2 - 1$ and the x-axis.

29. Find the centre of gravity of the region included between the x-axis and the curve $y = 2x - x^2$.

30. Find the centre of gravity of the region between the curve $y = x^2$, the y-axis and the line $y = 1$.

31. Find the centre of gravity of the triangle formed by the lines $3y = 4x$, $y = 0$ and $x = 3$.

32. Find the centre of gravity of the region included by the curve $y = x^2 - 4x$, the x-axis and the lines $x = 1$ and $x = 2$.

33. The curve $y = x^2$ between $x = 0$ and $x = 1$ is rotated through four right angles about the axis of x. Find the centre of gravity of the solid so formed.

34. Find the centre of gravity of the region included between $y = x^2$ and $y = x$.

35. Find the centre of gravity of the region included between $y^2 = 4x$ and $y = x$.

20 More Differentiation and Integration

The limit of sin x/x

To find the differential coefficient of sin x, it is necessary to know the limit of the fraction $\dfrac{\sin x}{x}$ as x tends to zero. We shall assume that x is measured in radians, that OAB in Fig. 20.1 is a sector of a circle of radius r whose angle is x radians and that the tangent at A to the circle meets OB produced at N.

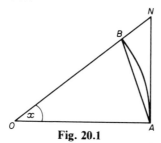

Fig. 20.1

Then, from the figure,

area of triangle OAB < area of sector OAB

< area of triangle OAN.

But the area of the triangle $OAB = \frac{1}{2}r^2 \sin x$ (using $\frac{1}{2}ab \sin C$);
the area of the sector OAB $= \frac{1}{2}r^2 x$ (sector formula);
the area of the triangle OAN

$$= \tfrac{1}{2}OA \cdot AN = \tfrac{1}{2}r \cdot r \tan x = \tfrac{1}{2}r^2 \tan x.$$
$$\therefore \tfrac{1}{2}r^2 \sin x < \tfrac{1}{2}r^2 x < \tfrac{1}{2}r^2 \tan x,$$

or $\qquad\qquad\qquad \sin x < x < \tan x.$

Dividing throughout by sin x (which is positive provided that x is acute),

$$1 < \frac{x}{\sin x} < \sec x.$$

But as $x \to 0$, sec $x \to 1$.

$$\therefore \frac{x}{\sin x} \to 1.$$

$$\therefore \frac{\sin x}{x} \to 1 \quad \text{as} \quad x \to 0.$$

The differential coefficient of sin x

To find the D.C. of sin x where x is measured in radians, suppose that $(x, \sin x)$ and $(x + h, \sin \overline{x + h})$ are two points on the curve $y = \sin x$.
The gradient of the line joining these two points is

$$\frac{\sin (x + h) - \sin x}{(x + h) - x}.$$

This expression $= \dfrac{2 \cos \dfrac{2x + h}{2} \sin \dfrac{h}{2}}{h} = \cos \left(\dfrac{2x + h}{2}\right) \dfrac{\sin \dfrac{h}{2}}{\dfrac{h}{2}}.$

The D.C. of sin x is the limit of this expression as $h \to 0$. As $h \to 0$, $\cos \left(\dfrac{2x + h}{2}\right) \to \cos x$, and we have just proved that $\dfrac{\sin h/2}{h/2} \to 1$ when h is measured in radians.

$$\therefore \frac{d}{dx}(\sin x) = \cos x.$$

The differential coefficient of cos x

Using a similar notation, the gradient of the chord joining the points $(x, \cos x)$ and $(x + h, \cos \overline{x + h})$ on the curve $y = \cos x$ is

$$\frac{\cos (x + h) - \cos x}{(x + h) - x} = \frac{-2 \sin \dfrac{2x + h}{2} \sin \dfrac{h}{2}}{h}$$

$$= -\sin \frac{2x + h}{2} \cdot \frac{\sin \dfrac{h}{2}}{\dfrac{h}{2}}.$$

The limit of this expression as $h \to 0$ is $-\sin x$ and therefore

$$\frac{d}{dx}(\cos x) = -\sin x.$$

The integrals of sin x and cos x

From definition, it follows that

$$\int \cos x \, dx = \sin x + c;$$

$$\int \sin x \, dx = -\cos x + c.$$

Products

We may be asked to differentiate a product such as $x \cos x$ or $x^2 \sin x$ and it is obviously unwieldy to work from first principles each time. Is it possible to find a method of differentiating a product, knowing the D.C. of each factor?

Suppose that u and v are functions of x and that we wish to differentiate the product uv with respect to x.

Let $y = uv$ and suppose that when x has a small increment δx, u has a small increment δu, v has a small increment δv and that the consequent increment in y is δy.

The new value of y will equal the product of the new values of u and v.

$$\therefore y + \delta y = (u + \delta u)(v + \delta v).$$

Substituting $y = uv$,

$$\delta y = (u + \delta u)(v + \delta v) - uv$$
$$= u \, \delta v + v \, \delta u + \delta u \cdot \delta v.$$

Dividing both sides by δx,

$$\frac{\delta y}{\delta x} = u \frac{\delta v}{\delta x} + v \frac{\delta u}{\delta x} + \frac{\delta u \cdot \delta v}{\delta x}.$$

When δx and consequently δu, δv, δy become small, this equation becomes

$$\frac{dy}{dx} = u \frac{dv}{dx} + v \frac{du}{dx}.$$

The last term disappears because $\frac{\delta u}{\delta x}$ which becomes $\frac{du}{dx}$ in the limit is multiplied by the quantity δv which tends to zero.

So to differentiate a product, keep each factor constant in turn and multiply it by the D.C. of the other factor; add the results.

This may be extended to the product of three factors, u, v and w. Each factor is differentiated in turn and multiplied by the other two factors unaltered. The three results are then added. The proof of this follows by a double application of the formula

$$\frac{d}{dx}(uv) = u\frac{dv}{dx} + v\frac{du}{dx}$$

and is left as an exercise for the reader.

Examples.

$$\frac{d}{dx}(x \cos x) = x\frac{d}{dx}(\cos x) + \cos x\frac{d}{dx}(x) = -x \sin x + \cos x.$$

$$\frac{d}{dx}(x^2 \sin x) = x^2\frac{d}{dx}(\sin x) + \sin x\frac{d}{dx}(x^2)$$

$$= x^2 \cos x + 2x \sin x.$$

$$\frac{d}{dx}(x \cos x \sin x) = \cos x \sin x\frac{d}{dx}(x) + x \sin x\frac{d}{dx}(\cos x) + x \cos x\frac{d}{dx}(\sin x)$$

$$= \cos x \sin x - x \sin^2 x + x \cos^2 x.$$

Quotients

Suppose we now wish to differentiate $\frac{u}{v}$ with respect to x. The appropriate formula may be deduced as was the product formula or it may be proved direct from the product formula as follows.

If $y = \frac{u}{v}$, where u and v are functions of x,

$$yv = u.$$

$$\therefore \frac{d}{dx}(yv) = \frac{du}{dx},$$

and

$$y\frac{dv}{dx} + v\frac{dy}{dx} = \frac{du}{dx}.$$

$$\therefore v\frac{dy}{dx} = \frac{du}{dx} - y\frac{dv}{dx} = \frac{du}{dx} - \frac{u}{v}\cdot\frac{dv}{dx}$$

$$= \frac{v\frac{du}{dx} - u\frac{dv}{dx}}{v}.$$

$$\therefore \frac{dy}{dx} = \frac{v\frac{du}{dx} - u\frac{dv}{dx}}{v^2}.$$

The quotient formula is frequently used when easier methods produce the same result. Remember the following points:

1. Never differentiate a product by the product formula if one of the factors is a constant.
2. Never use the quotient formula if the denominator is a constant, e.g.

$$\frac{d}{dx}\left(\frac{\sin x}{4}\right) = \frac{1}{4} \cdot \frac{d}{dx}(\sin x) = \frac{1}{4}\cos x.$$

3. It is sometimes easier to use the product rather than the quotient formula, or even to use neither.

Example 1. *Differentiate* $\dfrac{x+1}{x}$ *with respect to* x.

Using the quotient formula:

$$\frac{d}{dx}\left(\frac{x+1}{x}\right) = \frac{x(1) - (x+1)(1)}{x^2} = -\frac{1}{x^2}.$$

Using the product formula:

$$\frac{d}{dx}[(x+1)x^{-1}] = x^{-1} - \frac{x+1}{x^2} = \frac{x-(x+1)}{x^2} = -\frac{1}{x^2}.$$

The easiest method:

$$\frac{x+1}{x} = 1 + \frac{1}{x}.$$

$$\therefore \frac{d}{dx}\left(\frac{x+1}{x}\right) = \frac{d}{dx}\left(1 + \frac{1}{x}\right) = -\frac{1}{x^2}.$$

Example 2. *Differentiate* $\dfrac{x^2}{\sin x}$ *with respect to* x.

$$\frac{d}{dx}\left(\frac{x^2}{\sin x}\right) = \frac{\sin x \dfrac{d}{dx}(x^2) - x^2 \dfrac{d}{dx}(\sin x)}{\sin^2 x} = \frac{2x \sin x - x^2 \cos x}{\sin^2 x}.$$

Example 3. *Differentiate* $\dfrac{\cos x}{2x+3}$ *with respect to* x.

$$\frac{d}{dx}\left(\frac{\cos x}{2x+3}\right) = \frac{(2x+3)\dfrac{d}{dx}(\cos x) - \cos x \dfrac{d}{dx}(2x+3)}{(2x+3)^2}$$

$$= \frac{-(2x+3)\sin x - 2\cos x}{(2x+3)^2}.$$

Exercise 20a

Differentiate with respect to x:

1. $x \sin x$ **2.** $\sin x \cos x$.   **3.** $\dfrac{x-1}{x+1}$.

4. $\dfrac{x^2-1}{x+2}$.   **5.** $\dfrac{x+1}{\sin x}$.   **6.** $\dfrac{\cos x}{x+1}$.

7. $(x+2)(x^2+1)$.   **8.** $\dfrac{x^2}{x+1}$.   **9.** $\dfrac{1}{1+x}$.

10. $\dfrac{1}{x^2+1}$.   **11.** $x^3 \cos x$.   **12.** $\tan x \left(= \dfrac{\sin x}{\cos x} \right)$.

13. $\cot x$.   **14.** $\dfrac{1}{\sin x}$.   **15.** $\dfrac{1}{\cos x}$.

16. $\dfrac{1}{x^2+x+1}$.   **17.** $\dfrac{x}{x^2+x+1}$.   **18.** $\dfrac{x+1}{\sqrt{x}}$.

19. $\dfrac{\sqrt{x}}{x-1}$.   **20.** $\dfrac{\sqrt{x+1}}{\sqrt{x}}$.

Evaluate

21. $\displaystyle\int_0^{\pi/2} \sin x \, dx$.   **22.** $\displaystyle\int_0^{\pi/2} \cos y \, dy$.   **23.** $\displaystyle\int_0^{\pi/4} \sin t \, dt$.

24. $\displaystyle\int_{\pi/4}^{\pi/2} \cos u \, du$.   **25.** $\displaystyle\int_{\pi/6}^{\pi/3} \sin x \, dx$.

Function of a function

We are all familiar with the phrase 'rate of change' when we are thinking of time. In fact, rate of change will automatically bring to our minds the idea of time because it is generally used in that context. The rate of change of distance, for example, means the change of distance per unit of time. The rate of change of velocity similarly gives the acceleration. Generally speaking, by rate of change we mean the average rate of change over some definite time interval, but in the context of the calculus we are concerned with the limit which this average rate approaches as the time interval tends to zero. This is of course the differential coefficient with respect to the time.

The ideas behind calculus came to two men, Isaac Newton (1642–1727) and G. W. Leibnitz (1646–1716), independently. Leibnitz was responsible for the notation $\dfrac{dy}{dx}$ but Newton used $\dot{x}$ to denote the D.C. of x, a notation which has been adopted today to stand for differentiation with respect to time and is used almost exclusively in

Mechanics and Differential Geometry. In Mechanics, $\dot{x}$ stands for velocity and $\ddot{x}$ or $\dot{v}$ for acceleration.

The rate of change of one quantity with respect to another need not necessarily refer to time. The rate of change of y with respect to x means the ratio of the increase of y to the increase of x and is the D.C. of y with respect to x or the gradient of the tangent to the curve formed by plotting y against x.

Let us now consider the D.C. of $(x + 1)^2$. Expanding this as $x^2 + 2x + 1$, we see the D.C. is $2x + 2$ or $2(x + 1)$.

The D.C. of $(3x + 1)^2$, which equals $9x^2 + 6x + 1$, is $18x + 6$ or $6(3x + 1)$.

Why is it that the first formula agrees with what we should expect from $\dfrac{d}{dx}(x^2) = 2x$ while the second apparently does not?

The fact that $\dfrac{d}{dx}(x^2) = 2x$ means that the rate of increase of x^2 with respect to x is $2x$. From this it follows that:

(i) The rate of increase of $(x + 1)^2$ with respect to $(x + 1)$ is $2(x + 1)$;
(ii) The rate of increase of $(3x + 1)^2$ with respect to $(3x + 1)$ is $2(3x + 1)$.

Now if A runs twice as fast as B and B runs three times as fast as C then A runs six times as fast as C.

The rate of increase of $(x + 1)$ with respect to x is 1 (its D.C.).

The rate of increase of $(3x + 1)$ with respect to x is 3. Therefore the rate of increase of $(x + 1)^2$ with respect to x is still $2(x + 1)$ but the rate of increase of $(3x + 1)^2$ with respect to x is $3 \times 2(3x + 1)$ or $6(3x + 1)$.

Similarly we should expect the D.C. of $(x^2 + x + 1)^2$ to be $2(2x + 1)(x^2 + x + 1)$ which indeed it is.

This brings us to the idea of a function of a function. A simple function of x is an expression which depends on x directly and not on some function of x. Examples are x^2, $\sqrt{x}$, $\sin x$ etc. but $\sqrt{x^2 + 1}$, $\sin (x^2 + x + 1)$ are functions of a function of x and we need to formulate some rule for differentiating them.

The rate of increase of y with respect to z is $\dfrac{dy}{dz}$; the rate of increase of z with respect to x is $\dfrac{dz}{dx}$. The rate of increase of y with respect to x we should expect to be the product of these, which leads us to the equation

$$\frac{dy}{dx} = \frac{dy}{dz} \cdot \frac{dz}{dx}.$$

This formula may not be proved by cancelling the dz term as dz is not a finite quantity and cannot be treated as a multiplier or divisor.

However if δx, δy, δz are small corresponding increments in the quantities x, y, z so that δy and δz become small with δx,

$$\frac{\delta y}{\delta x} = \frac{\delta y}{\delta z} \cdot \frac{\delta z}{\delta x}$$

because since δx, δy and δz are finite quantities, however small, they may be cancelled.

If now δx, δy, δz are allowed to get smaller and smaller $\dfrac{\delta y}{\delta x}$ will tend to $\dfrac{dy}{dx}$ and so we get

$$\frac{dy}{dx} = \frac{dy}{dz} \cdot \frac{dz}{dx}$$

(assuming that the limit of a product is the product of the limits of its factors).

We should therefore get the same result in this case if we treated dx, dy and dz as algebraic quantities and it is true, for example, that

$$\frac{dy}{dx} = \frac{1}{dx/dy}.$$

It should be noted that these statements are only true for first-order differentials and that $\dfrac{d^2 y}{dx^2}$ is not equal to $\dfrac{d^2 y}{dz^2} \cdot \dfrac{d^2 z}{dx^2}$ nor is $\dfrac{d^2 y}{dx^2}$ the reciprocal of $\dfrac{d^2 x}{dy^2}$.

However it is a common convention that $\dfrac{dy}{dx} = x + 1$, for example, may be written $dy = (x + 1)\, dx$. In fact $2x$ is called the D.C. of x^2 with respect to x and $2x\, dx$ is called the differential of x^2.

The following worked examples show how to differentiate a function of a function.

Example 1. $\dfrac{d}{dx}(x^2 + x + 1)^3$.

Put $\qquad x^2 + x + 1 = z \quad$ and $\quad y = (x^2 + x + 1)^3$.

Then $\qquad\qquad\qquad y = z^3 \quad$ and so $\quad \dfrac{dy}{dz} = 3z^2$.

But $$\frac{dz}{dx} = 2x + 1 \quad \text{and} \quad \frac{dy}{dx} = \frac{dy}{dz} \cdot \frac{dz}{dx}.$$

$$\therefore \frac{dy}{dx} = 3z^2(2x + 1) = 3(2x + 1)(x^2 + x + 1)^2.$$

With practice, the student should be able to write down the D.C. directly by the following argument.

The D.C. of $(x^2 + x + 1)^3$ with respect to $(x^2 + x + 1)$ is $3(x^2 + x + 1)^2$.

The D.C. of $(x^2 + x + 1)$ with respect to x is $(2x + 1)$.

Therefore the D.C. of $(x^2 + x + 1)^3$ with respect to x is $3(2x + 1)(x^2 + x + 1)^2$.

Example 2. $\dfrac{d}{dx}(\sin 5x)$.

Put $$5x = z \quad \text{and} \quad y = \sin 5x.$$

Then $$y = \sin z \quad \text{and so} \quad \frac{dy}{dz} = \cos z. \quad \text{Also} \; \frac{dz}{dx} = 5.$$

$$\therefore \frac{dy}{dx} = \frac{dy}{dz} \cdot \frac{dz}{dx} = 5 \cos z = 5 \cos 5x.$$

The argument in words is:
    the D.C. of $\sin 5x$ with respect to $5x$ is $\cos 5x$;
    the D.C. of $5x$ with respect to x is 5;
    therefore the D.C. of $\sin 5x$ with respect to x is $5 \cos 5x$.

Example 3. $\dfrac{d}{dx}(\sin \sqrt{x^2 + 1})$.

First differentiate $\sqrt{x^2 + 1}$ with respect to x.

The D.C. of $\sqrt{x^2 + 1}$ with respect to $(x^2 + 1)$ is $\frac{1}{2}(x^2 + 1)^{-\frac{1}{2}}$ or $\dfrac{1}{2\sqrt{x^2 + 1}}$.

The D.C. of $(x^2 + 1)$ with respect to x is $2x$.

$\therefore$ the D.C. of $\sqrt{x^2 + 1}$ with respect to x is

$$2x \times \frac{1}{2\sqrt{x^2 + 1}} = \frac{x}{\sqrt{x^2 + 1}}.$$

The D.C. of $\sin \sqrt{x^2 + 1}$ with respect to $\sqrt{x^2 + 1}$ is $\cos \sqrt{x^2 + 1}$.

The D.C. of $\sqrt{x^2 + 1}$ with respect to x is $\dfrac{x}{\sqrt{x^2 + 1}}$.

∴ the D.C. of $\sin \sqrt{x^2 + 1}$ with respect to x is

$$\frac{x}{\sqrt{x^2 + 1}} \cos \sqrt{x^2 + 1}.$$

Exercise 20b

Find the D.C. of the following functions with respect to x:

1. $(x + 1)^3$. **2.** $(3x - 1)^3$. **3.** $(1 - 2x)^4$.

4. $(1 + x + x^2)^5$. **5.** $\sqrt{2x + 1}$. **6.** $\dfrac{1}{\sqrt{2x + 1}}$.

7. $\sqrt{1 - 4x}$. **8.** $\dfrac{1}{\sqrt{1 - 4x}}$. **9.** $(3x^2 - 4x + 1)^2$.

10. $\sin 7x$. **11.** $\cos 3x$. **12.** $\sin^3 x$.

13. $\cos^3 4x$. **14.** $\sin \sqrt{x}$. **15.** $\cos (x^2 + 1)$.

16. $\cos (x^2 + x + 1)$. **17.** $\cos \sqrt{x^2 + 1}$. **18.** $\cos (x^3)$.

19. $\sin^2 (x + 1)$. **20.** $\sin^3 (2x + 1)$.

Implicit functions

We are now able to deal with implicit functions. If y is given in terms of x, y is said to be an **explicit** function of x.

For example, $y = x^3 - 4x^2 + 5$ and $y = \sin 3x$ both express y explicitly in terms of x. If, however, y and x are connected by an equation (from which we may or may not be able to express y in terms of x) then y is said to be an **implicit** function of x.

For example, $x^2 + y^2 = 8$ and $y^2 + x \sin y = 1$ both express y implicitly in terms of x.

In order to find $\dfrac{dy}{dx}$ from an implicit function, we need to be able to differentiate expressions such as y^2, xy and xy^2 with respect to x.

(i) The D.C. of y^2 with respect to y is $2y$;

the D.C. of y with respect to x is $\dfrac{dy}{dx}$;

therefore the D.C. of y^2 with respect to x is $2y\dfrac{dy}{dx}$.

(ii) $\dfrac{d}{dx}(xy) = x\dfrac{dy}{dx} + y$,   using the product formula.

(iii) $\dfrac{d}{dx}(xy^2) = x\dfrac{d}{dx}(y^2) + y^2$ (using the product formula)

$$= 2xy\dfrac{dy}{dx} + y^2.$$

Example 1. *Find* $\dfrac{dy}{dx}$ *given that* $x^2 + 3xy + 5y^2 = 1$.

Differentiating the equation with respect to x,

$$2x + 3\left(x\dfrac{dy}{dx} + y\right) + 5\left(2y\dfrac{dy}{dx}\right) = 0.$$

$$\therefore \dfrac{dy}{dx}(3x + 10y) = -(2x + 3y),$$

and

$$\dfrac{dy}{dx} = -\dfrac{2x + 3y}{3x + 10y}.$$

Example 2. *Find* $\dfrac{dy}{dx}$ *when* $x = 2$ *and* $y = 3$, *given that*

$$xy^2 + y - xy = 15.$$

It is first essential to differentiate the equation before putting in the particular values of x and y.

$$\dfrac{d}{dx}(xy^2) + \dfrac{dy}{dx} - \dfrac{d}{dx}(xy) = 0.$$

$$\therefore y^2 + 2xy\dfrac{dy}{dx} + \dfrac{dy}{dx} - x\dfrac{dy}{dx} - y = 0.$$

Now put $x = 2$, $y = 3$.

$$9 + 12\dfrac{dy}{dx} + \dfrac{dy}{dx} - 2\dfrac{dy}{dx} - 3 = 0.$$

$$\therefore 11\dfrac{dy}{dx} = -6 \quad \text{and} \quad \dfrac{dy}{dx} = -\dfrac{6}{11}.$$

Exercise 20c.

Find $\dfrac{dy}{dx}$ from the following equations:

1. $x^2 + y^2 = 4$.   **2.** $y^3 = 8x$.   **3.** $x + y = 8$.

4. $\sqrt{x} + \sqrt{y} = \sqrt{a}$.   **5.** $x^2 + 2xy = 1$.   **6.** $x^3 + y^3 = a^3$.

7. $xy + y^2 = 1$.   **8.** $y \sin x + y^2 = 1$.   **9.** $x = \sin y$.

10. $x = \cos y$.

Find the numerical values of $\dfrac{dy}{dx}$ in the following:

11. $x^2 + y^2 = 17$ at $(1, 4)$.   **12.** $x = \sin 2y$ at $\left(1, \dfrac{\pi}{4}\right)$.

13. $x = \cos 3y$ at $\left(0, \dfrac{\pi}{6}\right)$.   **14.** $x^2 - y^2 = 15$ at $(4, 1)$.

15. $x^2 + xy^2 = 6$ at $(2, 1)$.

Integration of a function of a function

Suppose we wish to find the integral of $(3x - 4)^2$. The integral of x^2 with respect to x is $\dfrac{x^3}{3} + c$. The integral of $(3x - 4)^2$ with respect to $(3x - 4)$ is therefore $\dfrac{(3x - 4)^3}{3} + c$, i.e.

$$\int (3x - 4)^2 \, d(3x - 4) = \frac{(3x - 4)^3}{3} + c.$$

But $d(3x - 4) = 3dx$.

$$\therefore \int (3x - 4)^2 \cdot 3dx = \frac{(3x - 4)^3}{3} + c$$

or

$$3 \int (3x - 4)^2 \, dx = \frac{(3x - 4)^3}{3} + c.$$

$$\therefore \int (3x - 4)^2 \, dx = \frac{(3x - 4)^3}{9} + c.$$

(A different c but still an arbitrary constant.)

This corresponds very closely to the differentiation of a function of a function but note that the similarity applies only to linear functions such as $(3x - 4)$ or in general $(ax + b)$.

Suppose we try the same operation in an attempt to integrate $(x^2 + 1)^3$.

$$\int (x^2 + 1)^3 \, d(x^2 + 1) = \frac{(x^2 + 1)^4}{4} + c.$$

But $d(x^2 + 1) = 2x \, dx$ and therefore

$$\int (x^2 + 1)^3 \cdot 2x \, dx = \frac{(x^2 + 1)^4}{4} + c.$$

Since $2x$ is not a constant, it cannot be taken outside the integral sign.

All we can deduce is that

$$\int x(x^2 + 1)^3 \, dx = \frac{(x^2 + 1)^4}{8} + c.$$

So it is easier to find the integral of $x(x^2 + 1)^3$ than it is to integrate $(x^2 + 1)^3$.

An integral is often simple when one factor of the integrand (the expression to be integrated) is the D.C. of an expression which occurs elsewhere in the integrand. The easiest method often is to guess the answer and check by differentiation.

Example 1. $\int x^2(x^3 + 1)^2 \, dx.$

The D.C. of $(x^3 + 1)$ is $3x^2$. Try $(x^3 + 1)^3$.

$$\frac{d}{dx}(x^3 + 1)^3 = 3(x^3 + 1)^2 . 3x^2 = 9x^2(x^3 + 1)^2.$$

$$\therefore \int x^2(x^3 + 1)^2 \, dx = \tfrac{1}{9}(x^3 + 1)^3 + c.$$

Example 2. $\int x \sin(x^2 + 1) \, dx.$

The D.C. of $(x^2 + 1)$ is $2x$. Try $\cos(x^2 + 1)$.

$$\frac{d}{dx} \cos(x^2 + 1) = -2x \sin(x^2 + 1).$$

$$\therefore \int x \sin(x^2 + 1) \, dx = -\tfrac{1}{2} \cos(x^2 + 1) + c.$$

Integration by substitution

The integrals above may also be done by the method of substitution. The substitution is usually chosen to simplify the more difficult factor of the integrand. Remember to substitute for the x of the dx as well as in the expression to be integrated.

Example 1. $\int x^2(x^3 + 1)^2 \, dx.$

Put $y = x^3 + 1$ so that $dy = 3x^2 \, dx$ and $x^2 \, dx = \tfrac{1}{3} dy.$

Then

$$\int x^2(x^3 + 1)^2 \, dx = \int \frac{dy}{3} . y^2 = \frac{1}{3}\left(\frac{y^3}{3}\right) + c.$$

$$= \frac{(x^3 + 1)^3}{9} + c.$$

Example 2. $\int x \, sin \, (x^2 + 1) \, dx.$

Put $y = x^2 + 1$ so that $dy = 2x \, dx$ and $x \, dx = \frac{1}{2}dy.$

Then
$$\int x \, sin \, (x^2 + 1) \, dx = \int \frac{1}{2} \, sin \, y \, dy = -\frac{1}{2} \cos y + c.$$

$$= -\frac{1}{2} \cos \, (x^2 + 1) + c.$$

N.B. When the integrand contains the expression $\sqrt{a^2 - x^2}$, the substitution $x = a \, sin \, \theta$ is often successful. This reduces the expression $\sqrt{a^2 - x^2}$ to $\sqrt{a^2 - a^2 \, sin^2 \, \theta}$ or $a \cos \theta.$

Definite integrals by substitution

The limits in a definite integral are the limiting values of the variable, so when the variable is changed, the limits also should be changed. This is generally easier than changing the variable back as in indefinite integrals. Care should be taken in substituting that the substitution is one that is acceptable throughout the complete range of the limits. For example, in an indefinite integral where the limits for x are 2 and 0, the substitution $x = sin \, \theta$ is not valid because $sin \, \theta$ cannot be greater than 1.

The following examples illustrate the method.

Example 1. $\int_0^1 \frac{x^2}{(x^3 + 1)^2} \, dx.$

Put $y = x^3 + 1$. Then $dy = 3x^2 \, dx$ and $x^2 \, dx = \frac{1}{3} \, dy.$
When $x = 0$, $y = 1$; when $x = 1$, $y = 2.$

$$\therefore \int_0^1 \frac{x^2}{(x^3 + 1)^2} \, dx = \frac{1}{3} \int_1^2 \frac{dy}{y^2} = \frac{1}{3}\left[-\frac{1}{y} \right]_1^2 = \frac{1}{3}\left(-\frac{1}{2} + 1 \right) = \frac{1}{6}.$$

Example 2. $\int_{-1}^1 \frac{dx}{\sqrt{4 - x^2}}.$

Put $x = 2 \, sin \, \theta$. Then
$$dx = 2 \cos \theta \, d\theta \text{ and } \sqrt{4 - x^2} = \sqrt{4 - 4 \, sin^2 \, \theta} = 2 \cos \theta.$$

When $x = -1$, $sin \, \theta = -\frac{1}{2}$ and $\theta = -\frac{\pi}{6}$; when $x = 1$, $\theta = \frac{\pi}{6}.$

$$\therefore \int_{-1}^1 \frac{dx}{\sqrt{4 - x^2}} = \int_{-\pi/6}^{\pi/6} \frac{2 \cos \theta \, d\theta}{2 \cos \theta} = \int_{-\pi/6}^{\pi/6} d\theta$$

$$= \left[\theta \right]_{-\pi/6}^{\pi/6} = \frac{\pi}{3}.$$

Exercise 20d: Miscellaneous

1. Differentiate $x \sin 2x$ with respect to x.

2. Differentiate $\dfrac{x}{\sin 2x}$ with respect to x.

3. Evaluate $\displaystyle\int_{0}^{\pi/4} \cos 2x \, dx$.

4. Find the area under the curve $y = \sin x$ between $x = 0$ and $x = \pi/2$.

5. Evaluate $\displaystyle\int_{\pi/6}^{\pi/3} \sin 2x \, dx$.

6. Differentiate $x^4 \cos 2x$.

7. Calculate the values of $\dfrac{dy}{dx}$ when $x = 1$ for the curve given by the equation
$x^2 - 3xy + 4y^2 = 2$.

8. Evaluate $\displaystyle\int_{0}^{2} \dfrac{dx}{(2x + 5)^2}$.

9. Find the value of $\tan x$ which makes $2 \cos x + 3 \sin x$ a maximum or a minimum.

10. Evaluate $\displaystyle\int_{0}^{1} \dfrac{dx}{\sqrt{1 - x^2}}$.

11. Evaluate $\displaystyle\int_{1}^{2} \dfrac{x^2 \, dx}{(x^3 + 1)^2}$.

12. Differentiate $\dfrac{\cos x}{2 + 3 \sin x}$.

13. Find the equation of the tangent to the curve $y = \cos x + \sin x$ at the point where $x = \dfrac{\pi}{4}$.

14. Evaluate $\displaystyle\int_{0}^{3} \dfrac{x}{\sqrt{1 + x^2}} \, dx$ by substitution $y^2 = 1 + x^2$.

15. Differentiate $\dfrac{\cos x}{1 - \sin x}$.

16. Differentiate $x \sin x + \cos x$.

17. If $2x^2 + xy + y^2 = 8$, find the values of $\dfrac{dy}{dx}$ and $\dfrac{d^2y}{dx^2}$ at the point $(1, 2)$.

18. Differentiate $\sqrt{\sin 3x}$.

19. Differentiate $x^3 \sin 2x$.

20. Evaluate $\displaystyle\int_0^{\pi/3} \cos 2x \, dx$.

21. Differentiate $\dfrac{\cos 3x}{\sin x}$.

22. Evaluate $\displaystyle\int_0^4 \dfrac{x \, dx}{\sqrt{1 + 3x^2}}$.

23. Differentiate $x\sqrt{x^2 + a^2}$.

24. Evaluate $\displaystyle\int_0^3 \dfrac{dx}{\sqrt{9 - x^2}}$.

25. Differentiate $\dfrac{\cos x}{\cos 2x}$.

26. Evaluate $\displaystyle\int_0^1 \dfrac{x \, dx}{\sqrt{1 - x^2}}$.

27. Evaluate $\displaystyle\int_0^4 \dfrac{x \, dx}{\sqrt{2x + 1}}$ by the substitution $2x + 1 = y^2$.

28. If $\dfrac{dy}{dx} = xy + x + y + 1$, find the value of $\dfrac{d^2y}{dx^2}$ when $x = y = 1$.

29. Evaluate $\displaystyle\int_0^3 \dfrac{2x + 4}{\sqrt{x + 1}} \, dx$.

30. Integrate $\sin x \sin 2x$. (Hint: $\cos x - \cos 3x = 2 \sin x \sin 2x$.)

31. Integrate $\cos x \cos 3x$.

32. Evaluate $\displaystyle\int_0^{\pi/6} \cos x \cos 5x \, dx$.

33. Find the area under the curve $y = \cos x + \sin x$ between $x = 0$ and $x = \dfrac{\pi}{2}$.

34. Differentiate $\cos (px + q) \sin (px + q)$.

35. Integrate $\sin (ax + b)$.

36. Find $\dfrac{dy}{dx}$ given that $y = \dfrac{x^2 + x + 1}{x^2 - x + 1}$.

37. Evaluate $\displaystyle\int_5^8 \dfrac{x^2 \, dx}{\sqrt{x - 4}}$.

38. The curve $y^2 = \sin x$ between $x = 0$ and $x = \pi/2$ is rotated about the x-axis. Find the volume generated.

39. Given that $y(y + 1) = x(x - 1)$, find $\dfrac{dy}{dx}$.

40. Find the equation of the tangent to the curve $y(y + 1) = x(x - 2)$ at the point $(2, -1)$.

41. Use the formula $\cos 2x = 2 \cos^2 x - 1$ to integrate $\cos^2 x$.

42. Evaluate $\displaystyle\int_0^{\pi/2} \sin^2 x \, dx$.

43. Use the formula $\cos 3x = 4 \cos^3 x - 3 \cos x$ to integrate $\cos^2 x$.

44. Integrate $\tan^2 x$. (Use $\sec^2 x = 1 + \tan^2 x$.)

45. The area under the curve $y = \sin x$ between $x = 0$ and $x = \pi/2$ is rotated about the x-axis. Find the volume generated.

21 Approximate Methods of Integration

An approximation for the area under a curve may be found by several different methods. One of these, that of counting squares under the curve, has already been mentioned but this is crude and we shall now consider other and better methods. There are only two main cases in which an approximation might be required. The first is when y is not known or cannot be expressed as a function of x; the second is when the known function of x cannot be integrated.

Let us consider the curve $y = \dfrac{1}{x}$ between $x = 1$ and $x = 2$. Divide the interval between 1 and 2 into ten equal parts so that the length of each division is 0.1 and erect ordinates at each point of division including the end points. For ten divisions there will be eleven ordinates and we call these $y_1, y_2 \ldots y_{11}$ as shown in Fig. 21.1.

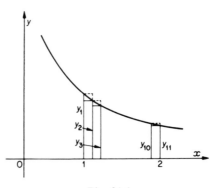

Fig. 21.1

By considering the areas of ten rectangles all drawn below the curve we see that

$$A \text{ (the area)} > \tfrac{1}{10}(y_2 + y_3 \ldots + y_{11})$$

and by considering rectangles drawn above the curve

$$A < \tfrac{1}{10}(y_1 + y_2 \ldots + y_{10}).$$
$$\therefore \tfrac{1}{10}(y_2 + y_3 + \ldots + y_{11}) < A < \tfrac{1}{10}(y_1 + y_2 + \ldots + y_{10}).$$

The value of y_1 is 1; the value of y_2 is $\dfrac{1}{1.1}$ or 0.9091 and so on. The results are tabulated below:

x	Upper approximation	Lower approximation
1.0	$y_1 = 1.0000$	
1.1	$y_2 = 0.9091$	$y_2 = 0.9091$
1.2	$y_3 = 0.8333$	$y_3 = 0.8333$
1.3	$y_4 = 0.7692$	$y_4 = 0.7692$
1.4	$y_5 = 0.7143$	$y_5 = 0.7143$
1.5	$y_6 = 0.6667$	$y_6 = 0.6667$
1.6	$y_7 = 0.6250$	$y_7 = 0.6250$
1.7	$y_8 = 0.5882$	$y_8 = 0.5882$
1.8	$y_9 = 0.5556$	$y_9 = 0.5556$
1.9	$y_{10} = 0.5263$	$y_{10} = 0.5263$
2.0		$y_{11} = 0.5000$
Sum	7.1877	6.6877

$$\therefore \ 0.66877 < A < 0.71877.$$

One would imagine that the average of these results should give a good approximation. The average is $\frac{1}{2}(0.66877 + 0.71877)$ or 0.6938.

The integral of $\dfrac{1}{x}$ is outside the scope of this book but is, in fact, $\log_e x$, abbreviated to ln x, and between the limits 1 and 2 the value of the integral is 0.6931 which agrees fairly well with our approximation.

The trapezoidal rule

By taking the average of the two limits in the last example, we have arrived at a rule for approximating to an area called the **trapezoidal rule**.

This is found by joining the ends of consecutive ordinates and by treating each trapezium so formed as an approximation for the area under the corresponding portion of the graph. In Fig. 21.2, the area of the first trapezium is $\dfrac{h}{2}(y_1 + y_2)$, of the second $\dfrac{h}{2}(y_2 + y_3)$ and so on where h is the distance between consecutive ordinates.

Therefore an approximation for the area is

$$\tfrac{1}{2}h(\overline{y_1 + y_2} + \overline{y_2 + y_3} + \ldots + \overline{y_{n-1} + y_n})$$

where y_n is the last ordinate.

$$\therefore \ A \doteqdot \tfrac{1}{2}h(y_1 + 2y_2 + 2y_3 + \ldots + 2y_{n-1} + y_n).$$

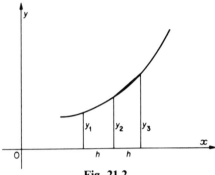

Fig. 21.2

This may be expressed as follows:

Take the average of the first and last ordinates and add to it the sum of all the intermediate ordinates. The product of this sum and the distance between consecutive ordinates is an approximation for the area.

Simpson's rule

The trapezoidal rule is obtained by treating each part of the curve as a straight line, namely as $y = bx + c$. Suppose that we now find the best approximation for each part of the curve in the form $y = ax^2 + bx + c$, which is a parabola. This must always (unless the curve is made up of straight lines) give a better approximation because, at the worse, we revert to the case when $a = 0$, which is the trapezoidal rule.

First, we shall find a formula for the area under the curve $y = ax^2 + bx + c$ in terms of the ordinates.

Consider the case of three ordinates, taking one at $x = 0$ and the others equally spaced at $x = -h$ and $x = h$ (see Fig. 21.3).

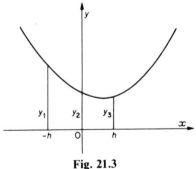

Fig. 21.3

The actual area under the curve is

$$\int_{-h}^{h} (ax^2 + bx + c)\, dx = \left[\frac{ax^3}{3} + \frac{bx^2}{2} + cx\right]_{-h}^{h} = \frac{2ah^3}{3} + 2ch.$$

The values of the ordinates are:

$$y_1 = ah^2 - bh + c,$$
$$y_2 = c,$$
$$y_3 = ah^2 + bh + c.$$
$$\therefore y_1 + y_3 = 2ah^2 + 2c = 2ah^2 + 2y_2,$$

and
$$2ah^2 = y_1 + y_3 - 2y_2.$$

$$\therefore A = h\left(\frac{2ah^2}{3} + 2c\right) = h\{\tfrac{1}{3}(y_1 + y_3 - 2y_2) + 2y_2\}$$

$$= \frac{h}{3}(y_1 + 4y_2 + y_3).$$

This is a geometrical formula expressing A in terms of consecutive ordinates and will be true for this curve wherever the axis of y. An approximation for the area based on this formula is called **Simpson's rule**. To apply this rule, the division must be made into an *even* number of parts. Suppose the corresponding ordinates are y_1, y_2, $y_3 \ldots y_n$, where n is odd. (The number of ordinates is one greater than the number of intervals.)

The area between y_1 and y_3 is approximately
$$\tfrac{1}{3}(y_1 + 4y_2 + y_3).$$

The area between y_3 and y_5 is approximately
$$\tfrac{1}{3}h(y_3 + 4y_4 + y_5).$$

The area between y_5 and y_7 is approximately
$$\tfrac{1}{3}h(y_5 + 4y_6 + y_7).$$

And finally the area between y_{n-2} and y_n is
$$\tfrac{1}{3}h(y_{n-2} + 4y_{n-1} + y_n) \qquad \text{(see Fig. 21.4).}$$

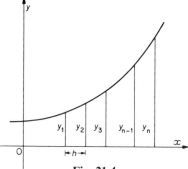

Fig. 21.4

The total area is therefore

$$\tfrac{1}{3}h(y_1 + 4y_2 + 2y_3 + 4y_4 \ldots + 4y_{n-1} + y_n) \text{ approximately.}$$

In words:

Divide the interval into an even number of equal parts. To the sum of the first and last ordinates, add twice the sum of the other odd ordinates and four times the sum of the other even ordinates; one-third of the product of this sum and the interval between consecutive ordinates is an approximation for the area under the curve.

Let us now apply Simpson's rule to the problem we have already solved by the trapezoidal rule, that of finding the area under the curve $y = \dfrac{1}{x}$ between $x = 1$ and $x = 2$.

The values of $y_1, y_2, \ldots y_{11}$ have already been found, and the results are tabulated below in appropriate columns.

$$\therefore A = \tfrac{1}{3}(\tfrac{1}{10})(1.5 + 2 \times 2.7282 + 4 \times 3.4595)$$

$$= \frac{20.7944}{30} = 0.69315.$$

	First and last	Even ordinates	Odd ordinates
$y_1 = 1$ $y_2 = 0.9091$ $y_3 = 0.8333$ $y_4 = 0.7692$ $y_5 = 0.7143$ $y_6 = 0.6667$ $y_7 = 0.6250$ $y_8 = 0.5882$ $y_9 = 0.5556$ $y_{10} = 0.5263$ $y_{11} = 0.5000$	$y_1 = 1.0000$ $y_{11} = 0.5000$	$y_2 = 0.9091$ $y_4 = 0.7692$ $y_6 = 0.6667$ $y_8 = 0.5882$ $y_{10} = 0.5263$	$y_3 = 0.8333$ $y_5 = 0.7143$ $y_7 = 0.6250$ $y_9 = 0.5556$
Sum	1.5000	3.4595	2.7282

As will be seen this gives a better approximation and is very close to the actual value.

The addition can be done quickly and accurately using a calculator, especially when the values are tabulated as above.

Example 1. *Find the area under the curve* $y = \dfrac{1}{1 + x^2}$ *between* $x = 0$ *and* $x = 1$ *using (i) the trapezoidal rule, (ii) Simpson's rule.*

Divide the base into ten equal parts and call the ordinates y_1 to y_{11}.

$$y_1 = \frac{1}{1 + 0^2} = 1.$$

$$y_2 = \frac{1}{1 + (0.1)^2} = \frac{1}{1.01} = 0.9901.$$

$$y_3 = \frac{1}{1 + (0.2)^2} = \frac{1}{1.04} = 0.9615 \quad \text{etc.}$$

The results are tabulated below.

(i) The trapezoidal rule gives

$$A \doteqdot \frac{1}{10}\left(\frac{1.5000}{2} + 3.9311 + 3.1687\right) = 0.78498.$$

(ii) Simpson's rule gives

$$A \doteqdot \frac{1}{10}\left(\tfrac{1}{3}\right)\{1.5000 + 2(3.1687) + 4(3.9311)\}$$
$$= 0.78539.$$

	First and last	Even ordinates	Odd ordinates
y_1	1.0000		
y_2		0.9901	
y_3			0.9615
y_4		0.9174	
y_5			0.8621
y_6		0.8000	
y_7			0.7353
y_8		0.6711	
y_9			0.6098
y_{10}		0.5525	
y_{11}	0.5000		
Sum	1.5000	3.9311	3.1687

[The integral of $\dfrac{1}{1 + x^2}$ is $\tan^{-1} x$ and the value of the integral between 0 and 1 is $\dfrac{\pi}{4}$. The Simpson's rule approximation therefore gives 3.14156 as the value of π.]

Example 2. *The velocity, v m s^{-1}, of a particle moving in a straight line is given at second intervals by the following table:*

v	0	6	11	15	16	14	12	7	0
t	0	1	2	3	4	5	6	7	8

Find the distance travelled in the 8 seconds.

Figure 21.5 shows the velocity–time graph. The area under the curve represents the distance travelled.

The sum of the first and last ordinates is 0.
The sum of the odd ordinates is 39.
The sum of the even ordinates is 42.

By Simpson's rule, the distance travelled is approximately
$$\tfrac{1}{3}\{0 + 2(39) + 4(42)\} \text{ metres}$$

or 82 m.

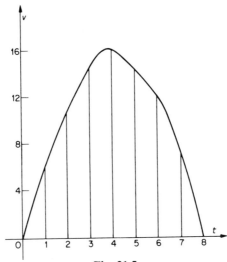

Fig. 21.5

Exercise 21

1. Find the area under the curve $y = x^2 + x$ between $x = 0$ and $x = 3$ using (i) the trapezoidal rule; (ii) Simpson's rule with 10 intervals; (iii) integration.

2. Find an approximate value for $\displaystyle\int_0^6 y \, dx$ given the following values:

x	0	1	2	3	4	5	6
y	8	12	14	11	9	3	1

3. Find the distance travelled in a straight line by a particle in 6 s given that the velocity, v m s^{-1}, is connected with the time t seconds as follows:

t	0	1	2	3	4	5	6
v	10	12	15	16	11	5	3

4. Find an approximate value for $\displaystyle\int_0^1 10^x \, dx$ using Simpson's rule with 10 intervals.

5. Evaluate $\int_0^{\frac{1}{2}} \dfrac{dx}{\sqrt{1-x^2}}$ by Simpson's rule with 10 intervals. Given that the value of the integral is $\dfrac{\pi}{6}$, find an approximation for π.

6. A car starts from rest and its acceleration in m s^{-2} over the first 10 s is given by the following table:

t	0	1	2	3	4	5	6	7	8	9	10
a	2	2.4	2.8	3.2	4.0	4.2·	4.5	6.0	6.2	6.3	6.4

Find the velocity at the end of the 10 seconds.

7. The area A square metres of the cross-section of a trunk of a tree at height h metres above the ground is given in the table:

h	0	1	2	3	4
A	0.35	0.33	0.3	0.25	0.24

Find approximately the volume of the first 4 m of the trunk.

8. Evaluate $\int_0^1 \dfrac{dx}{\sqrt{1+x^2}}$.

9. The cross-section of a sphere of radius r at a distance x from the centre is of area $\pi(r^2 - x^2)$. Find a formula for the volume of a sphere (i) by integration; (ii) using Simpson's rule. Does it matter in this example how many intervals are taken?

10. The length of the ordinate of a circle of radius 10 cm at a distance of x cm from the centre is equal to $\sqrt{100 - x^2}$ cm. Calculate the area of the circle using Simpson's rule.

COORDINATE GEOMETRY

22 Distances, Mid-points and Gradients

Coordinates

The coordinates of a point referred to perpendicular axes are written as an ordered pair with the x-coordinate always written first, followed, after a comma, by the y-coordinate. Thus the point (3, 2) represents the point whose x-coordinate or **abscissa** is 3 and whose y-coordinate or **ordinate** is 2. The x-coordinate is the distance of the point from the y-axis or the distance measured parallel to the axis of x; the y-coordinate is the distance from the x-axis or the distance measured parallel to the axis of y. It is conventional to draw the x-axis horizontal and the y-axis vertical and the point of intersection of these axes is called the origin, O. Either or both of the coordinates may be negative depending on the quadrant, in which the point lies. In the second and third quadrants the abscissa is negative; in the third and fourth quadrants, the ordinate is negative.

When we deal with a general point instead of a particular point such as $(-1, 3)$, we need letters for the coordinates. It is perfectly reasonable to call a point (h, k) for example but there is considerable advantage in naming it (x_1, y_1). In the first place, x_1 obviously stands for the x-coordinate and y_1 for the y-coordinate and in the second place further points may be named (x_2, y_2), (x_3, y_3), etc. There is no connection between x_1 and x_2 except that they both represent abscissae; the suffix in no way corresponds to an index.

The distance between two points

Suppose we wish to find the distance between the points A (1, 2) and B (4, 6) as shown in Fig. 22.1.

Through A draw the line parallel to the axis of x to meet the line through B parallel to the axis of y at N.

Then
$$AN = 4 - 1 = 3;$$
$$BN = 6 - 2 = 4.$$
$$\therefore AB^2 = AN^2 + BN^2 = 3^2 + 4^2 = 25.$$

Therefore $AB = 5$ and the distance between the points is 5.

Now let us generalize this example by finding the distance between the points P_1 (x_1, y_1) and P_2 (x_2, y_2).

In Fig. 22.2,
$$P_1N = x_2 - x_1 \quad \text{and} \quad P_2N = y_2 - y_1.$$

By Pythagoras, $P_1P_2 = \sqrt{(x_2 - x_1)^2 + (y_2 - y_1)^2}.$

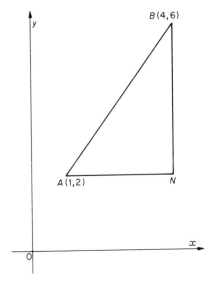

Fig. 22.1

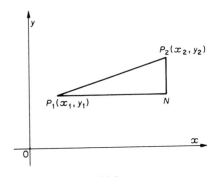

Fig. 22.2

Notice that this formula still holds if any one of the coordinates is negative. If the points given are $(-1, 5)$ and $(11, 10)$, the distance between the points measured parallel to the axis of x is $11 - (-1)$ or 12, i.e. the **difference** between the abscissae. The distance between the points measured parallel to the axis of y is $10 - 5$ or 5 and the distance between the points is $\sqrt{12^2 + 5^2}$ or 13.

The distance between P_1 and P_2 measured parallel to the x-axis is called the **projection** of $P_1 P_2$ on the x-axis.

The mid-point of a straight line

To find the mid-point of the line joining A $(1, 3)$ and B $(5, 7)$, draw parallels to the axes to meet in N as shown in Fig. 22.3.

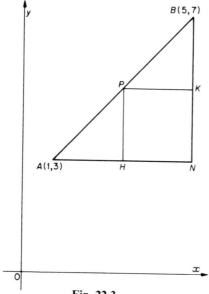

Fig. 22.3

If P is the mid-point of the line, let the line through P parallel to the x-axis meet BN at K and the line through P parallel to the y-axis meet AN at H. Then H and K are the mid-points of AN and BN respectively (mid-point theorem).

$$AN = 5 - 1 = 4. \quad \therefore AH = 2.$$

The x-coordinate of P is therefore $2 + 1$ or 3.

$$BN = 7 - 3 = 4. \quad \therefore NK = 2.$$

The y-coordinate of P is therefore $3 + 2$ or 5.

$$\therefore P \text{ is the point } (3, 5).$$

We shall now generalize to find the mid-point of the line joining P_1 (x_1, y_1) to P_2 $(x_2\ y_2)$.

In Fig. 22.4, let the mid-point of P_1P_2 be Q and let parallels through Q to the axes meet AN and BN at H and K as shown.

Then H and K are the mid-points of P_1N and P_2N respectively.

$$P_1N = x_2 - x_1. \quad \therefore P_1H = \tfrac{1}{2}(x_2 - x_1).$$

The x-coordinate of $Q = x_1 + P_1H$

$$= x_1 + \tfrac{1}{2}(x_2 - x_1) = \tfrac{1}{2}(x_1 + x_2).$$

Similarly the y-coordinate of Q is $\frac{1}{2}(y_1 + y_2)$ and the coordinates of Q are $\left(\dfrac{x_1 + x_2}{2}, \dfrac{y_1 + y_2}{2}\right)$.

Thus the mid-point of a line is found by taking the average of the coordinates separately.

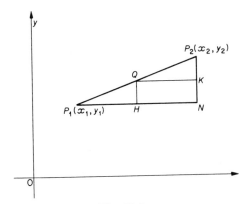

Fig. 22.4

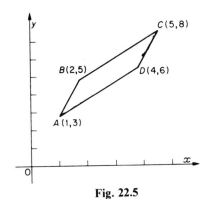

Fig. 22.5

Example 1. *Show that the points A (1, 3), B (2, 5), C (5, 8) and D (4, 6) form the vertices of a parallelogram.*

The points are shown in Fig. 22.5.

The mid-point of AC is $\left(\dfrac{1 + 5}{2}, \dfrac{3 + 8}{2}\right)$ or $(3, 5\frac{1}{2})$.

The mid-point of BD is $\left(\dfrac{2 + 4}{2}, \dfrac{5 + 6}{2}\right)$ or $(3, 5\frac{1}{2})$.

Since AC and BD have a common mid-point, these lines must bisect each other. Therefore $ABCD$ is a parallelogram (diagonals bisect).

Example 2. *Prove that the points $A\ (-1,\ 2)$, $B\ (3,\ 4)$ and $C\ (2,\ -4)$ form a right-angled triangle and find its area.*

$$AB^2 = (3 + 1)^2 + (4 - 2)^2 = 20.$$
$$AC^2 = (2 + 1)^2 + (-4 - 2)^2 = 45.$$
$$BC^2 = (3 - 2)^2 + (4 + 4)^2 = 65.$$

$\therefore BC^2 = AB^2 + AC^2$ and so the angle A is a right angle.
Since $A = 90°$,

$$\text{the area of the triangle} = \tfrac{1}{2}AB \cdot AC$$
$$= \tfrac{1}{2}\sqrt{20} \times \sqrt{45}$$
$$= \tfrac{1}{2}\sqrt{900}$$
$$= 15 \text{ square units.}$$

Exercise 22a

Find the distances between the pairs of points in questions 1 to 5:

1. $(3, 4)$ and $(2, 7)$.   **2.** $(5, 1)$ and $(-3, -4)$.

3. $(-3, 1)$ and $(2, -3)$.   **4.** $(-2, -3)$ and $(3, -1)$.

5. $(1, 2)$ and $(-1, 4)$.

Write down the mid-points of the lines joining the points given in questions 6 to 10:

6. $(1, 1)$ and $(4, 5)$.   **7.** $(5, 1)$ and $(3, -1)$.

8. $(4, 2)$ and $(-5, 0)$.   **9.** $(6, 1)$ and $(8, -2)$.

10. $(-1, -6)$ and $(-3, -2)$.

11. Prove that the points $(-1, 4)$, $(2, 2)$, $(2, 5)$ and $(5, 3)$ are the vertices of a parallelogram.

12. Prove that the points $(2, 0)$, $(-2, 0)$, $(0, 3)$ and $(0, -3)$ are the vertices of a rhombus.

13. Show that the points $A\ (1, 4)$, $B\ (2, 7)$ and $C\ (-5, 6)$ form a right-angled triangle and find its area.

14. Show that the points $(2, 0)$, $(4, 0)$, $(3, -5)$ and $(3, 5)$ form a rhombus and find its area.

15. A, B and C are three points such that B is the mid-point of AC. Given that A is $(-1, 6)$, B is $(2, 4)$, find the coordinates of C.

16. Show that the points $(5, 2)$, $(5, 8)$, $(3, 5)$ and $(7, 5)$ are the vertices of a rhombus.

17. Show that the points $(2, 4)$ and $(4, -2)$ subtend a right angle at the origin.

18. Show that the points $(1, m)$ and $(-m, 1)$ subtend a right angle at the origin.

19. Show that the lines $y = mx$ and $my + x = 0$ are perpendicular.

20. Find the point of intersection of the lines $2x + y = 3$ and $2y - x = 1$. By considering one particular point on each line, show that the lines are perpendicular.

The area of a triangle

In a previous example, we showed that the area of the triangle formed by joining the points $A(-1, 2)$, $B(3, 4)$ and $C(2, -4)$ is 15 square units. The fact that the area of the triangle was found as the product of two surds which gave a rational result suggests there might be an easier way of finding the area. Figure 22.6 shows the relative positions of the points A, B and C. Draw perpendiculars AP and BQ to the line through C parallel to the axis of x.

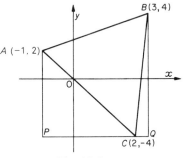

Fig. 22.6

The area of the trapezium $APQB$ is $\frac{1}{2}(6 + 8)4 = 28$.
The area of the triangle $APC = \frac{1}{2}(6)(3) = 9$.
The area of the triangle $BCQ = \frac{1}{2}(1)(8) = 4$.
$\therefore$ the area of the triangle $ABC = 28 - 9 - 4 = 15$.

It is possible to find a formula for the area of the triangle formed by the points (x_1, y_1), (x_2, y_2) and (x_3, y_3) but it is certainly not worth memorizing the result which is

$$\frac{1}{2}(x_1 y_2 - y_1 x_2 + x_2 y_3 - x_3 y_2 + x_3 y_1 - x_1 y_3).$$

The formula for the area of the triangle formed by the origin and the points $P_1(x_1, y_1)$ and $P_2(x_2, y_2)$ is however often useful and we shall now see how this formula may be obtained.

In Fig. 22.7, suppose that $P_1 N_1$ and $P_2 N_2$ are perpendiculars to the axis of x.

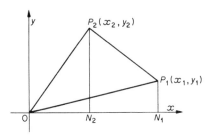

Fig. 22.7

The area of OP_1P_2 = (area of ON_2P_2) + (area of $N_2N_1P_1P_2$)
$\qquad\qquad\qquad$ − (area of ON_1P_1).
$\quad$ Area of triangle $ON_2P_2 = \frac{1}{2}x_2y_2$.
$\quad$ Area of trapezium $N_2N_1P_1P_2 = \frac{1}{2}(y_1 + y_2)(x_1 - x_2)$.
$\quad$ Area of triangle $ON_1P_1 = \frac{1}{2}x_1y_1$.
$\therefore$ area of triangle $OP_1P_2 = \frac{1}{2}x_2y_2 + \frac{1}{2}(y_1 + y_2)(x_1 - x_2) - \frac{1}{2}x_1y_1$
$\qquad\qquad\qquad\qquad\quad = \frac{1}{2}(y_2x_1 - y_1x_2)$.

Notice that the value of the area quoted for the triangle $P_1P_2P_3$ when $x_3 = y_3 = 0$ reduces to the same expression.

Gradient

In the first chapter of the calculus section, the gradient of a line was defined as $\dfrac{\text{increase in } y}{\text{increase in } x}$, where the increases are measured between two points of the line.

The gradient of a line is also equal to the tangent of the angle which the line makes with the positive direction of the x-axis (provided that the scales are equal on both axes). A line which makes an acute angle with the positive direction of the x-axis has a positive gradient; a line which makes an obtuse angle with this direction has a negative gradient.

If two lines are parallel, it is obvious by similar triangles that their gradients are equal and so parallel lines have equal gradients.

The line $y = mx + c$ is such that $\dfrac{dy}{dx} = m$ (if m and c are constants), i.e. the line has gradient m.

When $x = 0$, $y = c$ and so the line has gradient m and makes an intercept equal to c on the y-axis.

The line of gradient m through the origin is $y = mx$.

Now consider the lines $y = mx$ and $y = tx$ which both pass through the origin and are shown in Fig. 22.8.

Let P be the point on $y = mx$ where $x = 1$; i.e. P is $(1, m)$.
Let Q be the point on $y = tx$ where $x = 1$; i.e. Q is $(1, t)$.
Then $\quad OP^2 = 1 + m^2$; $\quad OQ^2 = 1 + t^2$; $\quad PQ = m - t$.
What is the condition that the lines should be perpendicular?
If the angle $POQ = 90°$,

$$PQ^2 = OP^2 + OQ^2.$$
$$\therefore (m - t)^2 = 1 + m^2 + 1 + t^2$$

or $\qquad\qquad\qquad\quad -2mt = 2$

i.e. $\qquad\qquad\qquad\quad mt = -1$.

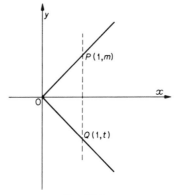

Fig. 22.8

Since parallel lines have equal gradients, this condition is true if two lines are perpendicular whether they pass through the origin or not.

The product of the gradients of perpendicular lines is -1.

The equation of a line

The equation of a line is a relation which exists between the coordinates of any and every point of the line. For example, the line $2x + 3y = 5$ passes through the point $(1, 1)$ because $2(1) + 3(1) = 5$. Conversely, the coordinates of a point which does not lie on the line cannot satisfy the equation. Since $2(2) + 3(3)$ is not equal to 5, the point $(2, 3)$ cannot lie on the line.

This line is shown in Fig. 22.9 and it is worth pointing out that for any point in the region above the lines (shaded horizontally)

$$2x + 3y > 5.$$

For any point below the line

$$2x + 3y < 5.$$

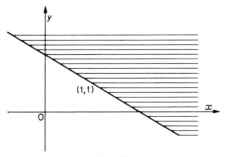

Fig. 22.9

Equation of the line of gradient m through (x_1, y_1)

To find the equation of the line of gradient 3 through the point $(1, 2)$, suppose that (x, y) is a point of the line.

The increase in y between $(1, 2)$ and (x, y) is $y - 2$.

The increase in x between $(1, 2)$ and (x, y) is $x - 1$.

$$\therefore \frac{y - 2}{x - 1} = 3 \quad \text{or} \quad y = 3x - 1.$$

Since this equation connects the coordinates of any point on the line, it is the equation of the line.

More generally, if the line joining (x_1, y_1) to (x, y) has gradient m, $\dfrac{y - y_1}{x - x_1} = m$ and therefore $y - y_1 = m(x - x_1)$ is the equation of the line.

Example 1. *Find the equation of the line through $(2, -1)$ parallel to $3x + y = 7$.*

The gradient of $3x + y = 7$ is -3.

The equation of the line is

$$\frac{y + 1}{x - 2} = -3$$

or $\qquad\qquad 3x + y = 5.$

It should be noted that any line parallel to $3x + y = 7$ is of the form $3x + y = c$, where c is a constant.

If $3x + y = c$ passes through $(2, -1)$,

$$3(2) - 1 = c \quad \text{and} \quad c = 5.$$

The line required is $3x + y = 5$.

Example 2. *Find the equation of the line through $(2, -1)$ perpendicular to $3x + y = 7$.*

The gradient of $3x + y = 7$ is -3.

The gradient of a perpendicular line is $+\frac{1}{3}$ (product -1).

The equation of the perpendicular through $(2, -1)$ is

$$\frac{y + 1}{x - 2} = \frac{1}{3}$$

or $\qquad\qquad 3y - x = -5.$

This may also be written down more quickly by noting that any line perpendicular to $3x + y = 7$ must be of the form $3y - x = k$, where k is a constant. The coefficients of x and y are interchanged and the sign between the terms altered.

Since the line passes through $(2, -1)$,

$$3(-1) - 2 = k$$

and the equation of the line is $3y - x = -5$.

Exercise 22b

In questions 1 to 10, ABC is the triangle formed by the points A (2, 3), B(4, 7) and C $(-2, 1)$.

1. Find the equation of the line BC.
2. Find the equation of the line through A parallel to BC.
3. Find the equation of the altitude through A.
4. Find the equation of the median through A.
5. Find the coordinates of the centre of gravity of the triangle ABC (i.e. the meet of the medians).
6. Find the coordinates of the orthocentre of the triangle ABC (i.e. the meet of the altitudes).
7. Find the equation of the perpendicular bisector of BC.
8. Find the coordinates of the circumcentre of the triangle ABC (i.e. the meet of the perpendicular bisectors of the sides).
9. Find the lengths of the sides of the triangle ABC.
10. Find cos A.
11. Find the equation of the line through (1, 2) parallel to $3x + 5y = 8$.
12. Find the equation of the line through $(1, -1)$ perpendicular to $2x - 3y = 4$.
13. Find the equation of the line joining (1, 2) and (2, 4).
14. Find the area of the triangle formed by the origin and the points (2, 3) and (3, 4).
15. Find the equation of the line through (1, 2) parallel to the join of (3, 4) and (5, 6).
16. Find the equation of the line through (2, 3) which makes equal positive intercepts on the axes.
17. Find the equation of the line through (1, 4) of gradient 2.
18. Find the equation of the line of gradient 2 through the point of intersection of $2x + 3y = 5$ and $3x - y = 2$.
19. Find the equation of the line joining the origin to the point of intersection of the lines $x + 2y = 4$ and $3x + y = 7$.
20. The point A lies on the x-axis and the point B on the y-axis. Show that the distance of the mid-point of AB from the origin is half the length of AB.

Determination of a linear law

The equation of a line in the form $y = mx + c$ is useful in determining a linear law between two variables when corresponding values of the variables are found by experiment.

Example. *In a system of pulleys an effort of P newtons is required to lift a mass of M kg. It is thought that a law of the form $P = aM + b$, where a and b are* *constants, exists between a and b. From the given experimental values, find if such a law does hold and determine suitable values for a and b.*

P	12	21	32	43	52
M	2	4	6	8	10

It is of course possible to take two pairs of values and substitute them in the equation $P = aM + b$. The values of a and b can then be found by solving the resulting simultaneous equations for a and b. The disadvantage of this method is that one of the measurements taken may be a faulty one and the best method is one which utilizes all the information given. Plot the points on a graph taking P as the vertical axis and M as the horizontal axis. The axes are chosen in this way so that the equation $P = aM + b$ corresponds to $y = mx + c$.

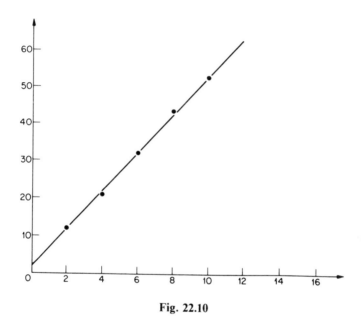

Fig. 22.10

If the points lie approximately on a straight line, a law such as $P = aM + b$ does hold. Draw the line which most nearly fits the points. If one of the points is clearly not on the best line suggested by other points, disregard it and consider the discrepancy to be due to an error in measurement. Produce the line to meet the axis of P as shown in Fig. 22.10. The value of a is the gradient of the line and b is equal to the intercept on the axis of P which is 2.

$$\text{The gradient} = \frac{52 - 2}{10} = 5.$$

$$\therefore a = 5, \quad b = 2.$$

The relation y = axⁿ

The same principle may also be used to find values of a and n when it is known that variables x and y satisfy an equation of the form $y = ax^n$, where a and n are constants.

If $y = ax^n$, then
$$\log y = \log a + n \log x.$$

So $\log y$ (vertically) is plotted against $\log x$ (horizontally). The points should lie on a straight line of which the gradient is n.

The intercept made by the line on the $\log y$ axis is equal to $\log a$, from which a may be found.

Example. *It is known that x and y obey the law $y = ax^n$. Find values of a and n from the corresponding values of x and y given in the table:*

y	178	617	1400	2820	4900
x	4	6	8	10	12

Find the values of $\log x$ and $\log y$.

log y	2.2504	2.7903	3.1461	3.4502	3.6902
log x	0.6021	0.7782	0.9031	1.0000	1.0792

Plot $\log y$ against $\log x$ as shown in Fig. 22.11. The points are seen to lie on a

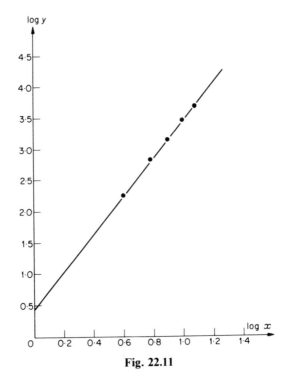

Fig. 22.11

straight line and this line when produced meets the axis of log y at the point where log $y = 0.45$.

$$\therefore \; \log a = 0.45 \quad \text{and} \quad a = 2.82.$$

The gradient of the line $= 3.45 - 0.45 = 3$.

$$y = 2.82x^3 \text{ approximately.}$$

Exercise 22c: Miscellaneous

1. Find the coordinates of the point which divides the join of $A\,(-2, 4)$ to B (6,12) in the ratio $3:1$.

2. Find the equation of the perpendicular bisector of the line joining $(-1, 5)$ to $(3, 7)$.

3. Find the equation of the line joining the intersection of $2x + 3y = 7$ and $3x - 2y = 1$ to the origin.

4. Find expressions for the intercepts made on the axes by the line of gradient m through the point $(1, 2)$.

5. Prove that the points $(2, 4)$, $(3, -1)$ and $(7, 3)$ form an isosceles triangle.

6. Find the area of the triangle formed by the points $(0, 0)$, $(4, 3)$ and $(1, 2)$.

7. Find the equation of the line through $(3, -3)$ parallel to $y + 2x = 8$.

8. Find the equation of the line through $(2, -1)$ perpendicular to $2x - 3y = 5$.

9. Find the equation of the line through $(1, 4)$ which makes equal positive intercepts on the axes.

10. If P is the point (x, y), A the point $(1, 2)$ and B the point $(3, 4)$, write down expressions for PA^2 and PB^2. Given that $PA = PB$, find an equation connecting x and y. What does this equation represent?

11. Show that the points $(1, 3)$, $(3, 5)$, $(8, 12)$ and $(6, 10)$ are the vertices of a parallelogram.

12. Show that the points $(1, 3)$, $(4, 5)$, $(6, 8)$ and $(3, 6)$ are the vertices of a rhombus. Find the area of the rhombus.

13. It is thought that the values of P and W given in the table are connected by an equation of the form $P = aW + b$. Find suitable values for a and b.

| P | 2.5 | 3.8 | 5.0 | 6.5 | 7.5 |
| W | 6 | 10 | 14 | 18 | 22 |

14. The following values of x and y are thought to satisfy $y = ax^n$. Find suitable values for a and n.

| y | 0.4 | 0.9 | 1.6 | 2.5 | 3.6 |
| x | 2 | 3 | 4 | 5 | 6 |

15. If corresponding values of p and v are given by

| p | 3.3 | 4.0 | 4.7 | 5.4 | 6.0 |
| v | 12 | 15 | 18 | 22 | 25 |

show that p and v satisfy $p = av + b$ and find the values of a and b.

16. The points whose coordinates are $(5, 11)$, $(6, 14)$ and $(8, 20)$ all lie on a straight line. Find its equation.

17. A law such as $p = av^n$ is thought to connect p and v. If experimental values are given below, find suitable values for a and n.

p	6.32	8.94	10	12.7	14.1
v	10	20	25	40	50

18. It is known that the quantities x and y are connected by a law such as $y = ax^n$. There is a misprint in the values of y given in the table. Find which value is wrong and state what it should be.

y	3.162	2.632	1.826	1.581	1.414
x	10	20	30	40	50

19. If A is the point $(1, 2)$ and B the point $(2, 5)$, find the coordinates of the point C which lies on AB produced and is such that $AC = 2BC$.

20. The line of gradient 2 is drawn through the point $(1, 0)$ and the line of gradient $-\frac{1}{2}$ through the point $(-1, 0)$. Find the point of intersection of these lines.

21. The line of gradient m is drawn through the point $(1, 0)$ and the line of gradient $-\dfrac{1}{m}$ through the point $(-1, 0)$. Show that the coordinates of the point of intersection of these lines satisfy the equation $x^2 + y^2 = 1$.

22. The triangle ABC is formed by the point A $(2, -1)$, the points B $(3, 2)$ and C $(4, 3)$. If D is the mid-point of BC, show that
$$AB^2 + AC^2 = 2AD^2 + 2BD^2.$$

23. If x and y are positive integers and it is given that $x + y < 4$ and $2x + y > 4$, find x and y by a graphical method.

24. Find the angle between the lines $y = 2x$ and $y = x$.

25. Find the gradient of the normal at the point $(1, 0)$ to the curve $y = x^2 - x$. (The normal is the perpendicular to the tangent at its point of contact.)

26. Find the gradient of the normal at the point $(2, 9)$ to the curve $y = 3x^2 - x - 1$.

27. Find the equation of the normal at the point $(2, 6)$ to the curve $y = x^2 + x$.

28. The tangent to the curve $y = x^2 + 2x$ at P $(1, 3)$ meets the axis of x at A and the axis of y at B. Show that $PA = 3AB$.

29. Find the equation of the tangent to the curve $x^2 + 4y^2 = 20$ at the point $(2, 2)$.

30. Find the equation of the straight line which is such that the axis of x bisects the angle between it and the line $3x + 2y = 7$.

23 The Straight Line

The point dividing *AB* in a given ratio

Suppose that A is $(2, 3)$ and B is $(8, 11)$. What are the coordinates of the point P on the line AB such that $AP:PB = 2:3$?

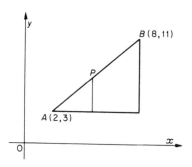

Fig. 23.1

In Fig. 23.1, the increase of x between A and B is $8 - 2$ or 6. Since $AP:PB = 2:3$, $AP = \frac{2}{5}AB$ and so the increase of x between A and P is $\frac{2}{5}(6)$ or 2.4. Therefore the x-coordinate of P is $2 + 2.4$ or 4.4.

Similarly the increase of y between A and B is 8. The increase of y between A and P is therefore $\frac{2}{5}(8)$ or 3.2. The y-coordinate of P is 6.2 and the point P has coordinates $(4.4, 6.2)$.

Let us now generalize this result by finding the coordinates of the point P which divides the line joining A (x_1, y_1) to B (x_2, y_2) in the ratio $\lambda:\mu$. Referring again to Fig. 23.1,

$$\frac{AP}{AB} = \frac{\lambda}{\lambda + \mu}.$$

The increase of x between A and P is

$$\frac{\lambda}{\lambda + \mu}(x_2 - x_1).$$

The x-coordinate of P is

$$x_1 + \frac{\lambda}{\lambda + \mu}(x_2 - x_1) \quad \text{or} \quad \frac{\lambda x_2 + \mu x_1}{\lambda + \mu}.$$

Similarly the y-coordinate of P is

$$\frac{\lambda y_2 + \mu y_1}{\lambda + \mu}$$

and P is the point

$$\left(\frac{\lambda x_2 + \mu x_1}{\lambda + \mu}, \frac{\lambda y_2 + \mu y_1}{\lambda + \mu}\right).$$

Applying this formula to the example at the beginning of the chapter, the coordinates of P are

$$\left(\frac{2 \times 8 + 3 \times 2}{5}, \frac{2 \times 11 + 3 \times 3}{5}\right) \quad \text{or} \quad (4.4, 6.2).$$

Provided we take account of sign, the coordinates of points on AB produced may be found from this formula.

If P is on AB produced and is such that $AP = \frac{5}{3}BP$, as shown in Fig. 23.2, then $\dfrac{AP}{PB} = -\frac{5}{3}$. In the formula, put $\lambda = 5$, $\mu = -3$.

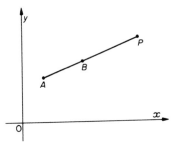

Fig. 23.2

Example. *Given that A is $(2, -1)$ and B is $(4, 3)$, find the coordinates of the point P on AB produced such that $BP = \frac{1}{2}AB$.*

Here $\dfrac{AP}{PB} = -\dfrac{3}{1}$. Take $\lambda = +3$, $\mu = -1$.

Then P is $\left(\dfrac{3 \times 4 + 2(-1)}{2}, \dfrac{3 \times 3 + (-1)(-1)}{2}\right)$, i.e. $(5, 5)$.

The centre of gravity of a triangle

The centre of gravity of a triangle is the meet of its medians. In Fig. 23.3, suppose that AD is a median and G the centre of gravity; we know that $\dfrac{AG}{GD} = \dfrac{2}{1}$.

If the coordinates of A, B and C are respectively (x_1, y_1), (x_2, y_2) and (x_3, y_3) then D is $\left(\dfrac{x_2 + x_3}{2}, \dfrac{y_2 + y_3}{2}\right)$.

G divides AD in the ratio $2:1$ and the coordinates of G are

$$\left(\frac{x_1 + x_2 + x_3}{3}, \frac{y_1 + y_2 + y_3}{3}\right).$$

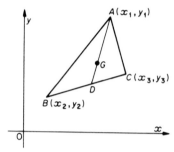

Fig. 23.3

The line joining (x_1, y_1) to (x_2, y_2)

We have already found in many examples the equation of a line joining two given points and this is so often wanted that a general formula for such a line is now found.

The line through (x_1, y_1) of gradient m is

$$\frac{y - y_1}{x - x_1} = m.$$

If the line also passes through the point (x_2, y_2), its gradient is

$$\frac{y_2 - y_1}{x_2 - x_1}.$$

Therefore the equation of the line is

$$\frac{y - y_1}{x - x_1} = \frac{y_2 - y_1}{x_2 - x_1}.$$

Example. *Find the equation of the line joining $(2, -1)$ to $(3, 2)$.*

Here,
$$\frac{y - (-1)}{x - 2} = \frac{2 - (-1)}{3 - 2} = \frac{3}{1}.$$

or
$$y = 3x - 7.$$

Intercept form

We now find the equation of the line which makes intercepts a and b on the axes. This line is the join of $(a, 0)$ to $(0, b)$ and so its equation is

$$\frac{y - 0}{x - a} = \frac{b - 0}{0 - a}$$

or

$$-ay = bx - ba$$

or

$$\frac{x}{a} + \frac{y}{b} = 1.$$

(p, α) form

To find the equation of the straight line such that the perpendicular from the origin is of length p and makes an angle α with the x-axis.

If (x, y) are the coordinates of any point on the line, from Fig. 23.4, we see that

$$ON = OS + PR.$$

Therefore

$$p = x \cos \alpha + y \sin \alpha.$$

Since this is an equation connecting the coordinates x and y of any point on the line, this is the equation of the line required.

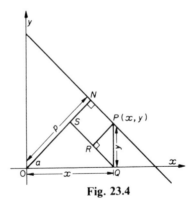

Fig. 23.4

Example 1. *Find the equation of the line which is at unit distance from the origin and makes an angle of 120° with the positive direction of the x-axis.*

Here

$$p = 1 \quad \text{and} \quad \alpha = 30°.$$

$$\sin \alpha = \tfrac{1}{2} \quad \text{and} \quad \cos \alpha = \frac{\sqrt{3}}{2}.$$

The equation of the line is

$$\frac{x\sqrt{3}}{2} + \frac{y}{2} = 1 \quad \text{or} \quad x\sqrt{3} + y = 2.$$

Example 2. *Find the length of the perpendicular from the origin to the line* $3x + 4y = 10$.

We wish to find an angle such that $\cos \alpha : \sin \alpha = 3:4$.

Draw a right-angled triangle as shown in Fig. 23.5 in which $\tan \alpha = \frac{4}{3}$.

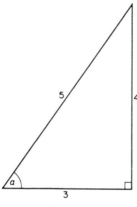

Fig. 23.5

The hypotenuse of the triangle is 5; $\cos \alpha = \frac{3}{5}$ and $\sin \alpha = \frac{4}{5}$.

Divide both sides of the given equation by 5. Then

$$\tfrac{3}{5}x + \tfrac{4}{5}y = 2.$$

This is in the form $x \cos \alpha + y \sin \alpha = p$ and so the length of the perpendicular from the origin to the line is 2.

More generally, the line $ax + by = c$ may be put in the form

$$\frac{a}{\sqrt{a^2 + b^2}}x + \frac{b}{\sqrt{a^2 + b^2}}y = \frac{c}{\sqrt{a^2 + b^2}}.$$

Since

$$\left(\frac{a}{\sqrt{a^2 + b^2}}\right)^2 + \left(\frac{b}{\sqrt{a^2 + b^2}}\right)^2 = 1,$$

it is possible to find an angle α such that

$$\cos \alpha = \frac{a}{\sqrt{a^2 + b^2}} \quad \text{and} \quad \sin \alpha = \frac{b}{\sqrt{a^2 + b^2}}.$$

Therefore the length of the perpendicular from the origin to the line $ax + by = c$ is

$$\frac{c}{\sqrt{a^2 + b^2}}.$$

(r, θ) form

Given that A is the point $(2, 3)$ and that a line is drawn through A making an angle θ with the x-axis, how may the coordinates of any point on the line be expressed?

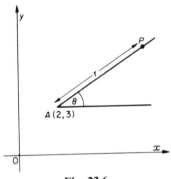

Fig. 23.6

In Fig. 23.6, suppose that P is the point on the line distant r from A. The increase of x between A and P is $r \cos \theta$; the increase of y between A and P is $r \sin \theta$. Therefore P is the point

$$(2 + r \cos \theta, 3 + r \sin \theta).$$

More generally, if A is the point (x_1, y_1), the coordinates of P are $(x_1 + r \cos \theta, y_1 + r \sin \theta)$.

The coordinates of a point on the line are thus expressed in terms of one variable r (θ is constant for a given line). When the coordinates of a point on a line or curve are expressed in terms of a third variable, this variable is called a **parameter** and the curve is said to be expressed in terms of parametric coordinates. The line may be put in the form

$$\frac{x - x_1}{\cos \theta} = \frac{y - y_1}{\sin \theta} = r,$$

which reduces to

$$x = x_1 + r \cos \theta, \quad y = y_1 + r \sin \theta.$$

Notice that it is necessary to consider r to be negative when the coordinates of a point on the other side of A are required. Two equal and opposite values of r give the points

$$(x_1 + r \cos \theta, y_1 + r \sin \theta) \quad \text{and} \quad (x_1 - r \cos \theta, y_1 - r \sin \theta)$$

and obviously $A (x_1, y_1)$ is the mid-point of the line joining them.

Example. *A line is drawn through the point $(2, 3)$ making an angle of $45°$ with the positive x-direction. Find the distance along the line from $(2, 3)$ to the point of intersection with the line $3x + 5y = 29$.*

Suppose the distance is r. Then the point of intersection is

$$(2 + r\cos 45°, 3 + r\sin 45°), \quad \text{or} \quad \left(2 + \frac{r}{\sqrt{2}}, 3 + \frac{r}{\sqrt{2}}\right).$$

This point lies on the line $3x + 5y = 29$ and therefore

$$3\left(2 + \frac{r}{\sqrt{2}}\right) + 5\left(3 + \frac{r}{\sqrt{2}}\right) = 29$$

or

$$6 + \frac{3r}{\sqrt{2}} + 15 + \frac{5r}{\sqrt{2}} = 29.$$

$$\therefore 21 + \frac{8r}{\sqrt{2}} = 29 \quad \text{and} \quad \frac{8r}{\sqrt{2}} = 8 \quad \text{or} \quad r = \sqrt{2}.$$

Exercise 23a

1. Find the coordinates of the point dividing the join of $(3, -1)$ and $(5, 2)$ in the ratio $3:4$.

2. The points A and B are $(2,3)$ and $(7, 10)$ respectively. Find the coordinates of the point P on AB produced such that $AP:BP = 7:4$.

3. Find the equation of the line joining $(4, 1)$ to $(5, -2)$.

4. Find the equation of the line which makes intercepts of 3 and 4 on the axes.

5. A line makes an angle of $135°$ with the positive x-direction. The length of the perpendicular from the origin to the line is $\sqrt{2}$. Find its equation.

6. Find the length of the perpendicular from the origin to the line $3x + 4y = 7$.

7. Write down the equation of the line through $(2, 3)$ parallel to the line $3x + 4y = 7$. Hence find the length of the perpendicular from the point $(2, 3)$ to the line $3x + 4y = 7$.

8. A line drawn through the point $(2, 1)$ makes an angle of $\cos^{-1}\frac{3}{5}$ with the axis of x. Find the distance from the point $(2, 1)$ along this line to its point of intersection with the line $4x + 3y = 23$.

9. Express the line $x + 3y - 2 = 0$ in the form $x\cos\alpha + y\sin\alpha = p$.

10. Find the coordinates of the centre of gravity of the triangle formed by the points $(1, 4)$, $(2, 7)$ and $(3, 10)$.

11. Write down the equation of the line perpendicular to $y = mx + \dfrac{a}{m}$ through the point $(a, 0)$.

12. Write down the equation of the line parallel to $xt + \dfrac{y}{t} = c$ through the point $\left(\dfrac{c}{t}, ct\right)$.

13. Calculate the area of the triangle formed by the line $3x + 4y = 12$ and the coordinate axes.

14. Find the coordinates of the foot of the perpendicular dropped from the point $(1, 1)$ to the line $5x + 12y = 30..$ Hence find the length of this perpendicular.

15. What angle does the perpendicular from the point $(1, 1)$ to the line $x + y = 4$ make with the x-axis? Use this angle to find the length of the perpendicular from the point to the line.

16. Find the equation of the line through the point $(a \cos \theta, a \sin \theta)$ perpendicular to $y = x \tan \theta$.

17. What are the coordinates of the points of trisection of the line joining $(2, 5)$ to $(8, 8)$?

18. Show that the point of trisection nearer the origin of the line joining O to the point $(3h, 3h^2)$ always lies on the curve $y = x^2$.

19. Write down the perpendicular distances of the origin from the lines $3x + 4y = 5$ and $3x + 4y + 5 = 0$. What does the sign of the perpendicular signify?

20. Find the equation of the line given parametrically by
$$x = t + 2, \quad y = 3t - 3.$$

Length of a perpendicular from a point to a line

In the last exercise two or three methods of finding the length of a perpendicular from a point to a line were indicated. Consider the perpendicular from the point $(1, 1)$ to the line $5x + 12y = 8$. The line through $(1, 1)$ parallel to $5x + 12y = 8$ is $5x + 12y = 17$. The line $5x + 12y = 8$ when expressed in the form $x \cos \alpha + y \sin \alpha = p$ is $\frac{5}{13}x + \frac{12}{13}y = \frac{8}{13}$ and the similar form for the parallel line is $\frac{5}{13}x + \frac{12}{13}y = \frac{17}{13}$. The perpendiculars from the origin to these lines are of length $\frac{8}{13}$ and $\frac{17}{13}$. The length of the perpendicular from the point to the line from Fig. 23.7 is therefore $\dfrac{17 - 8}{13}$ or $\dfrac{9}{13}$.

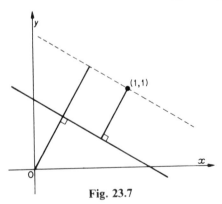

Fig. 23.7

If the line is $lx + my + n = 0$ and the point (x_1, y_1), the equation of the parallel is $lx + my = lx_1 + my_1$.

The length of the perpendicular from the origin to the line $lx + my + n = 0$ is $-\dfrac{n}{\sqrt{l^2 + m^2}}$ and the length of the perpendicular from the origin to $lx + my = lx_1 + my_1$ is $\dfrac{lx_1 + my_1}{\sqrt{l^2 + m^2}}$.

The length of the perpendicular from (x_1, y_1) to

$$lx + my + n = 0$$

is equal to the difference between these lengths and is

$$\frac{lx_1 + my_1 + n}{\sqrt{l^2 + m^2}}.$$

Alternative method

Suppose that the perpendicular from $P(x_1, y_1)$ to

$$lx + my + n = 0$$

makes an angle α with the x-axis.

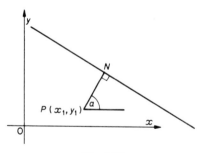

Fig. 23.8

Since the gradient of $lx + my + n = 0$ is $-\dfrac{l}{m}$, the gradient of the perpendicular is $\dfrac{m}{l}$ and $\tan \alpha = \dfrac{m}{l}$.

Therefore

$$\cos \alpha = \frac{l}{\sqrt{l^2 + m^2}} \quad \text{and} \quad \sin \alpha = \frac{m}{\sqrt{l^2 + m^2}} \quad \text{(see Fig. 23.9)}.$$

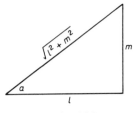

Fig. 23.9

If the length of the perpendicular is p, the coordinates of the foot of the perpendicular must be $(x_1 + p \cos \alpha, y_1 + p \sin \alpha)$. But this point lies on $lx + my + n = 0$ and therefore

$$l(x_1 + p \cos \alpha) + m(y_1 + p \sin \alpha) + n = 0.$$

From this equation,

$$p(l \cos \alpha + m \sin \alpha) = -(lx_1 + my_1 + n).$$

But

$$l \cos \alpha + m \sin \alpha = \frac{l^2}{\sqrt{l^2 + m^2}} + \frac{m^2}{\sqrt{l^2 + m^2}}$$

$$= \sqrt{l^2 + m^2}.$$

$$\therefore p = -\frac{lx_1 + my_1 + n}{\sqrt{l^2 + m^2}}.$$

The sign of the perpendicular will be considered in the next paragraph. When the length of a perpendicular is asked, it is usually sufficient to give its positive value.

Example 1. *Find the length of the perpendicular from the point $(2, -1)$ to the line $3x - 4y + 7 = 0$.*

Substitute the coordinates in the left-hand side of the equation and divide by the square root of the sum of the squares of the coefficients of x and y.

$$p = \frac{3(2) - 4(-1) + 7}{\sqrt{3^2 + 4^2}} = \frac{17}{5} = 3.4.$$

Example 2. *Find the length of the perpendicular from the point $(2, 0)$ to the line $y = mx + c$.*

$$p = \frac{2m + c - 0}{\sqrt{m^2 + (-1)^2}} = \frac{2m + c}{\sqrt{1 + m^2}}.$$

The sign of the perpendicular

Consider a line in the form $y = mx + c$ or $y - mx - c = 0$.

Suppose Q is any point above the line and draw a parallel through Q to the y-axis to meet the line at P. At P, the value of $y - mx - c$ is zero because P lies on the line. For the points P and Q the value of x is the same but the y-coordinate of Q is greater than that of P. So for Q, $y - mx - c > 0$. Thus for any point above the line $y - mx - c > 0$ and similarly for any point below the line $y - mx - c < 0$. Any equation except $x = k$ may be put in the form $y = mx + c$ and the line $x = k$ is parallel to the y-axis and so 'above the line' has no meaning. If the line is such that the coefficient of y is negative, e.g. $3x - 2y - 5 = 0$, the reverse applies. For points above the line $3x - 2y - 5 < 0$ and for points below the line $3x - 2y - 5 > 0$. The important thing to remember is that for all points on one side of the line $lx + my + n = 0$, the expression $lx + my + n$ has the same sign, and for all points on the other side of the line, the expression takes the opposite sign.

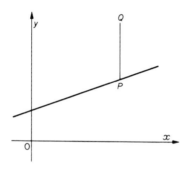

Fig. 23.10

The length of the perpendicular from (x_1, y_1) to $lx + my + n = 0$ is $\dfrac{lx_1 + my_1 + n}{\sqrt{l^2 + m^2}}$. If the positive sign of the square root is taken, the sign of the perpendicular has the same sign as $lx_1 + my_1 + n$ which depends on which side of the line the point is situated. So perpendiculars to a line from points on the same side of the line have the same sign; perpendiculars from points on opposite sides of the line have different signs.

The origin is often a good point with which to make comparison as is illustrated in Example 2 below.

Example 1. *Are the points $(1, 2)$ and $(2, -3)$ on the same side of the line $4x + 5y + 1 = 0$?*

$$4(1) + 5(2) + 1 = +15$$
$$4(2) + 5(-3) + 1 = -6.$$

Since the expressions are of opposite sign, the points are not on the same side of the line.

Example 2. *Indicate the position of the point* (2, 3) *with reference to the line*
$4x - y - 2 = 0$.

Express the line with the coefficient of y positive, i.e. as
$$y - 4x + 2 = 0.$$
The expression obtained by substituting $x = 2$, $y = 3$ is
$$3 - 4(2) + 2 = 5 - 8 \quad \text{or} \quad -3.$$
The point (2, 3) is therefore below the line.
When (0, 0) is substituted, we get $+2$. The origin is therefore above the line.

The line passes between the origin and the point (2, 3) so that the origin is above the line as shown in Fig. 23.11.

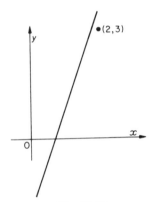

Fig. 23.11

The angle between two lines

If two lines make angles of θ and φ with the axis of x, the angle α between the lines is given by $\alpha = \theta - \varphi$ (see Fig. 23.12).

If the gradients of the lines are m and t so that $\tan \theta = m$ and $\tan \varphi = t$,

$$\tan \alpha = \tan(\theta - \varphi) = \frac{\tan \theta - \tan \varphi}{1 + \tan \theta \tan \varphi}$$

$$= \frac{m - t}{1 + mt}.$$

So the angle between the lines is $\tan^{-1} \frac{m - t}{1 + mt}$.

If the lines are parallel, the angle between them is zero.

$$\therefore \frac{m - t}{1 + mt} = 0 \quad \text{or} \quad m = t.$$

If the lines are perpendicular, the angle between them is 90°, and tan 90° is infinite. Therefore the denominator of the fraction must be zero and so $mt = -1$.

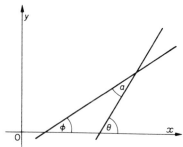

Fig. 23.12

Example. *Find the angle between the lines $3x - 4y = 2$ and $5x + 2y = 7$.*
The gradients of the lines are $\frac{3}{4}$ and $-\frac{5}{2}$.

$$\tan \alpha = \frac{m - t}{1 + mt} = \frac{\frac{3}{4} - (-\frac{5}{2})}{1 + \frac{3}{4}(-\frac{5}{2})} = \frac{6 + 20}{8 - 15} = -\frac{26}{7}.$$

The angle is $\tan^{-1}(-\frac{26}{7})$ or $\pi - \tan^{-1}\frac{26}{7}$.

The acute angle between the lines is $\tan^{-1}\frac{26}{7}$.

Exercise 23b

1. Find the lengths of the perpendiculars from the following points to the given lines:
   (a) $(1, 3)$ to $3x + 2y - 6 = 0$.
   (b) $(1, -1)$ to $y = 2x - 4$.
   (c) $(0, 4)$ to $3y = 1$.
   (d) $(5, 2)$ to $x - 2y - 7 = 0$.
   (e) $(4, 6)$ to $3x + 5y = 7$.

2. Say whether the points given are on the same or opposite sides of the line:
   (a) $(1, 4)$ and $(2, -2)$; $3x + y = 5$.
   (b) $(1, -1)$ and $(7, 2)$; $x + y + 6 = 0$.
   (c) $(1, -4)$ and $(1, 2)$; $3x - y - 2 = 0$.
   (d) $(3, 1)$ and $(4, -1)$; $y = 2x - 4$.
   (e) $(1, 4)$ and $(4, -2)$; $3x - y - 2 = 0$.

3. Is the origin above or below the line $3x - y - 2 = 0$?

4. Is the point $(2, 3)$ above or below the line $x + y = 6$?

5. Shade in a figure the region which includes the points defined by $y > 3$, $x > 2$, $x + y < 6$.

6. Find the angle between the lines $x + y = 8$ and $x - y = 2$.

7. Find the angle between the lines $3y - 4x = 2$ and $7y = x + 1$.

8. Find the tangent of the acute angle between the lines $x + 2y = 6$ and $y - 3x = 1$.

Lines through the intersection of two given lines

The lines $3x + 4y - 11 = 0$ and $4x - y - 2 = 0$ meet at the point $(1, 2)$. This may be shown either by substitution or by solving the simultaneous equations. Now consider the equation

$$3x + 4y - 11 + k(4x - y - 2) = 0,$$

where k is any constant. If the values $x = 1$, $y = 2$ are substituted in this equation, the left-hand side must become zero because both $(3x + 4y - 11)$ and $(4x - y - 2)$ become zero.

Therefore $3x + 4y - 11 + k(4x - y - 2) = 0$ must represent a line passing through the intersection of $3x + 4y - 11 = 0$ and $4x - y - 2 = 0$ whatever the value of k.

More generally, if $l = 0$ and $l' = 0$ represent two lines (i.e. l stands for an expression such as $ax + by + c$), then $l + kl' = 0$ represents a line passing through the point of intersection of l and l' for all values of k.

The value of k can be found when another piece of information fixing the line is given. For example, if we are told that the line passes through another fixed point, an equation giving k is found by substituting the coordinates of the point in $l + kl' = 0$.

Notice that if $S = 0$ and $S' = 0$ are the equations of two curves, $S + kS' = 0$ is the equation of a curve passing through all the points of intersection of S and S' whatever the value of k.

Example 1. *Find the equation of the line joining the point of intersection of the lines $3x - y - 2 = 0$ and $2x + y + 4 = 0$ to the origin.*

The line must be of the form

$$3x - y - 2 + k(2x + y + 4) = 0.$$

The equation of a line through the origin contains no constant term and therefore $\qquad -2 + 4k = 0$, i.e. $k = \tfrac{1}{2}$.
The equation of the line is

$$2(3x - y - 2) + (2x + y + 4) = 0$$
$$y = 8x.$$

or

Example 2. *Find the equation of the line through the intersection of $x - y - 2 = 0$ and $2x + y + 3 = 0$ perpendicular to $4x - 3y = 5$.*

The line must be of the form

$$(x - y - 2) + k(2x + y + 3) = 0.$$

The gradient of this line is $\dfrac{1 + 2k}{1 - k}$.

The gradient of $4x - 3y = 5$ is $\tfrac{4}{3}$.

Since the lines are perpendicular,

$$\left(\frac{1 + 2k}{1 - k}\right)\left(\frac{4}{3}\right) = -1.$$

$$\therefore 4 + 8k + 3 - 3k = 0$$
$$k = -\tfrac{7}{5}.$$

The equation of the line is

$$5(x - y - 2) - 7(2x + y + 3) = 0$$
or
$$9x + 12y + 31 = 0.$$

Exercise 23c: Miscellaneous

1. Find the equation of the line joining the origin to the point of intersection of $3x + y + 1 = 0$ and $4x - 2y - 1 = 0$.

2. Find the equation of the line joining the point $(1, 2)$ to the intersection of $3x + y = 7$ and $4x + 2y = 10$.

3. Find the equation of the line which passes through the intersection of $4x - y - 2 = 0$ and $5x + 3y + 1 = 0$ perpendicular to $4x - y - 2 = 0$.

4. Find the equation of the line joining the intersection of $x = y$ and $x + y + 2 = 0$ to the intersection of $x + y - 4 = 0$ and $2x + 3y - 10 = 0$.

5. Find the area of the triangle formed by the axis of x, the line $3x + 4y = 12$ and the perpendicular to this line from the origin.

6. Show that the points $(0, 12)$, $(4, 2)$ and $(14, 6)$ are three corners of a square. Find the coordinates of the fourth vertex.

7. Find the equation of the line which makes intercepts of 2 and 3 on the axes. Find the length of the perpendicular from the origin to this line and the coordinates of the foot of this perpendicular.

8. Find the coordinates of the point on the line $2x = y + 1$ which is equidistant from $(1, 2)$ and $(3, 2)$.

9. Show that the points $(-4, 2)$, $(4, 14)$ and $(10, 10)$ form three corners of a rectangle in which the length of one side is double the other. Find the coordinates of the fourth vertex.

10. A straight line makes equal intercepts a on the two axes. Find the length of the perpendicular from the point (a, a) to the line.

11. The points A $(1, 1)$, B $(3, -1)$ and C $(2, 2)$ form a triangle. The mid-point of AB is D and E is the point of trisection of BC nearer B. Find the equation of DE.

12. Find the angle A of the triangle ABC formed by the points A $(2, 5)$, B $(3, 6)$ and C $(4, 8)$.

13. Find the angle between the tangents to the curve $y = x^3$ at the points $(1, 1)$ and $(2, 8)$.

14. Find the equation of the tangent at the point $\left(ct, \dfrac{c}{t}\right)$ to the curve $xy = c^2$.

15. Find the image of the point $(1, 1)$ in the line $3x + 4y = 12$.

16. Find the equations of the two lines which pass through the point $(1, 3)$ and are at distance 3 from the origin.

17. The coordinates of A and B are $(a, 2a)$ and $(3a, a)$ respectively. Calculate the area of the triangle OAB. Hence find the length of the perpendicular from the origin to the line AB.

18. Find the equation of the line joining the points $\left(cm, \dfrac{c}{m}\right)$ and $\left(ct, \dfrac{c}{t}\right)$.

19. Find the equation of the line joining the point of intersection of the lines $ax + by = 1$ and $px + qy = 1$ to the origin.

20. Find the equation of the line joining the points $(am^2, 2am)$ and $(at^2, 2at)$.

21. Show that the points $(-2, -6)$, $(4, 2)$ and $(12, 8)$ form three vertices of a rhombus. Find the coordinates of the fourth vertex.

22. A triangle OAB is formed by the axes and the line $3x + 2y = 5$. Does the point $(1, 2)$ lie inside this triangle?

23. Two lines of gradients m and t include an angle of $45°$. Show that $1 + mt = \pm(m - t)$.

24. The line $3x + 4y = k$ forms with the axes a triangle whose area is four times as great as that formed by $3x + 4y = 12$ with the axes. Find the value of k.

25. If the product of the perpendiculars from the point (h, k) to the lines $x - y = 0$ and $x + y = 0$ is 2, show that
$$h^2 - k^2 = \pm 4.$$

24 Loci

The locus of a point is the path traced out by the point as it moves under certain conditions. In this chapter we are only concerned with a point moving in a plane, and we find the cartesian equation of the locus, that is, the relation between x and y, the coordinates of the point referred to perpendicular axes.

The idea of a locus involves two properties:

(a) All points satisfying the condition lie on the locus.

(b) No point, other than those on the locus, satisfies the conditions.

For example, a circle is the locus (in a plane) of a point which moves so that its distance from a fixed point is constant. Denote the fixed point by O, the constant distance by r (the radius of the circle). Then if P is any point in the plane,

(i) if $OP = r$, P lies on the circle;

(ii) unless P lies on the circle, OP does not equal r.

We sometimes define a locus as a set of points satisfying a given condition. A 'set' is a well-defined collection of elements, so that we can tell whether any one element does not belong to that set. The set of all points in a straight line contains all the points on that straight line. 'The locus of a point which moves so that it is equidistant from two fixed points A and B' means the same as 'The set of all points equidistant from A and B', and is the perpendicular bisector of AB. We can find the cartesian equation of the locus in the form $3x + 4y + 5 = 0$, or we may wish to use set notation, and write $\{(x, y): 3x + 4y + 5 = 0\}$. The former is shorter and just as precise.

There are two methods of finding a locus. The first, and neatest when it can be applied, is to name the point whose locus we wish to find (x_1, y_1). We then find a connection between x_1, y_1 and constants. Such a condition means that the point (x_1, y_1) must lie on another curve obtained by dropping the suffix 'one' and this equation is therefore the locus required.

The second method is to express the coordinates x and y of any point P satisfying the conditions in terms of a parameter which may be a distance, an angle or a gradient. This gives the parametric equation of the locus. The Cartesian equation may then be obtained by eliminating the parameter between the two equations. This method is more generally applicable than the first but the elimination can be tedious.

Examples are now given and solutions by both methods are included in some cases.

Example 1. *Find the equation of the perpendicular bisector of the line joining* (2, 3) *and* (4, 2).

Suppose (x_1, y_1) is a point of the locus. Then (x_1, y_1) is equidistant from (2, 3) and (4, 2).

$$\therefore (x_1 - 2)^2 + (y_1 - 3)^2 = (x_1 - 4)^2 + (y_1 - 2)^2.$$
$$\therefore x_1{}^2 + y_1{}^2 - 4x_1 - 6y_1 + 13 = x_1{}^2 + y_1{}^2 - 8x_1 - 4y_1 + 20,$$

i.e. $$4x_1 - 2y_1 = 7.$$

This condition implies that (x_1, y_1) must lie on the line $4x - 2y = 7$. This is the equation of the locus of points equidistant from the two given points and is therefore the perpendicular bisector of the line joining them.

Notice that the gradient of the line joining (2, 3) and (4, 2) is $\dfrac{3 - 2}{2 - 4}$ or $-\tfrac{1}{2}$,

so that the locus which is of gradient 2 is perpendicular to the join of the two points. The line also passes through the point $(3, 2\tfrac{1}{2})$, which is the mid-point of the join, because $(3, 2\tfrac{1}{2})$ satisfies the equation $4x - 2y = 7$.

Example 2. *Find the equation of the locus of points P such that PA is perpendicular to PB where A is* (1, 0) *and B is* (−1, 0).

Suppose one point of the locus is (x_1, y_1).

The gradient of PA is $\dfrac{y_1}{x_1 - 1}$.

The gradient of PB is $\dfrac{y_1}{x_1 + 1}$.

These lines are perpendicular and therefore

$$\left(\frac{y_1}{x_1 - 1}\right)\left(\frac{y_1}{x_1 + 1}\right) = -1$$

or $$x_1{}^2 + y_1{}^2 = 1.$$

The equation of the locus is $x^2 + y^2 = 1$, a circle on AB as diameter.

The parameter method

The equation of the line of gradient m through $(-1, 0)$ is $y = m(x + 1)$.

The equation of the line of gradient $-\dfrac{1}{m}$ through (1, 0) is $y = -\dfrac{1}{m}(x - 1)$.

These two equations when solved for x and y give the parametric equations of the curve. However it is not necessary to find x and y before eliminating. Eliminate m by multiplication:

$$y^2 = -(x + 1)(x - 1) \quad \text{or} \quad x^2 + y^2 = 1.$$

Example 3. *The point Q lies on the curve* $y^2 = 4x$. *Find the equation of the locus of the point P such that P is the mid-point of OQ.*

Suppose the coordinates of one point of the path are (x_1, y_1). Then $(2x_1, 2y_1)$ must lie on $y^2 = 4x$.

$$\therefore (2y_1)^2 = 4(2x_1)$$
or $$y_1{}^2 = 2x_1.$$

The equation of the locus is $y^2 = 2x$.

Parameter method

Suppose Q is (α, β). Then $\beta^2 = 4\alpha$.

The coordinates of P are $\left(\dfrac{\alpha}{2}, \dfrac{\beta}{2}\right)$.

Put $x = \dfrac{\alpha}{2}, y = \dfrac{\beta}{2}$ and eliminate α and β between these equations and $\beta^2 = 4\alpha$.

Since
$$\alpha = 2x \quad \text{and} \quad \beta = 2y,$$
$$4y^2 = 8x \quad \text{or} \quad y^2 = 2x.$$

Example 4. *Find the equation of the locus of points P such that $PA = 2$, where A is the point $(2, 1)$.*

If one point of the path is (x_1, y_1), the distance between (x_1, y_1) and $(2, 1)$ is 2.

$$\therefore (x_1 - 2)^2 + (y_1 - 1)^2 = 4$$
or
$$x_1^2 + y_1^2 - 4x_1 - 2y_1 + 1 = 0.$$

The equation of the locus is $x^2 + y^2 - 4x - 2y + 1 = 0$, which is the equation of the circle centre $(2, 1)$ and radius 2.

Parameter method

If a radius is drawn through $(2, 1)$ making an angle θ with the x-axis, the coordinates of the end point are given by
$$x = 2 + 2 \cos \theta, \quad y = 1 + 2 \sin \theta.$$
The equation of the locus is found by eliminating θ.
$$2 \cos \theta = x - 2 \quad \text{and} \quad 2 \sin \theta = y - 1.$$
Squaring and adding,
$$(x - 2)^2 + (y - 1)^2 = 4.$$

Example 5. *Find the equation of the locus of the point P such that $PA = 2PB$ where A is $(1, 2)$ and B is $(3, 1)$.*

If P is (x_1, y_1),
$$PA^2 = (x_1 - 1)^2 + (y_1 - 2)^2,$$
$$PB^2 = (x_1 - 3)^2 + (y_1 - 1)^2.$$
But
$$PA^2 = 4PB^2,$$
and so
$$x_1^2 + y_1^2 - 2x_1 - 4y_1 + 5 = 4(x_1^2 + y_1^2 - 6x_1 - 2y_1 + 10)$$
or
$$3x_1^2 + 3y_1^2 - 22x_1 - 4y_1 + 35 = 0.$$
The equation of the locus is
$$3x^2 + 3y^2 - 22x - 4y - 35 = 0.$$

Exercise 24: Miscellaneous

1. Find the equation of the perpendicular bisector of the line joining the points $(2, 0)$ and $(4, 6)$.
2. Find the equation of the perpendicular bisector of the line joining the points (a, a) and (b, b).

3. Find the equation of the circle, centre the origin and radius 4.

4. Find the equation of the circle, centre $(1, 3)$ and radius 4.

5. Find the equation of the circle on the line joining $(1, 1)$ and $(2, 2)$ as diameter.

6. Find the equation of the circle on the line joining (a, a) and (b, b) as diameter.

7. The point A has coordinates $(3, 2)$. Find the equation of the locus of the point P such that $PA = 2PO$.

8. The point Q lies on $x^2 + y^2 = 4$. Find the equation of the locus of the point P such that P is the mid-point of OQ.

9. The point Q lies on $3x^2 + 4y^2 = 8$. Find the equation of the locus of the point P such that P is the mid-point of OQ.

10. The point Q lies on $xy = 16$. Find the equation of the locus of the point P such that P is the mid-point of OQ.

11. The point Q lies on $y^2 = 4x$. Find the equation of the locus of the point P such that the mid-point of the line joining Q to $(1, 0)$ is P.

12. Find the equation of the locus of the point P such that P is equidistant from the point $(1, 0)$ and the y-axis.

13. Find the equation of the locus of the point P such that P is equidistant from the origin and the line $x = y$.

14. Find the equation of the locus of the point P such that $PA^2 + PB^2 = 7$, where A is $(2, 0)$ and B is $(1, 3)$.

15. Find the equation of the locus of the point P such that $PA + PB = 4$, where A is $(1, 0)$ and B is $(-1, 0)$.

16. A straight line cuts the axes at A and B and $AB = 8$ units. Find the locus of the mid-point of AB.

17. H and K are the feet of the perpendiculars to the axes from the point P. Find the equation of the locus P given that $HK = 4$ units.

18. The mid-point of the line joining P to $(1, 2)$ lies on the line $2x + 3y = 5$. Find the equation of the locus P.

19. A line of gradient m is drawn through the point $(1, 0)$ and a line of gradient $2m$ through the point $(2, 1)$. Find the locus of the point of intersection of the two lines.

20. OAP is a triangle of area 2 square units. The point A has coordinates $(3, 2)$. Find the equation of the locus of P.

21. The feet of the perpendiculars from a point P to the y-axis and the line $x = y$ are R and S. Given that $RS = 4$, find the equation of the locus of the point P.

22. Q is the point $(1, 1)$ and R a point on the line $y = 2x$. P is the mid-point of QR. Find the equation of the locus of the point P.

23. Q is the point $(1, 1)$ and R is a point on $y^2 = 4x$. P is the mid-point of QR. Find the equation of the locus of the point P.

24. Q is a point on $y^2 = 4x$. R and S are the feet of the perpendiculars from Q to the axes and P is the mid-point of RS. Find the equation of the locus of the point P.

25. The point Q (1, 2) lies on $y^2 = 4x$. Another point R is taken on $y^2 = 4x$ and P is the mid-point of QR. Find the equation of the locus of the point P.

26. A variable line through the point (2, 3) meets the axes at Q and R. The mid-point of QR is P. Find the equation of the locus of the point P.

27. A variable line through the fixed point (α, β) meets the axes at Q and R. The mid-point of QR is P. Find the equation of the locus of the point P.

28. Q lies on the x-axis and R is the point (2, 3). The perpendicular to QR through Q meets the axis of y at T and P is the mid-point of QT. Find the equation of the locus of the point P.

29. The product of the perpendiculars from a point P to the axes is 4 square units. Find the equation of the locus of the point P.

30. Q lies on the curve $xy = c^2$. The feet of the perpendiculars from Q to the axes are R and S. Find the locus of the mid-point of RS.

31. Q is a point on $y^2 = 4x$ and QN is perpendicular to the y-axis. If P is the mid-point of QN, and N lies on the y-axis, find the equation of the locus of the point P.

32. Find the equation of the circle on the line joining (x_1, y_1) and (x_2, y_2) as diameter.

33. Find the equation of the circle centre (α, β) and radius r.

34. P is the mid-point of a line of length 6 units which moves so that its ends lie on the axes. Find the equation of the locus of the point P.

35. P is the point of trisection nearer the x-axis of a line of length 6 units which moves so that its ends lie on the axes. Find the equation of the locus of the point P.

36. P is the point of intersection of the lines $m(y - 1) = (x + 1)$ and $y = mx$ where m is a variable. Find the equation of the locus of the point P.

25 The Circle

The equation of a circle

A circle is the locus of a point which moves in a plane so that its distance from a fixed point of that plane is constant. The fixed point is called the centre and the constant distance the radius.

If the circle has centre the origin and radius r, the distance of P (x_1, y_1), a point on the locus, from $(0, 0)$ is equal to r.

$$\therefore x_1{}^2 + y_1{}^2 = r^2$$

and so the locus is

$$x^2 + y^2 = r^2.$$

If the circle has centre (a, b) and radius r, the distance between (x_1, y_1) and (a, b) is r.

$$\therefore (x_1 - a)^2 + (y_1 - b)^2 = r^2$$

and so the equation of the circle is

$$(x - a)^2 + (y - b)^2 = r^2.$$

Example. *Find the equation of the circle which has centre $(2, -1)$ and passes through $(4, 2)$.*

The equation of the circle is $(x - 2)^2 + (y + 1)^2 = r^2$, where r is the radius. It passes through $(4, 2)$ and therefore by substitution:

$$2^2 + 3^3 = r^2 = 13.$$

The equation of the circle is

$$(x - 2)^2 + (y + 1)^2 = 13$$

or

$$x^2 + y^2 - 4x + 2y - 8 = 0.$$

The form $x^2 + y^2 + 2gx + 2fy + c = 0$

The most general equation of a circle is

$$(x - a)^2 + (y - b)^2 = r^2,$$

which is

$$x^2 + y^2 - 2ax - 2by + (a^2 + b^2 - r^2) = 0.$$

This suggests that every circle is of the form

$$x^2 + y^2 + 2gx + 2fy + c = 0.$$

This equation may be put in the form

$$(x + g)^2 + (y + f)^2 = g^2 + f^2 - c$$

which states that the square of the distance of the point $(-g, -f)$ from the point (x, y) is constant and equal to $\sqrt{g^2 + f^2 - c}$. The equation $x^2 + y^2 + 2gx + 2fy + c = 0$ therefore represents a circle of centre $(-g, -f)$ and radius $\sqrt{g^2 + f^2 - c}$.

Example. *Find the equation of the circle which passes through the points $(1, 1)$, $(2, 4)$ and $(3, 2)$.*

Suppose the circle is $x^2 + y^2 + 2gx + 2fy + c = 0$.
Substituting the coordinates of the three points in this equation:

$$2 + 2g + 2f + c = 0 \tag{i}$$
$$20 + 4g + 8f + c = 0 \tag{ii}$$
$$13 + 6g + 4f + c = 0 \tag{iii}$$

Subtracting (i) from (iii) and (iii) from (ii):

$$11 + 4g + 2f = 0 \tag{iv}$$
$$7 - 2g + 4f = 0 \tag{v}$$

Multiplying (v) by 2 and adding:

$$25 + 10f = 0. \quad \therefore 2f = -5.$$

By substitution, $\qquad 2g = -3 \quad$ and $\quad c = 6$.
The equation of the circle is $x^2 + y^2 - 3x - 5y + 6 = 0$.
The equation of the circle could also have been found by obtaining the intersection of two of the three perpendicular bisectors of the lines joining the points.

Exercise 25a

1. Find the equations of the circles with the following centres and radii:
   (i) $(1, 0)$; 3. (ii) $(2, -1)$; $\sqrt{5}$. (iii) $(3, 2)$; 4.
   (iv) $(-1, -1)$; $\sqrt{2}$. (v) $(1, 4)$; $\sqrt{5}$.
2. Find the centres and radii of the following circles:
   (i) $x^2 + y^2 + 2x + 4y + 4 = 0$.
   (ii) $x^2 + y^2 - 4x - 2y - 4 = 0$.
   (iii) $x^2 + y^2 - 3x = 12$.
   (iv) $x^2 + y^2 - 4y = 0$.
   (v) $4x^2 + 4y^2 - 7x - 8y = 2$.
3. Find the equation of the circle of centre $(2, -1)$ which passes through the point $(1, 2)$.
4. Find the equation of the circle which passes through the points $(4, 0)$, $(9, 0)$ and $(0, 6)$.
5. Find the equation of the circle which passes through the points $(2, 0)$, $(3, 3)$ and $(4, 1)$.
6. Find the equation of the circle which passes through the points $(2, 2)$, $(7, 2)$ and $(-2, 8)$.
7. Find the locus of the centres of circles which pass through the points $(2, 0)$ and $(4, 0)$.
8. Show that the point $(r \cos \theta, r \sin \theta)$ lies on the circle $x^2 + y^2 = r^2$ for all values of θ.

The equation of a tangent

Suppose we wish to find the equation of the tangent at the point $(2, 1)$ to the circle $x^2 + y^2 - 2x - 4y + 3 = 0$.

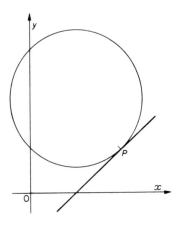

Fig. 25.1

For the circle,

$$2x + 2y\frac{dy}{dx} - 2 - 4\frac{dy}{dx} = 0.$$

$$\therefore \frac{dy}{dx} = \frac{1 - x}{y - 2}.$$

At the point $(2, 1)$,

$$\frac{dy}{dx} = \frac{1 - 2}{1 - 2} = 1.$$

This is the gradient of the tangent and its equation is

$$\frac{y - 1}{x - 2} = 1 \quad \text{or} \quad y = x - 1.$$

Now let us find the tangent at the point (x_1, y_1) to the circle $x^2 + y^2 + 2gx + 2fy + c = 0$.

For this circle

$$2x + 2y\frac{dy}{dx} + 2g + 2f\frac{dy}{dx} = 0.$$

$$\therefore \frac{dy}{dx} = -\frac{x + g}{y + f},$$

and at the point (x_1, y_1),

$$\frac{dy}{dx} = -\frac{x_1 + g}{y_1 + f}.$$

The equation of the tangent is

$$\frac{y - y_1}{x - x_1} = -\frac{x_1 + g}{y_1 + f},$$

or $\quad yy_1 + yf - y_1{}^2 - y_1 f + xx_1 + xg - x_1{}^2 - x_1 g = 0.$

This reduces to

$$xx_1 + yy_1 + gx + fy = x_1{}^2 + y_1{}^2 + gx_1 + fy_1.$$

But (x_1, y_1) lies on the circle and so

$$x_1{}^2 + y_1{}^2 + 2gx_1 + 2fy_1 + c = 0$$

or $\quad x_1{}^2 + y_1{}^2 + gx_1 + fy_1 = -(gx_1 + fy_1 + c).$

The equation of the tangent may therefore be put in the form

$$xx_1 + yy_1 + gx + fy = -(gx_1 + fy_1 + c)$$

or $\quad xx_1 + yy_1 + g(x + x_1) + f(y + y_1) + c = 0.$

This equation could be found without using calculus by using the geometrical property that the tangent is perpendicular to the radius at the point of contact.

Condition for a line to be a tangent

The line $y = 4x + c$ has gradient 4. For what values of c is it a tangent to the circle $x^2 + y^2 = 17$?

The easiest method is to use the property that if the line is a tangent to the circle, the length of the perpendicular to it from the centre must be equal to the radius.

The length of the perpendicular from $(0, 0)$ to $y = 4x + c$ is

$$\pm \frac{c}{\sqrt{1^2 + 4^2}} \text{ or } \pm \frac{c}{\sqrt{17}}.$$

$$\therefore \pm \frac{c}{\sqrt{17}} = \sqrt{17} \text{ (the radius)} \quad \text{and} \quad c = \pm 17.$$

The lines $y = 4x \pm 17$ are tangents of gradient 4.

Another method is now given. This is not so neat but is applicable to many other curves when no simple geometrical property exists.

Substitute $y = 4x + c$ in the equation $x^2 + y^2 = 17$.

$$x^2 + (4x + c)^2 = 17$$

or $\quad 17x^2 + 8cx + (c^2 - 17) = 0.$

This equation is a quadratic giving the x-coordinates of the points in which the line $y = 4x + c$ meets the circle. If the line is a tangent, the two points of intersection are coincident and the equation has equal roots. The left-hand side is a perfect square. The condition that $ax^2 + bx + c$ is a perfect square is $b^2 = 4ac$ and therefore

$$64c^2 = 4(17)(c^2 - 17).$$
$$\therefore 16c^2 = 17c^2 - 17^2$$
or $$c^2 = 17^2.$$
$$\therefore c = \pm 17.$$

Equations of tangents of gradient m

Suppose that $y = mx + c$ is a tangent to the circle

$$x^2 + y^2 = a^2.$$

The length of the perpendicular from $(0, 0)$ to $y = mx + c$ is then of length a.

$$\therefore \pm \frac{c}{\sqrt{1 + m^2}} = a,$$

or $$c = \pm a\sqrt{1 + m^2}.$$

The tangents of gradient m are $y = mx \pm a\sqrt{1 + m^2}$.

Exercise 25b

1. Write down the equations of the tangents at the given points to the following circles:
   (i) $(2, 2)$; $x^2 + y^2 = 8$.
   (ii) $(6, 0)$; $x^2 + y^2 - 6x = 0$.
   (iii) $(1, 1)$; $x^2 + y^2 + 2x + 4y = 8$.
   (iv) $(1, 2)$; $x^2 + y^2 - 4x - 6y + 11 = 0$.

2. Find the equations of the tangents of gradient $\frac{3}{4}$ to the circle $x^2 + y^2 = 4$.

3. Find the condition that $y = mx$ is a tangent to
   $$x^2 + y^2 - 10x + 16 = 0$$
   and hence find the equations of the tangents from the origin to the circle.

4. Find the length of the chord made by the line $x + y = 4$ on the circle $x^2 + y^2 = 25$.

5. Find the point at which the following lines touch the given circles:
   (i) $2x + 3y = 13$; $x^2 + y^2 = 13$.
   (ii) $x + y = 3$; $x^2 + y^2 + 2x = 7$.
   (iii) $x + 3y = 4$; $x^2 + y^2 + 4y = 6$.
   (iv) $y = -1$; $x^2 + y^2 - 2x - 2y = 2$.
   (v) $x - 4y - 3 = 0$; $x^2 + y^2 - 4x - 8y + 3 = 0$.

Length of tangent from a point to a circle

To find the length of a tangent from the point $(4, 5)$ to the circle $x^2 + y^2 - 2x - 2y = 14$, first find the centre and radius of the circle.

The circle is $(x - 1)^2 + (y - 1)^2 = 16$ and so its centre is $(1, 1)$ and its radius 4.

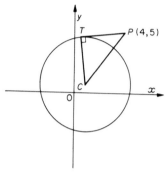

Fig. 25.2

In Fig. 25.2, since the angle $PTC = 90°$,
$$PC^2 = PT^2 + TC^2.$$
P is $(4, 5)$ and C is $(1, 1)$.
$$\therefore PC^2 = 3^2 + 4^2 = 25.$$
Also $CT = 4$.
$$\therefore 25 = PT^2 + 16 \quad \text{and} \quad PT = 3.$$

The length of the tangent is 3 units.

Now suppose the circle is $x^2 + y^2 + 2gx + 2fy + c = 0$ and the point (α, β).

The centre of the circle is $(-g, -f)$ and its radius is
$$\sqrt{g^2 + f^2 - c}.$$

Referring again to Fig. 25.2,
$$PC^2 = (\alpha + g)^2 + (\beta + f)^2$$
and
$$CT^2 = g^2 + f^2 - c.$$
$$\therefore PT^2 = (\alpha + g)^2 + (\beta + f)^2 - (g^2 + f^2 - c)$$
$$= \alpha^2 + \beta^2 + 2g\alpha + 2f\beta + c.$$

So the square of the length of the tangent from (α, β) is given by substituting the coordinates α, β in the left-hand side of the equation of the circle.

To check the numerical example given earlier in this section by the rule, substitute $(4, 5)$ in $x^2 + y^2 - 2x - 2y - 14$.
$$PT^2 = 4^2 + 5^2 - 8 - 10 - 14 = 9.$$
$$\therefore PT = 3.$$

Remember however that this rule can only be applied directly when the equation of the circle begins with the terms $x^2 + y^2$. An example to illustrate this point is given below.

Example. *Find the length of a tangent to the circle*
$$4x^2 + 4y^2 - 8x - 2y + 15 = 0$$
from the point (2, 3).
The circle is put in the form $x^2 + y^2 - 2x - \frac{1}{2}y + \frac{15}{4} = 0$.
$$PT^2 = 2^2 + 3^2 - 4 - 1\frac{1}{2} + \frac{15}{4}$$
$$= 11\frac{1}{4}.$$
$$\therefore PT = \frac{1}{2}\sqrt{45}.$$

The intersections of two circles

Suppose that S stands for $x^2 + y^2 + 2gx + 2fy + c$ and S' stands for $x^2 + y^2 + 2g'x + 2f'y + c'$ so that $S = 0$ and $S' = 0$ represent two circles. We have already pointed out when dealing with the intersection of two curves that $S + kS' = 0$ must always represent a curve passing through all the points of intersection of $S = 0$ and $S' = 0$. If $S = 0$ and $S' = 0$ are circles, the coefficients of x^2 and y^2 in $S + kS'$ are both $(1 + k)$ and there is no xy term. The equation $S + kS' = 0$ must therefore represent a circle or a system of circles as k varies and, for all values of k, each circle of the system must pass through both points of intersection of $S = 0$ and $S' = 0$.

It should be noted here that whether the circles S and S' intersect or not, the circle $S + kS' = 0$ is a member of the coaxial system defined by $S = 0$ and $S' = 0$.

In the particular case when $k = -1$, the equation $S - S' = 0$ becomes $2(g - g')x + 2(f - f')y + (c - c') = 0$ which is a straight line. This line must therefore be the common chord of the circles. What does the line represent if the circles do not intersect?

Suppose that (x_1, y_1) is a point from which tangents to the circles $S = 0$ and $S' = 0$ are equal. Then
$$x_1{}^2 + y_1{}^2 + 2gx_1 + 2fy_1 + c = x_1{}^2 + y_1{}^2 + 2g'x_1 + 2f'y_1 + c',$$
i.e. $S = S'$ is the locus of such points (x_1, y_1). So when the circles do not meet, the line $S = S'$ is the locus of points from which the tangents to S and S' are equal. This line is called the **radical axis** of the two circles. The reader should prove geometrically for himself that this equal tangent property also holds for any point on the common chord produced of two intersecting circles.

Example. *Find the points of intersection of the circles*
$$x^2 + y^2 - 2x - 4y + 4 = 0 \quad and \quad x^2 + y^2 - 5x - 8y + 11 = 0.$$
The equation of the common chord of the circles is
$$x^2 + y^2 - 2x - 4y + 4 - (x^2 + y^2 - 5x - 8y + 11) = 0$$
$$3x + 4y = 7.$$
or

We find where this line meets the first circle and substitute $x = \dfrac{7 - 4y}{3}$ in the first equation.

$$\left(\frac{7 - 4y}{3}\right)^2 + y^2 - 2\left(\frac{7 - 4y}{3}\right) - 4y + 4 = 0.$$

$$\therefore\ 49 - 56y + 16y^2 + 9y^2 - 42 + 24y - 36y + 36 = 0$$

or

$$25y^2 - 68y + 43 = 0.$$

$$\therefore\ (y - 1)(25y - 43) = 0$$

and

$$y = 1 \text{ or } \tfrac{43}{25}.$$

When $y = 1$, $x = 1$.

When $y = \tfrac{43}{25}$, $x = \tfrac{1}{3}(7 - \tfrac{172}{25}) = \tfrac{1}{25}$.

The points of intersection are $(1, 1)$ and $(\tfrac{1}{25}, \tfrac{43}{25})$.

Exercise 25c

1. Find the lengths of the tangents from the given points to the following circles:
   (i) $(1, -2)$; $x^2 + y^2 - 3x - 4y - 6 = 0$.
   (ii) $(1, -1)$; $x^2 + y^2 + 8x + 2y + 8 = 0$.
   (iii) $(0, 1)$; $x^2 + y^2 + 4x + 5y + 3 = 0$.
   (iv) $(-1, 2)$; $4x^2 + 4y^2 + 5x + 6y - 18 = 0$.

2. Find the equation of the common chord of the following pairs of circles:
   (i) $x^2 + y^2 + 6x + 5y - 13 = 0$; $x^2 + y^2 + 2x - 4y = 0$.
   (ii) $x^2 + y^2 - 2x - 3y + 3 = 0$; $x^2 + y^2 - 8x - 5y + 11 = 0$.
   (iii) $x^2 + y^2 - 4x + 7y + 3 = 0$; $x^2 + y^2 - 5x + 8y + 4 = 0$.
   (iv) $x^2 + y^2 + 5x - 2y + 1 = 0$; $x^2 + y^2 + 11x - 3y + 2 = 0$.
   (v) $x^2 + y^2 + x + y - 4 = 0$; $3x^2 + 3y^2 + 5x + 2y - 13 = 0$.

3. Find the locus of a point which moves so that the length of a tangent from it to $x^2 + y^2 - 8x - 2y + 8 = 0$ is equal to its distance from the point $(1, 2)$.

4. Find the locus of a point which moves so that the length of the tangent from it to $x^2 + y^2 = 4$ is double the length of the tangent from it to $x^2 + y^2 = 2$.

5. Find the points of intersection of the circles $x^2 + y^2 = 1$ and $x^2 + y^2 + 2x - 4y + 3 = 0$.

Exercise 25d: Miscellaneous

1. Find the equation of the circle through the point $(1, 1)$ concentric with $x^2 + y^2 - 2x - 4y = 9$.

2. Find the point of intersection of the tangents at the points $(2, -4)$ and $(-3, 1)$ to the circle $x^2 + y^2 - 4x - 2y - 20 = 0$.

3. Find the equation of the circle centre $(2, 3)$ which touches the axis of x.

4. Show that the point $(1 + 3 \cos \theta, 2 + 3 \sin \theta)$ lies on the circle $x^2 + y^2 - 2x - 4y - 4 = 0$ for all values of θ.

5. Find the condition that the circle $x^2 + y^2 + 2gx + 2fy + c = 0$ should touch the y-axis.

6. Find the equation of the circle on the line joining $(1, 1)$ and $(2, -2)$ as diameter.

7. Show that the line $x \cos \alpha + y \sin \alpha = 1$ touches the circle $x^2 + y^2 = 1$ for all values of α.

8. Find the equation of the circle which touches the y-axis and is concentric with $x^2 + y^2 - 6x + 1 = 0$.

9. Find the equation of the circle which has centre $(1, 0)$ and touches the line $3x + 4y = 8$.

10. Find the equation of the circle which passes through the origin and the points $(2, 4)$ and $(0, 8)$.

11. Find the equations of the circles through the point $(4, 2)$ which touch both axes.

12. Find the locus of a point P which moves so that its distance from the origin is double its distance from the point $(1, 2)$.

13. Show that the tangents from the point $(1, 1)$ to the circles
$$x^2 + y^2 + 5x + 8y - 12 = 0 \quad \text{and} \quad x^2 + y^2 - 2x + 4y - 1 = 0$$
are equal in length.

14. Find the equation of the common chord of the circles
$$x^2 + y^2 + 3x + 5y - 10 = 0 \quad \text{and} \quad x^2 + y^2 - 2x + 4y - 4 = 0.$$

15. A circle has its centre at the point $(1, 3)$ and passes through the point $(4, 2)$. Find the equation of the tangent at the point $(4, 2)$.

16. Find the point of intersection of the tangents to the circle $x^2 + y^2 = 5y$ at the origin and at the point $(2, 1)$.

17. Find the equation of the locus of a point P which moves so that the tangents from P to the circle $x^2 + y^2 = 4$ are perpendicular.

18. A circle passes through the origin and through the point $(1, 1)$. The equation of the tangent at the origin is $y = 2x$. Find the equation of the circle.

19. Show that the point $(2, 3)$ lies inside the circle
$$x^2 + y^2 - 2x - 2y - 20 = 0.$$

20. Find the equation of the circle which touches the circle
$$x^2 + y^2 - 2x - 5y = 0$$
at the origin and passes through the point $(2, 1)$.

21. The circles $x^2 + y^2 - 2x - 4y + 4 = 0$ and $x^2 + y^2 - x - y = 0$ meet at the point $(1, 1)$. Find the acute angle between the tangents to the circles at this point.

22. Find the equation of the chord of the circle $x^2 + y^2 - 2x - 4y = 18$ which is bisected at the point $(2, 2)$.

23. Find the equation of the circle which has centre $(3, 2)$ and touches the line $5x + 12y = 13$.

24. Find the values of c for which $x + y = c$ is a tangent to the circle $x^2 + y^2 - 2x - 4y = 3$.

25. A point P moves so that its distance from the origin is equal to the length of a tangent from the point to the circle
$$x^2 + y^2 + 4x + 5y = 11.$$
Find the equation of its locus.

26. Find the equation of the circle which passes through the points $(0, 8)$, $(0, 2)$ and $(4, 0)$.

27. P is any point on the circle $x^2 + y^2 = 2x$. Show that the square of the distance of P from the line $x = y$ is equal to the product of the distances of P from the lines $x = 0$ and $y = 1$.

28. Find the equations of the tangents to the circle
$$x^2 + y^2 - 3x - 4y + 2 = 0$$
at the points where it cuts the x-axis. If θ is the acute angle between the tangents, show that $\tan \theta = \frac{8}{15}$.

29. Find the equation of the circle which passes through both points of intersection of
$$x^2 + y^2 - 2x - 3y + 3 = 0 \quad \text{and} \quad x^2 + y^2 - 3x + 2y - 1 = 0$$
and also passes through the point $(2, 3)$.

30. A point moves so that its distance from the point $(-1, 0)$ is k times its distance from the point $(1, 0)$. Prove that its locus is a circle and find its centre and radius in terms of k, given that $k > 1$.

26 Parameters

The coordinates of the point distant r from $(1, 2)$ on the line through this point making an angle of $30°$ with the axis of x are

$$1 + r \cos 30° \quad \text{and} \quad 2 + r \sin 30°$$

It has already been pointed out in the chapter on the straight line that, in such a case, r is called a **parameter**. The equation of a curve is said to be expressed parametrically when each of the coordinates x and y of a point on the curve are given in terms of a third variable, which is called the parameter. The parameter is any convenient variable and may be for example a distance, an angle or a gradient. The Cartesian equation of the curve may be found by eliminating the parameter between the equations for x and y. In the parametric representation of a line, $x = 1 + r \cos 30°$, $y = 2 + r \sin 30°$, r may be expressed either

as $\dfrac{x - 1}{\cos 30°}$ or $\dfrac{y - 2}{\sin 30°}$. Therefore $\dfrac{x - 1}{\cos 30°} = \dfrac{y - 2}{\sin 30°}$ is the Cartesian

equation of the line.

Parametric equations of a circle

Consider a circle centre the origin and radius 2. Draw a radius which makes an angle θ with the x-axis, as shown in Fig. 26.1.

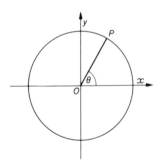

Fig. 26.1

The coordinates of the end point of the radius are $2 \cos \theta$ and $2 \sin \theta$. Therefore $x = 2 \cos \theta$ and $y = 2 \sin \theta$ are the parametric equations of the circle. The Cartesian equation is found by eliminating θ between these two equations.

Since $\cos^2 \theta + \sin^2 \theta = 1$,

$$\left(\frac{x}{2}\right)^2 + \left(\frac{y}{2}\right)^2 = 1 \quad \text{or} \quad x^2 + y^2 = 4.$$

If the circle has centre $(1, 3)$ and radius 2, the corresponding equations are

$$x = 1 + 2 \cos \theta, \quad y = 3 + 2 \sin \theta.$$

Now $\cos \theta = \dfrac{x-1}{2}$ and $\sin \theta = \dfrac{y-3}{2}$ and the equation of the circle is

$$(x-1)^2 + (y-3)^2 = 4.$$

In the most general case, when the circle has centre (a, b) and radius r, the coordinates of a point on the circle may be expressed as

$$a + r \cos \theta, \quad b + r \sin \theta.$$

To find the parametric equations of a curve through the origin

A curve which passes through the origin may often be expressed parametrically by using the substitution $y = mx$.

Consider the curve $y^2 = 4ax$. If we put $y = mx$,

$$m^2 x^2 = 4ax, \quad \text{from which provided } x \neq 0,$$

$$x = \frac{4a}{m^2}.$$

Since $y = mx$, $y = \dfrac{4a}{m}$

and the parametric equations are

$$x = \frac{4a}{m^2}, \quad y = \frac{4a}{m}.$$

The curve $y^2 = 4ax$ is a parabola and a better-known parametric form is $(am^2, 2am)$ and the reader should verify that this point lies on the curve for all values of m.

Example. *Express a point on the curve $x^3 + y^3 = 3xy$ parametrically.*

Put $y = mx$ in the equation.

Then

$$x^3 + m^3 x^3 = 3x(mx) = 3mx^2.$$

Provided $x \neq 0$,

$$x = \frac{3m}{1 + m^3},$$

and therefore

$$y = \frac{3m^2}{1 + m^3}.$$

Parametric forms of well-known curves

The curve $\dfrac{x^2}{a^2} + \dfrac{y^2}{b^2} = 1$ is called an ellipse and its shape is shown in Fig. 26.2.

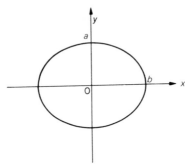

Fig. 26.2

The coordinates of a point on this ellipse may be taken as $(a \cos \theta,\ b \sin \theta)$.

The curve $\dfrac{x^2}{a^2} - \dfrac{y^2}{b^2} = 1$ is called a hyperbola and its shape is shown in Fig. 26.3.

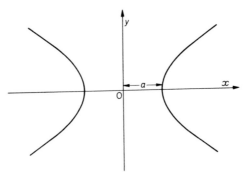

Fig. 26.3

The coordinates of a point on this hyperbola may be taken as $(a \sec \theta,\ b \tan \theta)$ or as $\left\{ \dfrac{a}{2}\left(t + \dfrac{1}{t}\right),\ \dfrac{b}{2}\left(t - \dfrac{1}{t}\right) \right\}$.

The curve $xy = c^2$ is called a rectangular hyperbola and its shape is shown in Fig. 26.4.

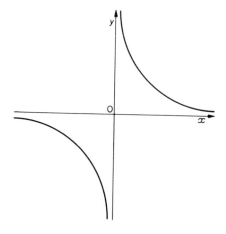

Fig. 26.4

This curve may be expressed very simply as $x = ct$, $y = \dfrac{c}{t}$.

The reader should satisfy himself that the points do in fact lie on the appropriate curves for all values of the variables.

Curve tracing

If a curve is given as parametric equations, it is usually easier to sketch the curve from these equations rather than from the Cartesian equation. The values of the parameter, t, which make x or y zero or infinite should be specially considered.

Example 1. *Sketch the curve given by the equations*
$$x = 1 - t^2, \quad y = t(1 - t^2).$$
The values of t which make x or y zero are -1, 0 and $+1$.
Consider first the positive values of t.
When $t = 0$, $x = 1$ and $y = 0$.
When t lies between 0 and 1, x and y are both positive.
When $t = 1$, $x = y = 0$.
When t is greater than 1, x and y are both negative and both become very large and negative when t becomes very large.
The shape of the part of the curve for which t is positive is shown in Fig. 26.5. The process could be repeated for negative values of t but it is simpler to note that if the sign of t is changed, x remains unaltered and y is changed in sign. The part of the curve for negative values of t is therefore the reflection in the x-axis of the part drawn. The complete curve is shown in Fig. 26.6.

If the curve is to be plotted exactly, values of t should be chosen and the corresponding values of x and y calculated as in the table below:

t	-2	-1	0	$\frac{1}{2}$	1	2
x	-3	0	1	$\frac{3}{4}$	0	-3
y	6	0	0	$\frac{3}{8}$	0	-6

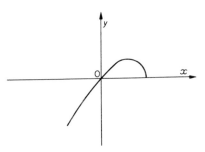

Fig. 26.5

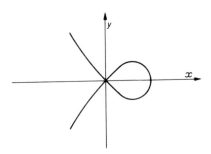

Fig. 26.6

Example 2. *Sketch the curve given by the equations*

$$x = \cos^3 t, \quad y = \sin^3 t.$$

The values of t which make x and y zero are $0°$ and $90°$.
Consider values of t from $0°$ to $90°$.

t	$0°$	$30°$	$45°$	$60°$	$90°$
x	1	0.65	0.35	0.125	0
y	0	0.125	0.35	0.65	1

The part of the curve in the positive quadrant is shown in Fig. 26.7.
When t lies between $90°$ and $180°$, x is negative and y positive.
When t lies between $180°$ and $270°$, x and y are both negative.
When t lies between $270°$ and $360°$, x is positive and y negative.
The complete curve which is called an astroid is shown in Fig. 26.8.

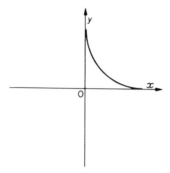

Fig. 26.7

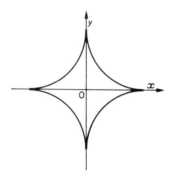

Fig. 26.8

Equations of tangent and normal

Consider the curve $x = 1 - t^2$, $y = t(1 - t^2)$.
   For this curve,

$$\frac{dx}{dt} = -2t \quad \text{and} \quad \frac{dy}{dt} = 1 - 3t^2.$$

Now
$$\frac{dy}{dx} = \frac{dy}{dt} \div \frac{dx}{dt}.$$

$$\therefore \frac{dy}{dx} = \frac{3t^2 - 1}{2t}.$$

This gives the gradient of the tangent to the curve for the corresponding value of t and is useful information in the sketching of the curve. For example, when $t = 0$, $x = 1$ and $y = 0$, and the gradient of the curve at $(1, 0)$ is infinite which agrees with the sketch in Fig. 26.6. The equation of the tangent at the point 't' is

$$\frac{y - t(1 - t^2)}{x - (1 - t^2)} = \frac{3t^2 - 1}{2t}$$

or $2ty - 2t^2(1 - t^2) = (3t^2 - 1)x - (3t^2 - 1)(1 - t^2)$
or $2ty - (3t^2 - 1)x = (1 - t^2)^2.$

The gradient of the normal is $\dfrac{2t}{1 - 3t^2}$ (product of gradients -1)

and its equation is

$$\frac{y - t(1 - t^2)}{x - (1 - t^2)} = \frac{2t}{1 - 3t^2}$$

or $y(1 - 3t^2) - t(1 - t^2)(1 - 3t^2) = 2tx - 2t(1 - t^2)$
or $y(1 - 3t^2) - 2tx = t(t^2 - 1)(3t^2 + 1).$

Exercise 26: Miscellaneous

1. Find the Cartesian equations of the following curves:
   (i) $x = 2 \cos \theta, \ y = 2 \sin \theta.$ (ii) $x = 2 \cos \theta, \ y = 3 \sin \theta.$
   (iii) $x = 3 \sec \theta, \ y = 2 \tan \theta.$ (iv) $x = am^2, \ y = 4am.$

   (v) $x = \dfrac{1}{t}, \ y = t + 1.$

2. Express the following curves parametrically:
   (a) $\dfrac{x^2}{16} + y^2 = 1.$ (b) $\dfrac{x^2}{16} - \dfrac{y^2}{9} = 1.$

   (c) $x^2 = 8y$
   (e) $x^3 + y^3 = x^2.$ (d) $x^3 = y^2.$

3. Find the equations of the tangents to the following curves at the points given:
   (a) $x = ct, \ y = \dfrac{c}{t}.$ (b) $x = 2 \cos \theta, \ y = 2 \sin \theta.$

   (c) $x = am^2, \ y = 2am.$ (d) $x = 2 \cos \theta, \ y = 3 \sin \theta.$
   (e) $x = 3 \sec \theta, \ y = 2 \tan \theta.$

4. Find the equations of the normals to the curves in question 3 at the points given.

5. Find the equation of the chord joining the points $(am^2, 2am)$ and $(at^2, 2at)$ on the parabola $y^2 = 4ax$. By putting $t = m$, find the equation of the tangent at the point $(am^2, 2am)$.

6. Find the equation of the chord joining the points $\left(cm, \dfrac{c}{m}\right)$ and $\left(ct, \dfrac{c}{t}\right)$ on the rectangular hyperbola $xy = c^2$. By putting $t = m$, find the equation of the tangent at the point $\left(cm, \dfrac{c}{m}\right)$.

7. Find the coordinates of the points where the line $x - y + a = 0$ meets the curve given by $x = am^2$, $y = 2am$. What do you deduce about the line and the curve?

8. Find the gradient of the tangent at the point θ to the curve $x = a(\theta + \sin \theta)$, $y = a(1 + \cos \theta)$.

9. Sketch the cycloid $x = a(\theta + \sin \theta)$, $y = a(1 + \cos \theta)$ between $\theta = 0$ and $\theta = \pi$.

10. Find the equation of the tangent at the point m to the curve $x = a \cos^3 m$, $y = a \sin^3 m$.

11. Find the equation of the tangent at the point m to the curve $x = am^3$, $y = am^2$.

12. Find the equation of the normal at the point t to the curve $x = at^3$, $y = at^2$.

13. Show that the line $x - 4y + 4 = 0$ touches the curve $y = m$, $x = m^2$.

14. Find the Cartesian equation of the curve $x = \dfrac{t}{1+t}$, $y = \dfrac{t^2}{1+t}$.

15. Sketch the curve $x = \dfrac{t}{1+t}$, $y = \dfrac{t^2}{1+t}$.

16. Find the equation of the tangent at the point t to the curve $x = \dfrac{t}{1+t}$, $y = \dfrac{t^2}{1+t}$.

17. Find the equation of the normal at the point t to the curve $x = \dfrac{t}{1+t}$, $y = \dfrac{t^2}{1+t}$.

18. Find the tangent at the point t to the curve $x = t(1 - t)$, $y = t^2(1 - t)$.

19. Sketch the curve $x = t(1 - t)$, $y = t^2(1 - t)$.

20. A tangent to the astroid $x = a \cos^3 t$, $y = a \sin^3 t$ meets the axes at P and Q. Show that the length of PQ is constant.

21. Find the Cartesian equation to the curve given by $x = \dfrac{t}{t+1}$, $y = \dfrac{t}{t+2}$.

22. Find the equation of the tangent at the point t to the curve $x = \dfrac{t}{t+1}$, $y = \dfrac{t}{t+2}$.

23. Express $x = (t - 1)^2$, $y = t^2 - 1$ as a Cartesian equation.

24. Find the equation of the tangent at the point t to the curve $x = (t - 1)^2$, $y = t^2 - 1$.

25. Sketch the curve $x = (t - 1)^2$, $y = t^2 - 1$.

TRIGONOMETRY

27 Circular Measure

An angle may be measured either in degrees or radians. The degree is already familiar but in more advanced trigonometry and in calculus the radian is used almost invariably. At first sight it may appear more artificial than the degree but, in fact, it is much more closely associated with the circle and generally leads to simpler formulae than when the degree is the unit of measurement. The radian also emphasizes the fact that an angle is merely a number and has no dimensions.

Definition of a radian

A radian is the angle subtended at the centre of a circle by an arc equal to the radius.

If AB is an arc of a circle, equal in length to the radius OA, then the angle AOB is said to equal 1 radian (abbreviated as 1 rad).

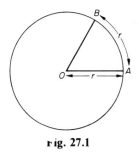

Fig. 27.1

If $A\hat{O}B = x°$, then

$$\frac{x}{360} = \frac{\text{arc } AB}{\text{circumference of circle}} = \frac{r}{2\pi r} = \frac{1}{2\pi}.$$

Therefore $x = \dfrac{180}{\pi}$.

So $\qquad$ 1 radian $= \dfrac{180}{\pi}$ degrees $\quad$ or $\quad 57° \, 18'$ approximately.

$\dfrac{\pi}{2}$ radians $= 90°$ $\quad$ and $\quad \pi$ radians $= 180°$.

By the symmetrical properties of a circle, the size of the radian is obviously independent of the radius of the circle and of the position of the arc on its circumference. Since a radian is the ratio of two lengths, we see that it is merely a number. The number of radians in an angle is called the circular measure of that angle. It should perhaps be pointed out that a trigonometrical ratio of an angle is the same whether the angle is measured in degrees or in radians, e.g. $\sin \dfrac{\pi}{2} = \sin 90°$.

The number of degrees in an angle may be changed to circular measure by multiplying by the factor $\dfrac{\pi}{180}$; e.g. $x° = \dfrac{x\pi}{180}$ radians.

Radians may be changed to degrees by multiplying by the factor $\dfrac{180}{\pi}$; e.g. y rad $= \dfrac{180y}{\pi}$ degrees.

When an angle is given without units (e.g. $\sin x$), assume that the angle is measured in radians.

To save computation, tables will be found in any well-known book of tables which convert degrees to radians and vice versa.

Length of arc

If PQ is an arc of a circle of radius r which subtends an angle of x radians at the centre O,

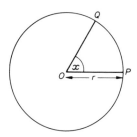

Fig. 27.2

$$\frac{\text{arc } PQ}{\text{circumference}} = \frac{x}{2\pi}$$

(there are 2π radians in a complete revolution).

Therefore

$$\frac{\text{arc } PQ}{2\pi r} = \frac{x}{2\pi}$$

and

$$\text{arc } PQ = rx.$$

Area of sector

Using Fig. 27.2,

$$\frac{\text{area of sector } POQ}{\text{area of circle}} = \frac{x}{2\pi}$$

$$\therefore \frac{\text{area of sector } POQ}{\pi r^2} = \frac{x}{2\pi}$$

and the area of the sector $POQ = \frac{1}{2}r^2 x$.

Proof that x must lie between sin x and tan x

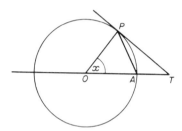

Fig. 27.3

Given that O is the centre of a circle of radius r, and P a point on the circle, the tangent at P meets a radius OA produced at T. Let the angle AOP equal x radians and consider the areas of the triangle AOP, the sector AOP and the triangle TOP.

The area of triangle $AOP = \frac{1}{2}r^2 \sin x$ $(\Delta = \frac{1}{2}ab \sin C)$.

The area of sector $AOP = \frac{1}{2}r^2 x$.

The area of triangle $POT = \frac{1}{2}OP \cdot PT = \frac{1}{2}r^2 \tan x$.

Area of triangle $AOP <$ area of sector $AOP <$ area of triangle TOP.

$$\therefore \frac{1}{2}r^2 \sin x < \frac{1}{2}r^2 x < \frac{1}{2}r^2 \tan x,$$

and $\sin x < x < \tan x$.

Dividing throughout by $\sin x$,

$$1 < \frac{x}{\sin x} < \frac{1}{\cos x}.$$

If the angle x is very small, $\dfrac{1}{\cos x}$ is very nearly equal to 1 and therefore $\dfrac{x}{\sin x}$ must also be nearly equal to 1 when x is small.

This may be written

$$\frac{x}{\sin x} \to 1 \quad \text{as} \quad x \to 0 \quad \text{or} \quad \underset{x \to 0}{\text{Lt}} \frac{x}{\sin x} = 1.$$

Obviously also

$$\underset{x \to 0}{\text{Lt}} \frac{\sin x}{x} = 1$$

Again

$$\frac{\tan x}{x} = \frac{1}{\cos x} \left(\frac{\sin x}{x} \right)$$

and since $\dfrac{1}{\cos x} \to 1$ as $x \to 0$,

$$\underset{x \to 0}{Lt} \frac{\tan x}{x} = 1.$$

These limits may be written in the following way:

$$\text{Lt}_{n \to \infty} \frac{\pi/n}{\sin \pi/n} = 1; \quad Lt_{n \to \infty} \frac{\pi/n}{\tan \pi/n} = 1.$$

So $\sin x$ and $\tan x$ are both approximately equal to x measured in circular measure when x is small.

For example, 1 degree is equal to $\dfrac{\pi}{180}$ rad $= 0.01745$ rad.

From tables, $\sin 1° = 0.0175$ and $\tan 1° = 0.0175$.

(sin $1°$ should of course be less than the circular measure of $1°$ but the tables are not accurate enough to show this.)

Example 1. *Express (a) 14° 18′ in radians and (b) 2.5 radians in degrees.*

(a)
$$14° \, 18' = 14.3° = \frac{14.3 \times \pi}{180} \text{ rad}$$

$$= 0.2496 \text{ rad}.$$

(b)
$$2.5 \text{ radians} = \frac{2.5 \times 180}{\pi} \text{ degrees}$$

$$= 143.2 \text{ degrees} \quad \text{or} \quad 143° \, 12'.$$

Example 2. *An arc AB of a circle of radius 10 cm subtends an angle of 14° 18′ at the centre O. Find (i) the length of the arc AB; (ii) the area of the sector AOB.*
$$14° \, 18' = 0.2496 \text{ rad} \quad \text{(see Example 1).}$$

(i) Length of arc $AB = rx = 10 \times 0.2496 = 2.496$ or 2.50 cm (to 3 s.f.).
(ii) Area of sector $= \frac{1}{2}r^2x = 50 \times 0.2496 = 12.48$ or 12.5 cm^2 (to 3 s.f.).

Example 3. *An arc AB of a circle of radius 8 cm is of length 6 cm. Find (i) the angle AB subtends at the centre O; (ii) the length of the chord AB.*

(i) If $A\hat{O}B = x$ radians, the length of arc $AB = rx$.

∴ $8x = 6$. So $x = 0.75$ rad or $42°$ $58'$.

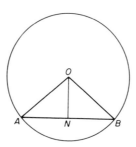

Fig. 27.4

(ii) If N is the foot of the perpendicular from O to AB,

$$N\hat{O}A = 21° 29'.$$

∴ $AN = 8 \sin 21° 29' = 8 \times 0.3662 = 2.9296.$

$$AB = 2AN = 5.8592 \text{ or } 5.86 \text{ cm (to3 s.f.).}$$

(N.B. This must of course be shorter than the arc.)

Exercise 27: Miscellaneous

1. Convert the following angles to degrees:

(i) 1.5 rad; (ii) 2.43 rad; (iii) 0.68 rad; (iv) 0.07 rad; (v) 1.34 rad.

2. Convert the following angles to radians:

(i) $32° 8'$; (ii) $106°$; (iii) $1° 4'$; (iv) $222°$; (v) $180°$.

3. Express in degrees the following angles measured in circular measure:

(i) $\dfrac{\pi}{2}$; (ii) $\dfrac{\pi}{3}$; (iii) $\dfrac{3\pi}{2}$; (iv) $\dfrac{4\pi}{3}$; (v) $\dfrac{5\pi}{8}$.

4. Express the following angles in radians as fractions of π:

(i) $210°$; (ii) $270°$; (iii) $36°$; (iv) $75°$; (v) $135°$.

5. In a circle of radius 8 cm, find the angle subtended at the centre by a chord of length (i) 4 cm, (ii) 3 cm, (iii) 1.6 cm. Give your answers in radians.

6. An arc PQ of a circle subtends an angle of $42°$ at the centre O. If the radius of the circle is 12 cm, calculate the length of the chord PQ and the length of the arc PQ.

7. Find the length of arc of the equator which subtends an angle of 1 minute at the centre of the earth. Take the earth to be a sphere of radius 6400 km.

8. An arc of a circle of radius 10 cm subtends an angle of (i) $\dfrac{\pi}{2}$; (ii) $\dfrac{\pi}{3}$; (iii) $\dfrac{\pi}{4}$

at the centre. Find in each case the length of the arc.

9. Write down without using tables the values of

(i) $\tan \dfrac{\pi}{4}$; (ii) $\sin \dfrac{\pi}{2}$; (iii) $\cos \dfrac{3\pi}{4}$; (iv) $\cos \pi$; (v) $\sin \dfrac{\pi}{3}$; (vi) $\cos \dfrac{\pi}{3}$;

(vii) $\cos \dfrac{2\pi}{3}$; (viii) $\cos \dfrac{\pi}{4}$.

10. Express the following as simply as possible:

(i) $\sin \left(\dfrac{\pi}{2} - x \right)$; (ii) $\cos \left(\dfrac{\pi}{2} + x \right)$; (iii) $\cos (\pi - x)$;

(iv) $\sin (\pi - x)$; (v) $\tan (\pi - x)$.

11. A wire in the shape of a square of side 4 cm is bent to form an arc of a circle which subtends an angle of 60° at the centre. Find the radius of the arc.

12. A wheel is turning at the rate of 120 revolutions per minute. Find the number of radians per second at which it is turning.

13. Express 20 revolutions per minute in radians per second.

14. Two of the angles of a triangle are $\dfrac{\pi}{4}$ and $\dfrac{\pi}{3}$. Find the third angle.

15. Write down (i) the supplement of $\dfrac{\pi}{3}$; (ii) the complement of $\dfrac{\pi}{6}$.

16. A cone of slant edge 8 cm and semi-vertical angle 30° is cut along a generator and opened to form a sector of a circle. Find the angle of the sector.

17. A wheel of radius 2 m is rotating at 80 revolutions per minute. Find the speed of a point on the rim of the wheel.

18. If a wheel of radius 1.5 m is turning so that every point of the rim is moving at 60 km h^{-1}, find the turning speed in radians per second.

19. An arc PQ of a circle of radius 8 cm subtends an angle of 42° at the centre O. Calculate the area of the sector POQ, the area of the triangle POQ and the area of the minor segment cut off by the chord PQ.

20. PQ is an arc of length 7 cm in a circle of radius 6 cm. Find the area of the sector on the arc PQ.

21. In a circle of radius 12 cm, a sector has an area of 75 cm^2. Find the length of arc of the sector.

22. Draw the graph of sin x (where x is in radians) from $x = 0$ to $x = \pi$.

23. Draw the graph of $y = \cos x$ (where x is in radians) from $x = 0$ to $x = \pi$.

24. Find the limit as x tends to 0 of $\dfrac{\sin 2x}{x}$.

25. Find the limit as x tends to 0 of $\dfrac{\sin 2x}{\sin 3x}$.

26. Find the limit as x tends to 0 of $\dfrac{\tan x}{\tan 2x}$.

27. Find the limit as n tends to infinity of $n \tan \dfrac{2\pi}{n}$.

28. Find the limit as n tends to infinity of $\dfrac{\sin 2\pi/n}{\sin 3\pi/n}$.

29. Find the limit as x tends to 0 of $\dfrac{\sin x°}{x}$.

30. The angles of a triangle are in arithmetic progression with common difference $\dfrac{\pi}{6}$. Find the angles.

31. Find the length of belt necessary to go round two wheels of radii 3 cm and 5 cm, whose centres are 10 cm apart.

32. A piece of cardboard is cut in the shape of a sector of a circle of radius 10 cm. If the area of the sector is 100 cm², find the perimeter of the piece of cardboard.

33. A chord PQ subtends an angle $2x$ rad at the circumference of a circle. If PQ divides the circle into two segments, one of which is twice the other in area, find an equation for x.

34. TP, TQ are tangents from a point T to a circle of radius 2 cm. If the angle between the tangents is 40°, find the area enclosed between TP, TQ and the circle.

35. Given the data of the previous question, calculate the length of the minor arc PQ.

36. If h is small, $\tan(x + h) = \tan x + h \sec^2 x$ approximately, where h is in radians. Use this formula to calculate $\tan 45° \, 1'$.

28 The General Angle

Draw a circle of radius 1 unit with centre O and take two perpendicular axes, Ox and Oy. Suppose that a radius rotates anticlockwise about O from the position OA in which it lies along OX.

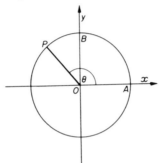

Fig. 28.1

If the radius rotates through an angle θ (however large) to arrive at the position OP, the coordinates of P are ($\cos \theta$, $\sin \theta$). This defines both $\cos \theta$ and $\sin \theta$ for any positive angle θ. The tangent is defined by the equation $\tan \theta = \dfrac{\sin \theta}{\cos \theta}$. If OP rotates through an angle ($360° + \theta$), the position of P is unchanged and so are the cosine, sine and tangent. The same properties hold for the addition of any complete number of revolutions.

$$\therefore \cos (720° + \theta) = \cos (360° + \theta) = \cos \theta$$

and
$$\sin (720° + \theta) = \sin (360° + \theta) = \sin \theta.$$

An angle of 360° is equal to 2π radians and the above equations may be more neatly expressed in radians in the following ways:

$$\sin (2n\pi + x) = \sin x;$$
$$\cos (2n\pi + x) = \cos x.$$

Negative angles

A similar definition holds for the cosine and sine of a negative angle. The radius OP rotates about O from OA in a clockwise direction through the given angle and the coordinates of P are respectively equal to the cosine and sine of the angle. As before,

$$\tan \theta = \frac{\sin \theta}{\cos \theta}.$$

If the angle θ is acute, the angle $-\theta$ will be in the fourth quadrant.

A ratio of any angle is equal numerically to the same ratio of the *acute* angle the radius makes with the *x*-axis. (This may easily be proved by congruent triangles.)

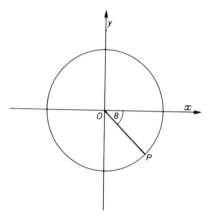

Fig. 28.2

$$\therefore \cos(-\theta) = \cos\theta;$$
$$\sin(-\theta) = -\sin\theta;$$
$$\tan(-\theta) = -\tan\theta.$$

If θ is obtuse, the angle $(-\theta)$ will be in the third quadrant. Figure 28.2 shows the relative positions of P and P' for the angles θ and $-\theta$.

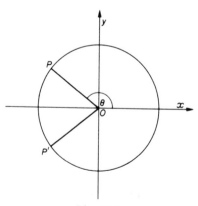

Fig. 28.3

Obviously the *x*-coordinates of P and P' are equal.

$$\therefore \cos(-\theta) = \cos\theta.$$

The y-coordinates of P and P' are obviously equal in magnitude but opposite in sign.

$$\therefore \sin(-\theta) = -\sin\theta$$

and by division

$$\tan(-\theta) = -\tan\theta.$$

The student should satisfy himself that these equations are always true whatever the magnitude of θ.

Ratios of $(90° + \theta)$

If θ is acute, $(90° + \theta)$ will be in the second quadrant. The acute angle made by the radius with the x-axis is $(90° - \theta)$.

$$\therefore \cos(90° + \theta) = -\cos(90° - \theta) = -\sin\theta;$$
$$\sin(90° + \theta) = \sin(90° - \theta) = \cos\theta;$$
$$\tan(90° + \theta) = -\tan(90° - \theta) = -\cot\theta.$$

The student should satisfy himself that these formulae hold even when the angle θ is not acute.

If the angle is measured in radians

$$\cos\left(\frac{\pi}{2} + x\right) = -\sin x; \quad \sin\left(\frac{\pi}{2} + x\right) = \cos x;$$

$$\tan\left(\frac{\pi}{2} + x\right) = -\cot x.$$

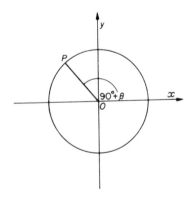

Fig. 28.4

Ratios of (180° − θ)

If θ is acute, $(180° - \theta)$ lies in the second quadrant. The acute angle made by the radius with the axis of x is θ.

$$\therefore \cos (180° - \theta) = -\cos \theta;$$
$$\sin (180° - \theta) = \sin \theta;$$
$$\tan (180° - \theta) = -\tan \theta.$$

These results proved when θ is acute are universally true. In radians,

$$\cos (\pi - x) = -\cos x; \quad \sin (\pi - x) = \sin x;$$
$$\tan (\pi - x) = -\tan x.$$

Ratios of (180° + θ)

If θ is acute, $(180° + \theta)$ lies in the third quadrant. The acute angle made by the radius with the axis of x is θ.

$$\therefore \cos (180° + \theta) = -\cos \theta;$$
$$\sin (180° + \theta) = -\sin \theta;$$
$$\tan (180° + \theta) = \tan \theta.$$

These results are true whatever the value of θ. In radians,

$$\cos (\pi + x) = -\cos x; \quad \sin (\pi + x) = -\sin x;$$
$$\tan (\pi + x) = \tan x.$$

Ratios of (270° − θ)

If θ is acute, $(270° - \theta)$ lies in the third quadrant. The acute angle made by the radius with the axis of x is $(90° - \theta)$.

$$\therefore \cos (270° - \theta) = -\cos (90° - \theta) = -\sin \theta;$$
$$\sin (270° - \theta) = -\sin (90° - \theta) = -\cos \theta;$$
$$\tan (270° - \theta) = \tan (90° - \theta) = \cot \theta.$$

These results are true whatever the value of θ. In radians,

$$\cos \left(\frac{3\pi}{2} - x\right) = -\sin x; \quad \sin \left(\frac{3\pi}{2} - x\right) = -\cos x;$$

$$\tan \left(\frac{3\pi}{2} - x\right) = \cot x.$$

Ratios of $(270° + \theta)$

If θ is acute, $(270° + \theta)$ lies in the fourth quadrant. The acute angle made by the radius with the axis of x is $(90° - \theta)$.

$$\therefore \cos (270° + \theta) = \cos (90° - \theta) = \sin \theta;$$
$$\sin (270° + \theta) = -\sin (90° - \theta) = -\cos \theta;$$
$$\tan (270° + \theta) = -\tan (90° - \theta) = -\cot \theta.$$

These results are true whatever the value of θ.

In radians,

$$\cos \left(\frac{3\pi}{2} + x\right) = \sin x; \quad \sin \left(\frac{3\pi}{2} + x\right) = -\cos x;$$

$$\tan \left(\frac{3\pi}{2} + x\right) = -\cot x.$$

Ratios of $(360° - \theta)$

If θ is acute, $(360° - \theta)$ lies in the fourth quadrant. The acute angle made by the radius with the axis of x is θ.

$$\therefore \cos (360° - \theta) = \cos \theta;$$
$$\sin (360° - \theta) = -\sin \theta;$$
$$\tan (360° - \theta) = -\tan \theta.$$

These results, which are true for all values of θ, are obviously the same as for the angle $(-\theta)$ since the position of the radius is unchanged by a complete revolution.

In radians,

$$\cos (2\pi - x) = \cos x; \quad \sin (2\pi - x) = -\sin x;$$
$$\tan (2\pi - x) = -\tan x.$$

Ratios of any angle

To find a ratio of any angle, first write down the sign according to the position of the radius in the circle. The result may then be written down using either

(i) any ratio of $(n\pi \pm x)$ is numerically equal to the same ratio of x;

or (ii) any ratio of $\left(\dfrac{(2n + 1)\pi}{2} \pm x\right)$ is numerically equal to the co-ratio of x.

Examples.

1. $\sin (270° - \theta)$ is in the third quadrant and is of form (ii).

$$\therefore \sin (270° - \theta) = -\cos \theta.$$

2. $\cos\left(\dfrac{3\pi}{2} + x\right)$ is in the fourth quadrant and is of form (ii).

$$\therefore \cos\left(\dfrac{3\pi}{2} + x\right) = \sin x.$$

3. $\tan(x - \pi)$ is of form (i) and is in the third quadrant when x is acute. In this quadrant the tangent is positive.

$$\therefore \tan(x - \pi) = +\tan x.$$

This result may also be proved using the formulae

$$\tan(\pi - x) = -\tan x \quad \text{and} \quad \tan(-x) = -\tan x.$$

Therefore $\tan(x - \pi) = -\tan(\pi - x) = +\tan x.$

4. If x lies between $\dfrac{3\pi}{2}$ and 2π and $\cos x = \tfrac{3}{5}$, find the values of $\sin x$ and $\tan x$.

If y is acute and $\cos y = \tfrac{3}{5}$, from the triangle in Fig. 28.5,

$$\sin y = \tfrac{4}{5} \quad \text{and} \quad \tan y = \tfrac{4}{3}.$$

Since x is in the fourth quadrant,

$$\sin x = -\tfrac{4}{5} \quad \text{and} \quad \tan x = -\tfrac{4}{3}.$$

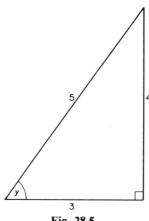

Fig. 28.5

Exercise 28: Miscellaneous

1. In which quadrant must x lie if $\cos x = \tfrac{1}{2}$ and $\sin x = -\dfrac{\sqrt{3}}{2}$?

2. In which quadrant must x lie if $\cos x = -\tfrac{4}{5}$ and $\sin x = \tfrac{3}{5}$?

3. In which quadrant must x lie if $\tan x = -1$ and $\sec x = \sqrt{2}$?

4. In which quadrant must x lie if $\sin x = -\tfrac{5}{13}$ and $\cos x = \tfrac{12}{13}$?

5. In which quadrant must x lie if $\sec x = -\sqrt{2}$ and $\operatorname{cosec} x = -\sqrt{2}$?

6. In which quadrant must x lie if $\tan x = \frac{1}{2}$ and $\sin x$ is negative?

7. In which quadrant must x lie if $\sec x = 3$ and $\tan x$ is negative?

8. In which quadrant must x lie if $\operatorname{cosec} x = 2$ and $\cot x$ is positive?

9. In which quadrant must x lie if $\sin x = \frac{3}{4}$ and $\cos x$ is negative?

10. In which quadrant must x lie if $\cos x = \frac{1}{2}$ and $\sin x$ is negative?

11. Write down the value of $\cos \dfrac{5\pi}{6}$.

12. Write down the value of $\tan \dfrac{4\pi}{3}$.

13. Write down the value of $\cot \dfrac{13\pi}{4}$.

14. Write down the value of $\sec \dfrac{8\pi}{3}$.

15. Write down the value of $\tan \dfrac{31\pi}{6}$.

16. Simplify $\cos \left(\dfrac{5\pi}{2} + \alpha \right)$.

17. Simplify $\cot \left(\dfrac{5\pi}{2} - \alpha \right)$.

18. Simplify $\tan (3\pi + \alpha)$.

19. Simplify $\sin \left(\dfrac{7\pi}{2} - \alpha \right)$.

20. Simplify $\operatorname{cosec} (3\pi - \alpha)$.

21. Simplify $\tan (\alpha - 2\pi)$.

22. Simplify $\cot \left(\alpha - \dfrac{3\pi}{2} \right)$.

23. Simplify $\cos (2\alpha - 2\pi)$.

24. Simplify $\sin (3\alpha - 2\pi)$.

25. Simplify $\operatorname{cosec} \left(\alpha - \dfrac{5\pi}{2} \right)$.

26. Evaluate exactly $\dfrac{\sin 135° + \cos 60°}{\sin 135° - \cos 60°}$.

27. Evaluate exactly $\dfrac{\tan 135° + \tan 240°}{\tan 135° - \tan 240°}$.

28. Evaluate exactly $\sin 60° + \sin 120° + \sin 240°$.

29. Evaluate exactly $\cos 60° + \cos 120° + \cos 240°$.

30. Evaluate exactly $\cos 135° + \cos 225° + \cos 315°$.

31. If $\sin x = \frac{7}{25}$, find the values of $\cos x$ and $\tan x$ when x is in the second quadrant.

32. If $\sin x = -\dfrac{1}{\sqrt{2}}$, find the values of $\sec x$ and $\cot x$ when x lies in the third quadrant.

33. If $\tan x = -1$, find the values of cosec x and cot x when x lies in the second quadrant.

34. If $\cot x = -\sqrt{2}$, find the values of cosec x and sec x when x lies in the fourth quadrant.

35. If cosec $x = \sqrt{2}$, find the values of sec x and cot x when x lies in the second quadrant.

36. Simplify $\cos^2\left(\dfrac{\pi}{2} + \alpha\right) + \sin^2\left(\dfrac{\pi}{2} - \alpha\right)$.

37. Simplify $\cos^2\left(\dfrac{\pi}{2} - \alpha\right) + \cos^2\alpha$.

38. Simplify $\sec^2(2\pi - \alpha) - \cot^2\left(\dfrac{\pi}{2} + \alpha\right)$.

39. Simplify $\tan\left(\dfrac{\pi}{2} + \alpha\right)\tan\alpha$.

40. Simplify $\operatorname{cosec}\left(\dfrac{3\pi}{2} + \alpha\right)\cos\alpha$.

41. Write down all the angles between $360°$ and $720°$ which have the same tangent as $35°$.

42. Write down all the angles between $400°$ and $600°$ which have the same sine as $30°$.

43. If cosec $x = -2$, what are all the possible values of x between $360°$ and $720°$?

44. If $\sin\alpha = \frac{4}{5}$ and α lies in the second quadrant, find the value of $(\tan\alpha - \cos\alpha)$.

45. If $\sin\alpha = \frac{5}{13}$, find the value of $(\cos\alpha + \tan\alpha)$, (i) when α is acute, (ii) when α is obtuse.

46. If $\sin\alpha$ is negative and $\cos\alpha$ is positive and $\sin\alpha$ is numerically greater than $\cos\alpha$, between what values must α lie if it is positive and less than $360°$?

47. If $\tan\alpha = \dfrac{1}{2}\left(c - \dfrac{1}{c}\right)$, find the possible values of sec α.

48. If $\cos\alpha + \sec\alpha = z$, find the possible values of $\cos\alpha$.

49. If cosec $\alpha - \cot\alpha = u$, find the possible values of cosec α.

50. If $\tan\alpha = 1 + \varepsilon$, where ε is small, prove that

$$\sec\alpha = \pm\sqrt{2}\left(1 + \dfrac{\varepsilon}{2}\right) \text{ approximately.}$$

29 Graphs of the Circular Functions

In this chapter, the graphs of sin x, cos x and tan x are considered and the functions are plotted against the angle measured in radians. The graphs are plotted for values of x between 0 and 2π. The extension to angles of any magnitude is not a difficult one, because, in any complete revolution, a circular function completes its cycle. For this reason the circular functions are called **periodic** functions and their graphs are repetitions of the part between 0 and 2π.

The graph of sin x

The graph of sin x may be plotted very simply by taking the measurements from a circle of unit radius as shown in Fig. 29.1. To find the plot corresponding to $x = \dfrac{\pi}{4}$, an angle of $\dfrac{\pi}{4}$ is drawn at the centre of the circle and the ordinate at the end of the radius is the value of y at the point where $x = \dfrac{\pi}{4}$. Angles of $\dfrac{\pi}{8}, \dfrac{\pi}{4}, \dfrac{3\pi}{8}, \dfrac{3\pi}{4}, \dfrac{5\pi}{4}$ and $\dfrac{7\pi}{4}$ are indicated in the diagram but the reader should draw the graph for himself and take more values than those given in the diagram.

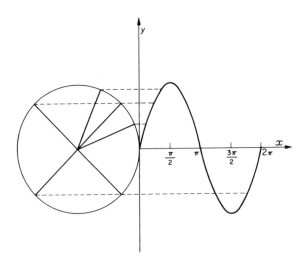

Fig. 29.1

The following points are worthy of notice:

(i) The curve from $x = 0$ to $x = \pi$ is symmetrical about the line $x = \dfrac{\pi}{2}$. This follows from the identity

$$\sin\left(\frac{\pi}{2} + x\right) = \sin\left(\frac{\pi}{2} - x\right).$$

(ii) The curve lies between the limits ± 1 for y.

(iii) The curve from π to 2π is below the axis of x and is the reflection (translated along the x-axis) of that part of the curve between 0 and π. This follows from the identity

$$\sin\left(\pi + x\right) = -\sin x.$$

(iv) The portion of the curve from 2π to 4π is the same as that portion between 0 and 2π. This follows from the identity

$$\sin\left(2\pi + x\right) = \sin x.$$

(v) A line $y = k$ where k lies between ± 1 cuts the curve in an infinite number of points. This shows that for a given value of $\sin x$ there is an infinite number of possible values for x. If α is the smallest possible value, from the shape of the curve it follows that the other positive values are $\pi - \alpha$, $2\pi + \alpha$, $3\pi - \alpha$, $4\pi + \alpha$, etc. In fact, the general value is $2n\pi + \alpha$, or $(2n + 1)\pi - \alpha$, where n may take any integral value. These two results may be combined to give the single formula

$$x = n\pi + (-1)^n \alpha.$$

This is the general solution of the equation $\sin x = \sin \alpha$.

The graph of cos x

The graph of cos x is shown in Fig. 29.2.

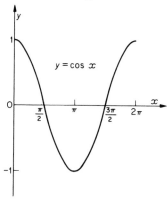

$$y = \cos x$$

Fig. 29.2

The following points are worthy of note:

(i) Since $\sin\left(\dfrac{\pi}{2} + x\right) = \cos x$, the graph of $\cos x$ is the same as that of $\sin x$ with the axis of y shifted.

(ii) The value of $\cos x$ lies between ± 1.

(iii) The portion of the curve between $x = \dfrac{\pi}{2}$ and $x = -\dfrac{\pi}{2}$ is symmetrical about the axis of y.

(iv) The portion of the curve between $x = 0$ and $x = 2\pi$ is the same as that from $x = 2\pi$ to $x = 4\pi$.

(v) A line $y = k$ where k lies between ± 1 cuts the curve in an infinite number of points. This shows that for a given value of $\cos x$, there is an infinite number of possible values for x. If α is the smallest positive value of x, from the shape of the curve, it follows that the other positive values are $2\pi - \alpha$, $2\pi + \alpha$, $4\pi - \alpha$, etc. In fact, the general value is $2n\pi \pm \alpha$, where n is any integer. The general solution of the equation $\cos x = \cos \alpha$ is $2n\pi \pm \alpha$.

The graph of tan x

The graph of $\tan x$ may be drawn in a way similar to that in which $\sin x$ was drawn. The ordinate corresponding to $\dfrac{\pi}{4}$, for example, is now

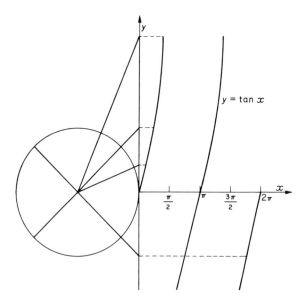

Fig. 29.3

found by producing the radius making an angle $\frac{\pi}{4}$ with the x-axis to meet the tangent to the circle at the point shown in Fig. 29.3. The angles shown in the diagram are $\frac{\pi}{8}, \frac{\pi}{4}, \frac{3\pi}{8}, \frac{3\pi}{4}, \frac{5\pi}{4}$, and $\frac{7\pi}{4}$, but the reader should draw the graph for himself and plot many more points than are shown in the figure. As the angle gets near 90°, the ordinate increases very rapidly and it is not possible to plot points for angles much larger than 70°.

The following points are worthy of note:

(i) In the graph for tan x, discontinuities occur at odd multiples of $\frac{\pi}{2}$. At these points, the value of the tangent of the angle changes from $+\infty$ to $-\infty$.

(ii) The value of tan x is not restricted as are the values of sin x and cos x which must both lie between ± 1.

(iii) The portion of the curve between 0 and 2π is the same as that between 2π and 4π.

(iv) A line $y = k$ cuts the curve in an infinite number of points for all values of k. This shows that for a given value of tan x, there is an infinite number of possible values for x. If α is the smallest positive value, from the shape of the curve it follows that the other positive values are $\pi + \alpha$, $2\pi + \alpha$, $3\pi + \alpha$, etc. In fact, the general solution is $n\pi + \alpha$, where n is any integer. The general solution of the equation tan x = tan α is

$$x = n\pi + \alpha.$$

(v) Since $\underset{x \to 0}{\mathrm{Lt}} \dfrac{\tan x}{x} = 1$, the gradient of the curve at the origin or at any point where it crosses the axis of x is 1. The curve therefore cuts the axis of x at an angle of $\frac{\pi}{4}$ (assuming that equal scales are chosen for x and y).

Solution of equations by graphs

An example is given to illustrate how trigonometrical equations are solved by graphical methods.

Example. *Find the solutions of the equation cos x = cos 2x between 0 and 2π. Deduce all the solutions of the equation.*

The shape of the graph of cos x between x = 0 and x = 2π has already been considered. The graph of cos 2x between x = 0 and x = π is the same as that of cos x between x = 0 and x = 2π. The graphs are shown in Fig. 29.4.

The solutions in the given range are $0, \dfrac{2\pi}{3}, \dfrac{4\pi}{3}$ and 2π. From the symmetry of the curves, the general solutions are $2n\pi$ and $\dfrac{2n\pi}{3}$, where n is any integer. These solutions are both included in the one expression $\dfrac{2n\pi}{3}$ because the value 4π for example may be obtained from $\dfrac{2n\pi}{3}$ by putting n equal to 6.

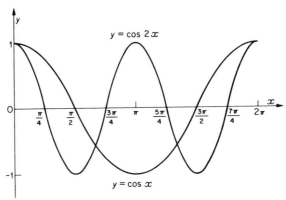

Fig. 29.4

Exercise 29: Miscellaneous

Draw the graphs from $x = 0$ to $x = 2\pi$ of

1. $\sin 2x$. **2.** $\cos \frac{1}{2}x$. **3.** $2 \sin x$.

4. $\sin^2 x$. **5.** $\tan 2x$. **6.** $\cot \frac{1}{2}x$.

7. $\sin x + \cos x$. **8.** $\sin x - \cos x$. **9.** $\tan x + \cot x$.

10. $\tan x - \cot x$.

11. Find graphically the smallest positive value of θ for which $\tan \theta = \cos^2 \theta$.

12. Find a value of x for which $\cos x + \frac{1}{2} \cos 2x = 0$.

13. Find an angle less than $180°$ satisfying the equation
$$3 \cos \theta + 4 \sin \theta = 2.$$

14. Find graphically a solution other than $90°$ of the equation
$$2 \cos 3x + 3 \cos x = 0.$$

15. Find a solution of the equation $3 \cos x - \sin x = \frac{1}{2}$.

16. Find a solution of the equation $1 + \cos 2x = 3 \sin x$.

17. Find two values of x between 0 and 90 which satisfy
$$3 \sin 3x° - \cos 2x° = 1.$$

18. Draw the graphs of $1 + \cos x$ and of $\sin x$ between $0°$ and $180°$. For what values of x between these limits is $\sin x > 1 + \cos x$?

19. Find the solution of the equation $\tan x° = \cos x°$ between 0 and 45.

20. Find the value of θ between $0°$ and $60°$ for which $\sin 3\theta = \tan \theta$.

30 Identities and Equations;
the Sine and Cosine Formulae

Identities

If, in a circle which has unit radius and centre at the origin, a radius is drawn making an angle θ with the positive x-direction, the coordinates of the end point of the radius are $(\cos \theta, \sin \theta)$. By the general definitions of sine and cosine, these are the coordinates for all values of θ.

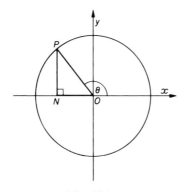

Fig. 30.1

From Fig. 30.1, it then follows, since

$$ON^2 + NP^2 = OP^2,$$

that
$$\cos^2 \theta + \sin^2 \theta = 1 \tag{i}$$

We also know by definition, that

$$\tan \theta = \frac{\sin \theta}{\cos \theta} \tag{ii}$$

Dividing both sides of equation (i) by $\cos^2 \theta$,

$$1 + \tan^2 \theta = \frac{1}{\cos^2 \theta} = \sec^2 \theta \tag{iii}$$

Dividing both sides of equation (i) by $\sin^2 \theta$,

$$\cot^2 \theta + 1 = \frac{1}{\sin^2 \theta} = \operatorname{cosec}^2 \theta \tag{iv}$$

These four identities are true for all values of θ and are so important that they are grouped together below.

$$\cos^2 \theta + \sin^2 \theta = 1 \qquad \text{(i)}$$

$$\tan \theta = \frac{\sin \theta}{\cos \theta} \qquad \text{(ii)}$$

$$\sec^2 \theta = 1 + \tan^2 \theta \qquad \text{(iii)}$$

$$\operatorname{cosec}^2 \theta = 1 + \cot^2 \theta \qquad \text{(iv)}$$

Example 1. *Show that* $(\cos \theta + \sin \theta)^2 + (\cos \theta - \sin \theta)^2 = 2.$

$$(\cos \theta + \sin \theta)^2 = \cos^2 \theta + 2 \cos \theta \sin \theta + \sin^2 \theta.$$
$$(\cos \theta - \sin \theta)^2 = \cos^2 \theta - 2 \cos \theta \sin \theta + \sin^2 \theta.$$
$$\therefore (\cos \theta + \sin \theta)^2 + (\cos \theta - \sin \theta)^2 = 2(\cos^2 \theta + \sin^2 \theta) = 2.$$

Example 2. *Show that* $\dfrac{1}{\sec \theta + 1} + \dfrac{1}{\sec \theta - 1} = 2 \operatorname{cosec} \theta \cot \theta.$

$$\frac{1}{\sec \theta + 1} + \frac{1}{\sec \theta - 1} = \frac{(\sec \theta - 1) + (\sec \theta + 1)}{(\sec \theta + 1)(\sec \theta - 1)} = \frac{2 \sec \theta}{\sec^2 \theta - 1}$$

$$= \frac{2 \sec \theta}{\tan^2 \theta} = 2 \cot \theta \left(\frac{\sec \theta}{\tan \theta} \right)$$

$$= 2 \cot \theta \left(\frac{1}{\sin \theta} \right) = 2 \operatorname{cosec} \theta \cot \theta.$$

Exercise 30a

Prove the following identities:

1. $\dfrac{1}{1 - \sin \theta} + \dfrac{1}{1 + \sin \theta} = 2 \sec^2 \theta.$

2. $\dfrac{\cos^2 \theta}{1 - \sin \theta} - \dfrac{\cos^2 \theta}{1 + \sin \theta} = 2 \sin \theta.$

3. $\dfrac{1}{\operatorname{cosec} \theta - \cot \theta} + \dfrac{1}{\operatorname{cosec} \theta + \cot \theta} = 2 \operatorname{cosec} \theta.$

4. $\dfrac{1}{\operatorname{cosec} \theta - \cot \theta} - \dfrac{1}{\operatorname{cosec} \theta + \cot \theta} = 2 \cot \theta.$

5. $(\cos x + \sin x)^3 - (\cos x - \sin x)^3 = 2(3 \sin x - 2 \sin^3 x).$

6. $\dfrac{\cos x}{1 - \sin x} - \tan x = \sec x.$

7. $\cos^4 x - \sin^4 x = \cos^2 x - \sin^2 x.$

8. $\cos^6 x + \sin^6 x = 1 - 3 \cos^2 x \sin^2 x.$

9. $\sqrt{\dfrac{1 - \cos^2 \theta}{1 - \sin^2 \theta}} = \tan \theta.$

10. $\left(\dfrac{1}{\sin^2 \theta} - 1\right)\left(\dfrac{1}{\cos^2 \theta} - 1\right) = 1.$

Elimination

The four identities proved in the last paragraph are also useful in eliminating an unknown between two trigonometrical equations. If $\cos x$ can be found from one equation and $\sin x$ from the other, the eliminant may be obtained by squaring and adding.

Example 1. *Eliminate θ between $\cos \theta + \sin \theta = a$, $\cos \theta - \sin \theta = b$.*

By adding the two equations, $2 \cos \theta = a + b.$
By subtracting the equations, $2 \sin \theta = a - b.$
Squaring and adding:
$$4 = (a + b)^2 + (a - b)^2$$
or
$$a^2 + b^2 = 2.$$

Example 2. *Eliminate θ between $\sec \theta + \tan \theta = a$,*
$$\sec \theta - \tan \theta = b.$$

This may be done in a similar way to the first example by finding $\sec \theta$ and $\tan \theta$ but it is simpler to multiply the equations as they stand.
$$(\sec \theta + \tan \theta)(\sec \theta - \tan \theta) = ab$$
and therefore $\sec^2 \theta - \tan^2 \theta = ab$ or $ab = 1.$

Example 3. *Eliminate x between $a \sin x = b$ and $\tan x = c$.*
$\tan^2 x = c^2$ and therefore $\sec^2 x - 1 = c^2$ and $\sec^2 x = 1 + c^2.$

$$\sin^2 x = 1 - \cos^2 x = 1 - \frac{1}{\sec^2 x} = 1 - \frac{1}{1 + c^2} = \frac{c^2}{1 + c^2}.$$

But $\sin^2 x = \dfrac{b^2}{a^2}$ and so $\dfrac{b^2}{a^2} = \dfrac{c^2}{1 + c^2}$ or $a^2 c^2 = b^2(c^2 + 1).$

Another method illustrates a useful device whenever one ratio can be found. Here we are given $\tan x = c$. Draw a right-angled triangle as shown in Fig. 30.2 with the sides containing the right angle of lengths c and 1. The angle

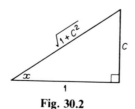

Fig. 30.2

opposite the side of length c is x and, since the length of the hypotenuse is $\sqrt{c^2 + 1}$, the other ratios of x may be written down. From the triangle

$$\sin x = \frac{c}{\sqrt{c^2 + 1}}, \text{ and since } \sin x = \frac{b}{a},$$

$$\frac{c}{\sqrt{c^2 + 1}} = \frac{b}{a} \quad \text{or} \quad a^2 c^2 = b^2(c^2 + 1).$$

Exercise 30b

Eliminate x between:

1. $\cos x = a$, $\tan x = b$.
2. $\cos x + \sin x = a$, $\tan x = b$.
3. $a \cos x + b \sin x = c$, $\cos x + \sin x = d$.
4. $\sec x = a$, $\sec x + \tan x = b$.
5. $a = 1 + \cos x$, $b = 1 + \sin x$.
6. $a = \operatorname{cosec} x - \cot x$, $b = \operatorname{cosec} x + \cot x$.
7. $a = \sin x$, $b = \operatorname{cosec} x + \cot x$.
8. $a = \tan x + \cos x$, $b = \tan x - \cos x$.

Solution of equations

In this section we shall consider the solutions of some trigonometrical equations for values of the variable between $0°$ and $180°$. The equation $\sin x = \frac{1}{2}$ has an obvious solution $x = 30°$. Is there any other solution between $0°$ and $180°$? Since $\sin x$ is positive in the second quadrant, the value $(180° - 30°)$ or $150°$ is also a solution.

On the other hand, the equation $\cos x = \frac{1}{2}$ has only one solution less than $180°$, namely $60°$. Since $\cos x$ is negative in the second quadrant, no solution can exist between $90°$ and $180°$.

The equation $2 \sin^2 x - 3 \sin x + 1 = 0$ can be factorized to give $(2 \sin x - 1)(\sin x - 1) = 0$. From this, $\sin x = 1$ or $\sin x = \frac{1}{2}$.

The solutions between the given limits are $x = 30°$, $90°$ and $150°$.

One general method of solution of a trigonometrical equation is to express the equation in terms of one ratio only and examples are given to show how our identities may be used to do this.

Example 1. *Solve the equation $1 - \sin x = \cos^2 x$.*

Since $\cos^2 x + \sin^2 x = 1$, the equation may be rewritten as

$$1 - \sin x = 1 - \sin^2 x = (1 + \sin x)(1 - \sin x).$$

Either $\sin x = 1$ or $1 = 1 + \sin x$, i.e. $\sin x = 1$ or 0.
The solutions are $0°$, $90°$ and $180°$.

Example 2. *Solve the equation* $\tan^4 x + 7 = 4\sec^2 x$.

Since $\sec^2 x = 1 + \tan^2 x$, the equations may be rewritten as

$$\tan^4 x + 7 = 4(1 + \tan^2 x)$$

or
$$\tan^4 x - 4\tan^2 x + 3 = 0.$$

$$\therefore (\tan^2 x - 3)(\tan^2 x - 1) = 0$$

or
$$\tan x = \pm\sqrt{3}, \quad \tan x = \pm 1.$$

The solutions are 45°, 60°, 120° and 135°.

Exercise 30c

Find all the solutions between 0° and 180° of the following equations:

1. $3(1 - \cos x) = 2\sin^2 x$.
2. $2\sin x \cos x = \sin x$.
3. $\sec^2 x + 1 = 3\tan x$.
4. $\tan^3 x + 4 = \sec^2 x + 3\tan x$.
5. $\cot^2 x + 3 = 3\operatorname{cosec} x$.
6. $(\tan x - 1)(2\sin x - 1) = 0$.
7. $\cos^2 x = 1 + \sin x$.
8. $3(1 + \cos x) = 2\sin^2 x$.

The sine formula

The reader should be familiar with the sine and cosine formulae. For the sake of completeness proofs of these formulae are given but examples on the solution of triangles by these formulae are not included. The sine formula is given in the form

$$\frac{a}{\sin A} = \frac{b}{\sin B} = \frac{c}{\sin C} = 2R,$$

where R is the radius of the circumcircle of the triangle ABC.

(i) *The angle A acute*

In Fig. 30.3, join B to the centre O of the circle and produce BO to meet the circle again at X.

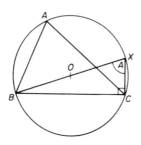

Fig. 30.3

Since BX is a diameter, the angle $BCX = 90°$ and the angle BXC equals the angle A, both standing on the arc BC.

In the triangle BXC,

$$\sin A = \frac{BC}{BX} = \frac{a}{2R}.$$

$$\therefore \frac{a}{\sin A} = 2R.$$

Similarly it may be proved that $\dfrac{b}{\sin B}$ and $\dfrac{c}{\sin C}$ both equal $2R$.

$$\therefore \frac{a}{\sin A} = \frac{b}{\sin B} = \frac{c}{\sin C} = 2R.$$

(ii) *The angle A obtuse*

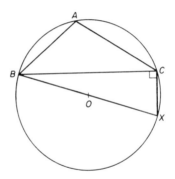

Fig. 30.4

The construction in Fig. 30.4 is the same as when A is acute.
The angle $BXC = 180° - A$ (opp. angles of cyclic quad.).
In the triangle BXC,

$$\sin (180° - A) = \frac{BC}{BX} = \frac{a}{2R}.$$

But $\sin (180° - A) = \sin A$ and therefore $\dfrac{a}{\sin A} = 2R$.

The proofs that $\dfrac{b}{\sin B} = \dfrac{c}{\sin C} = 2R$ follow as in case (i).

$$\therefore \frac{a}{\sin A} = \frac{b}{\sin B} = \frac{c}{\sin C} = 2R \quad \text{for all triangles.}$$

The cosine formula*

The projection of a line AB on a line L is the length $A'B'$ where A' and B' are the feet of the perpendiculars from A and B to the line L.

Fig. 30.5 shows the projection of AB on a line L in three different positions of AB. If the angle between AB and the line L is θ, the length of the projection is $AB \cos \theta$.

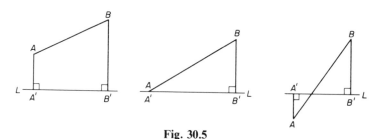

Fig. 30.5

If a line of length r through the origin makes an angle θ with the positive x-direction, the projection of the line on the x-axis is $r \cos \theta$. This is true for all values of θ provided that distances to the left of O are counted as negative. For example, as shown in Fig. 30.6, the projection of OP on the x-axis when θ is obtuse is ON. The value of $r \cos \theta$ is negative and numerically equal to ON.

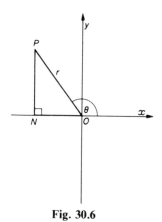

Fig. 30.6

Similarly the projection of OP on the y-axis is $r \sin \theta$ for all values of θ.

Now suppose that N is the foot of the perpendicular from B to the side AC of the triangle ABC. The projection of AB on AC is $c \cos A$ whether A is acute or obtuse.

* See also page 120.

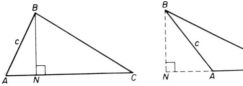

Fig. 30.7

In both cases of Fig. 30.7,

$$NC = AC - AN = b - c \cos A.$$

(This is true for either figure, because in the obtuse case $b - c \cos A$ is greater than b which fits with the figure.)

For both diagrams,

$$BN = c \sin A.$$

But $\qquad BC^2 = BN^2 + NC^2.$

$$\therefore a^2 = (c \sin A)^2 + (b - c \cos A)^2$$
$$= c^2 \sin^2 A + b^2 - 2bc \cos A + c^2 \cos^2 A$$
$$= b^2 + c^2 - 2bc \cos A.$$
$$\therefore a^2 = b^2 + c^2 - 2bc \cos A \quad \text{for all triangles.}$$

Example 1. *A line AB of length 3 cm is given. The point P moves on one side of AB so that the angle APB = 30°. Show that the locus of P is an arc of a circle and find its radius.*

Since $APB = 30°$, AB subtends a constant angle at a point on one side of it. The locus of P is therefore an arc of a circle through A and B.

Since $\qquad \dfrac{a}{\sin A} = \dfrac{b}{\sin B} = \dfrac{c}{\sin C} = 2R,$

$$2R = \frac{AB}{\sin 30°} = \frac{3}{\frac{1}{2}} = 6.$$

$$\therefore R = 3.$$

The radius of the arc is 3 cm.

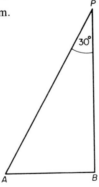

Fig. 30.8

Example 2. *In the triangle ABC, the angle $A = 110°$, $AB = 3$ cm and $AC = 2$ cm. Calculate the length of the circum-radius of the triangle.*

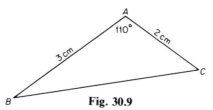

Fig. 30.9

From Fig. 30.9,

$$BC^2 = AB^2 + AC^2 - 2AB \cdot AC \cos 110°$$
$$= 9 + 4 - 2(3)(2) \cos 110°$$
$$= 13 + 12 \cos 70°$$
$$= 13 + 4.104$$
$$= 17.104.$$
$$\therefore BC = 4.135.$$

But $\dfrac{a}{\sin A} = 2R.$

$$\therefore 2R = \frac{4.135}{\sin 110°} = \frac{4.135}{\sin 70°}$$

$$= 4.4.$$

0.6165
$\bar{1}$.9730
0.6435

The length of the circum-radius is 2.2 cm.

Identities using the sine and cosine formulae

The sine and cosine formulae may be used to prove properties of a triangle. The sine formula is especially useful in converting properties about length of sides into connections between angles as in Example 1 below.

Example 1. *Show that $\sin^2 C = \sin^2 A + \sin^2 B - 2 \sin A \sin B \cos C$.*
We know that $c^2 = a^2 + b^2 - 2ab \cos C$ and also that
$$a = 2R \sin A, \quad b = 2R \sin B \quad \text{and} \quad c = 2R \sin C.$$
Substituting these values, we have
$$4R^2 \sin^2 C = 4R^2 \sin^2 A + 4R^2 \sin^2 B - 2(2R \sin A)(2R \sin B) \cos C$$
or $\quad \sin^2 C = \sin^2 A + \sin^2 B - 2 \sin A \sin B \cos C.$

Example 2. *Show that $b \cos C + c \cos B = a$.*

$$b \cos C + c \cos B = b\frac{a^2 + b^2 - c^2}{2ab} + c\frac{a^2 + c^2 - b^2}{2ac}$$
$$= \frac{1}{2a}(a^2 + b^2 - c^2 + a^2 + c^2 - b^2)$$
$$= \frac{1}{2a}(2a^2) = a.$$

Exercise 30d: Miscellaneous

1. In the triangle ABC, the angle $B = 60°$, the angle $C = 45°$ and $a = 2.5$ cm. Calculate the radius of the circum-circle.

2. In the triangle ABC, it is given that $BC = 8.5$ cm, $CA = 4.5$ cm and $AB = 5$ cm. Without using tables, calculate the values of cos A and sin A.

3. Given that $3 \sin^2 \theta - \cos^2 \theta = 2$, find the possible values of θ between $0°$ and $180°$.

4. Show that $\cos^4 \theta + \cos^2 \theta \sin^2 \theta + \sin^2 \theta = 1$.

5. Find all values of θ between $0°$ and $360°$ which satisfy the equation $2 \sin^2 \theta - 7 \sin \theta + 3 = 0$.

6. Prove that cosec $x - \sin x = \cos x \cot x$.

7. Find all the values of x between $0°$ and $180°$ which satisfy $\sec^4 x - 2 \sec^2 x = 4 (\tan^2 x - 1)$.

8. Show that $a \cos B - b \cos A = \dfrac{a^2 - b^2}{c}$.

9. By projecting the sides of an equilateral triangle, show that cos $x + \cos (120° + x) + \cos (240° + x) = 0$.

10. In the triangle ABC, the angle $A = 40°$, $b = 8$ cm and $c = 7$ cm. Calculate the radius of the circum-circle.

11. Find all values of x between $0°$ and $180°$ which satisfy $\sin 2x = \frac{1}{2}$.

12. Show that $(\cos A + \cos B)(\cos A - \cos B)$
$\qquad + (\sin A + \sin B)(\sin A - \sin B) = 0$.

13. In the triangle ABC, prove that
    (i) $a \cos B + b \cos A = c$,
    (ii) $\sin A \cos B + \sin B \cos A = \sin C$.

14. Using the result of question 13, show that for any acute angles A and B, $\sin (A + B) = \sin A \cos B + \sin B \cos A$.

15. In the triangle ABC, show that
    $ab \sin C = ac \sin B = bc \sin A = 4R^2 \sin A \sin B \sin C$.

16. Eliminate x between the equations
    $\cos x = a, \; b(\sec x + \text{cosec } x) = 1$.

17. If $\tan x = p$, find the value of cosec x when x is acute.

18. If $\sin x = s$, express $\dfrac{\sqrt{1 - s^2}}{s}$ in terms of x.

19. If $t = \tan x$, express $\dfrac{t}{\sqrt{1 + t^2}}$ in terms of x.

20. If the angle C of the triangle ABC is $120°$, show that
    $c^2 = a^2 + ab + b^2$.

21. Eliminate x between the equations
    $a + \cos x = b, \quad p + \sin x = q$.

22. Find all values of x between $0°$ and $360°$ which satisfy
    $6 - 9 \sin x = 2 \cos^2 x$.

23. The points P, Q and F lie in a straight line on level ground. A tower FT whose foot is at F is such that the angles of elevation of T from P and Q are $30°$ and $45°$ respectively. If the distance $PQ = 20$ m, calculate the height of the tower.

24. The points P, Q and F lie in a straight line on level ground. A tower FT whose foot is at F is such that the angles of elevation of T from P and Q are α and β respectively. If the distance $PQ = x$ metres, prove that the height of the tower is $\dfrac{x \sin \alpha \sin \beta}{\sin (\beta - \alpha)}$ metres.

25. Two points P, Q are on level ground. P lies due south of a tower FT and Q lies south-east of the tower. If the angles of elevation of the top of the tower from P and Q are α and β respectively, show that
$$PQ^2 = h^2(\cot^2 \alpha + \cot^2 \beta - \sqrt{2} \cot \alpha \cot \beta),$$
where h is the height of the tower.

26. In the triangle ABC, $b = 6$ cm, $c = 10$ cm and the angle $B = 30°$. If the angle C is obtuse, find the obtuse angle and the remaining side.

27. By considering an isosceles triangle in which $A = 45°$ and $C = 90°$ and finding where the bisector of the angle A cuts BC, show that
$$\tan 22° 30' = \sqrt{2} - 1.$$
Hence show that $\tan 45° = \dfrac{2 \tan 22° 30'}{1 - \tan^2 22° 30'}$.

28. By considering a triangle ABC in which $A = 30°$ and $B = 90°$ and finding the point in which the bisector of A meets BC, show that $\tan 15° = 2 - \sqrt{3}$. Hence show that
$$\tan (45° - 30°) = \frac{\tan 45° - \tan 30°}{1 + \tan 45° \tan 30°}$$

29. Solve the equation $\cos (x + 30°) = \cos 60°$, giving all solutions between $0°$ and $360°$.

30. Show that $bc \cos A + ca \cos B + ab \cos C = \frac{1}{2}(a^2 + b^2 + c^2)$.

31. Show that $\dfrac{\cos A}{a} + \dfrac{\cos B}{b} + \dfrac{\cos C}{c} = \dfrac{a^2 + b^2 + c^2}{2abc}$.

32. If $a = 123$, $b = 125$ and $c = 62$, find $\sin A$.

33. Show that the largest angle of the triangle whose sides are $(n^2 + n + 1)$, $(2n + 1)$ and $(n^2 - 1)$ is $120°$.

34. Show that $(a^2 + b^2 - c^2) \tan C = (b^2 + c^2 - a^2) \tan A$.

35. Show that $4R^2 (\cos^2 B - \cos^2 C) = c^2 - b^2$.

31 Compound Angles

Given the values of the angles A and B, the sine of their sum may be found from tables. It is possible to find the value of $\sin (A + B)$ without using tables and, if so, what ratios of A and B must be known to evaluate the expression? In this chapter formulae are found which express the sin, cos and tan of the sum and difference of two angles in terms of the ratios of the separate angles.

The formula for cos $(A - B)$

Draw two radii OP, OQ of a circle of unit radius making angles A and B respectively with the x-axis, as shown in Fig. 31.1.

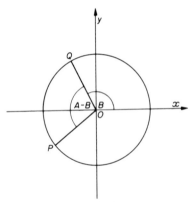

Fig. 31.1

Whatever the values of A and B, the coordinates of P and Q are $(\cos A, \sin A)$ and $(\cos B, \sin B)$.

The distance between these points, PQ, is given by
$$
\begin{aligned}
PQ^2 &= (\cos A - \cos B)^2 + (\sin A - \sin B)^2 \\
&= (\cos^2 A + \sin^2 A) + (\cos^2 B + \sin^2 B) \\
&\qquad - 2(\cos A \cos B + \sin A \sin B) \\
&= 2(1 - \cos A \cos B - \sin A \sin B).
\end{aligned}
$$

The cosine formula holds for angles of all magnitudes and therefore
$$
\begin{aligned}
PQ^2 &= OP^2 + OQ^2 - 2OP.OQ \cos (A - B) \\
&= 1 + 1 - 2 \cos (A - B).
\end{aligned}
$$
$\therefore\ 2 - 2 \cos (A - B) = 2 - 2 \cos A \cos B - 2 \sin A \sin B$

or $\qquad \cos (A - B) = \cos A \cos B + \sin A \sin B.$ $\qquad$ (i)

This equation holds for all values of A and B

If $(-B)$ is written for B, equation (i) becomes:

$$\cos (A + B) = \cos A \cos (-B) + \sin A \sin (-B)$$

or $\qquad \cos (A + B) = \cos A \cos B - \sin A \sin B \qquad$ (ii)

Again, writing $\left(\dfrac{\pi}{2} - A \right)$ for A in equation (i),

$$\cos \left(\frac{\pi}{2} - A - B \right) = \cos \left(\frac{\pi}{2} - A \right) \cos B + \sin \left(\frac{\pi}{2} - A \right) \sin B$$

or $\qquad \sin (A + B) = \sin A \cos B + \cos A \sin B \qquad$ (iii)

Finally, writing $\left(\dfrac{\pi}{2} + A \right)$ for A in equation (i),

$$\cos \left(\frac{\pi}{2} + A - B \right) = \cos \left(\frac{\pi}{2} + A \right) \cos B + \sin \left(\frac{\pi}{2} + A \right) \sin B$$

or $\qquad -\sin (A - B) = -\sin A \cos B + \cos A \sin B$

which leads to our fourth equation

$$\sin (A - B) = \sin A \cos B - \cos A \sin B \qquad \text{(iv)}$$

These four identities are very important and have many applications in all branches of mathematics. They should therefore be memorized.

Example. *Find values in surd form for* $\cos 15°$ *and* $\cos 75°$.

Since $\qquad \cos (A - B) = \cos A \cos B + \sin A \sin B,$

$\qquad \cos (45° - 30°) = \cos 45° \cos 30° + \sin 45° \sin 30°.$

$$\therefore \cos 15° = \frac{1}{\sqrt{2}} \cdot \frac{\sqrt{3}}{2} + \frac{1}{\sqrt{2}} \cdot \frac{1}{2} = \frac{\sqrt{3} + 1}{2\sqrt{2}}.$$

Since $\qquad \cos (A + B) = \cos A \cos B - \sin A \sin B,$

$\qquad \cos (45° + 30°) = \cos 45° \cos 30° - \sin 45° \sin 30°.$

$$\therefore \cos 75° = \frac{1}{\sqrt{2}} \cdot \frac{\sqrt{3}}{2} - \frac{1}{\sqrt{2}} \cdot \frac{1}{2} = \frac{\sqrt{3} - 1}{2\sqrt{2}}.$$

Exercise 31a

1. Find the value of $\sin A \cos B + \cos A \sin B$ when $A = B = 45°$.
2. Find the value of $\cos A \cos B - \sin A \sin B$ when $A = B = 45°$.
3. Find the value of $\sin A \cos B + \cos A \sin B$ when $A = 60°$, $B = 30°$.
4. Find the value of $\cos A \cos B - \sin A \sin B$ when $A = 60°$, $B = 30°$.
5. Find the value of $\sin 105°$ in surd form.
6. Find the value of $\cos 105°$ in surd form.
7. Find the value of $\cos 165°$ in surd form.
8. Find the value of $\sin 165°$ in surd form.

The tangent formulae

Formulae for tan $(A + B)$ and tan $(A - B)$ may be deduced from the four identities

$$\tan (A + B) = \frac{\sin (A + B)}{\cos (A + B)} = \frac{\sin A \cos B + \cos A \sin B}{\cos A \cos B - \sin A \sin B}.$$

Dividing both numerator and denominator by $\cos A \cos B$:

$$\tan (A + B) = \frac{\tan A + \tan B}{1 - \tan A \tan B}.$$

Similarly,

$$\tan (A - B) = \frac{\sin (A - B)}{\cos (A - B)} = \frac{\sin A \cos B - \cos A \sin B}{\cos A \cos B + \sin A \sin B}$$

$$= \frac{\tan A - \tan B}{1 + \tan A \tan B}.$$

The subsidiary angle

The expression $a \cos \theta + b \sin \theta$ may always be put in either of the forms $\sqrt{a^2 + b^2} \cos (\theta - A)$ or $\sqrt{a^2 + b^2} \sin (\theta + B)$.

This is called the subsidiary angle form of the expression $a \cos \theta + b \sin \theta$.

To prove that this can always be done, draw a right-angled triangle as shown in Fig. 31.2, in which $C = 90°$, $AC = a$ and $BC = b$.

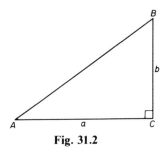

Fig. 31.2

Then $AB = \sqrt{a^2 + b^2}$.

From the triangle ABC,

$$\sqrt{a^2 + b^2} \cos A = a \quad \text{and} \quad \sqrt{a^2 + b^2} \sin A = b.$$

$\therefore a \cos \theta + b \sin \theta$

$$= \sqrt{a^2 + b^2} \cos A \cos \theta + \sqrt{a^2 + b^2} \sin A \sin \theta$$

$$= \sqrt{a^2 + b^2} (\cos \theta \cos A + \sin \theta \sin A)$$

$$= \sqrt{a^2 + b^2} \cos (\theta - A).$$

From the triangle ABC, a and b may also be expressed as follows:

$$a = \sqrt{a^2 + b^2} \sin B, \quad b = \sqrt{a^2 + b^2} \cos B.$$

$$\therefore a \cos \theta + b \sin \theta = \sqrt{a^2 + b^2}(\cos \theta \sin B + \sin \theta \cos B)$$
$$= \sqrt{a^2 + b^2} \sin (\theta + B).$$

Example 1. *Prove that* $\tan (A + 45°) = \dfrac{\cos A + \sin A}{\cos A - \sin A}$

$$\tan (A + 45°) = \frac{\tan A + \tan 45°}{1 - \tan A \tan 45°}$$

$$= \frac{\tan A + 1}{1 - \tan A} \qquad \text{(multiplying both numerator and denominator by } \cos A).$$

$$= \frac{\sin A + \cos A}{\cos A - \sin A}$$

Example 2. *Find the greatest value of* $3 \cos \theta + 4 \sin \theta$.

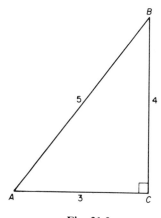

Fig. 31.3

Draw a triangle with sides 3, 4 and 5 as shown in Fig. 31.3.

$$3 = 5 \cos A \quad \text{and} \quad 4 = 5 \sin A.$$
$$\therefore 3 \cos \theta + 4 \sin \theta = 5(\cos \theta \cos A + \sin \theta \sin A)$$
$$= 5 \cos (\theta - A).$$

The greatest value of $\cos (\theta - A)$ is 1 when $\theta = A$.

So the greatest value of $3 \cos \theta + 4 \sin \theta$ is 5 and this greatest value occurs when $\theta = A$, i.e. the angle whose tangent is $\frac{4}{3}$.

Example 3. *Find a value of θ which satisfies the equation*
$$3 \cos \theta + 4 \sin \theta = 2.$$

Express $3 \cos \theta + 4 \sin \theta$ in the form $5 \cos (\theta - A)$ where $\tan A = \frac{4}{3}$.
$$\therefore 5 \cos (\theta - A) = 2$$
or
$$\cos (\theta - A) = 0.4.$$

One value of $(\theta - A)$ is $66° \ 25'$.

Also from tables $A = 53° \ 8'$.

By addition, one value of θ is $119° \ 33'$.

Notice that other solutions of the equation $\cos (\theta - A) = 0.4$ are $-66° \ 25'$ and $293° \ 35'$.

Other solutions of the original equation are $-13° \ 17'$ and $346° \ 43'$.

Summary

For convenience, the six identities are listed below.

$$\sin (A + B) = \sin A \cos B + \cos A \sin B.$$
$$\sin (A - B) = \sin A \cos B - \cos A \sin B.$$
$$\cos (A + B) = \cos A \cos B - \sin A \sin B.$$
$$\cos (A - B) = \cos A \cos B + \sin A \sin B.$$

$$\tan (A + B) = \frac{\tan A + \tan B}{1 - \tan A \tan B}.$$

$$\tan (A - B) = \frac{\tan A - \tan B}{1 + \tan A \tan B}.$$

Exercise 31b

Simplify (questions 1 to 10):

1. $\sin 10° \cos 20° + \sin 20° \cos 10°$.
2. $\cos 40° \cos 20° - \sin 40° \sin 20°$.
3. $\cos \theta \cos 2\varphi + \sin \theta \sin 2\varphi$.
4. $\sin A \cos 2A + \cos A \sin 2A$.
5. $\cos 50° \cos 20° + \sin 50° \sin 20°$.
6. $\cos A \sin 2B + \sin A \cos 2B$.
7. $\cos 2\varphi \cos 3\varphi - \sin 2\varphi \sin 3\varphi$.
8. $\cos (A + B) \cos A + \sin (A + B) \sin A$.
9. $\sin (A + 2B) \cos 2B - \cos (A + 2B) \sin 2B$.
10. $\sin \theta \cos 45° - \cos \theta \sin 45°$.
11. Find the greatest value of $5 \cos \theta + 12 \sin \theta$.
12. Find the greatest value of $a \cos \theta + b \sin \theta$.
13. Find the least value of $7 \sin \theta + 24 \cos \theta$.
14. Find the least value of $a \sin \theta + b \cos \theta$.
15. Find a solution of the equation $\cos \theta + \sin \theta = \sqrt{2}$.

16. Find a solution of the equation $2(\cos \theta + \sin \theta) = \sqrt{3} + 1$.

17. If $\sin A = \frac{3}{5}$ and A is acute, write down the values of $\cos A$ and $\tan A$.

18. If $\sin A = \frac{3}{5}$ and A is acute, find the value of $\tan (A + 45°)$.

19. If $\cos A = \frac{7}{25}$ and A is acute, find the value of $\tan (A + 45°)$.

20. If $\sin A = \frac{3}{5}$ and $\cos B = \frac{7}{25}$ and A and B are acute, find the value of $\tan (A + B)$.

Exercise 31c: Miscellaneous

1. Simplify $\sin (A + B) + \sin (A - B)$.

2. Simplify $\cos (A + 2B) + \cos (A - 2B)$.

3. Expand $\sin (\theta + \theta)$. Hence find an expression for $\sin 2\theta$.

4. Expand $\cos (\theta + \theta)$. Hence find an expression for $\cos 2\theta$.

5. Expand $\tan (\theta + \theta)$. Hence find an expression for $\tan 2\theta$.

6. Show that $\cot (A + B) = \dfrac{\cot A \cot B - 1}{\cot A + \cot B}$.

7. If $\sin A = \frac{4}{5}$ and $\cos B = \frac{12}{13}$ and A and B are acute, find the value of $\sin (A + B)$.

8. If $\tan A = \frac{1}{2}$ and $\tan B = \frac{1}{3}$, find $\tan (A + B)$.

9. If $\tan A = \frac{1}{2}$, $\tan B = \frac{1}{3}$ and $\tan C = 1$, find $\cot (A + B + C)$.

10. If $5 \cos \theta + 2 \sin \theta = r \cos (\theta - \alpha)$, find the values of r and $\tan \alpha$.

11. If $8 \sin \theta - 3 \cos \theta = r \sin (\theta - \alpha)$, find the values of r and $\tan \alpha$.

12. Find, without using tables, the value of
$$\cos 80° \cos 20° + \sin 80° \sin 20°.$$

13. Find, without using tables, the value of $\dfrac{\tan 70° - \tan 10°}{1 + \tan 70° \tan 10°}$.

14. Simplify $a \cos \theta + b \sin \theta$ if $\tan \theta = \dfrac{b}{a}$, given that θ is acute and a and b are positive.

15. If $\tan \theta = \cot \varphi$ and θ and φ are acute, show that $\theta + \varphi = 90°$.

16. If $\tan (\theta + \varphi) = \tan \theta$, show that $\tan \varphi = 0$.

17. A box is placed on horizontal ground. A vertical face of the box is $ABCD$ where AB is horizontal and $AB = 24$ cm, $BC = 10$ cm. The box is turned through an angle of $45°$ so that B is above A and A remains on the ground. Find the height of C above the ground.

18. Find the values of x between $0°$ and $180°$ which satisfy the equation $\cos x + 2 \sin x = 1.5$.

19. Find the values of x between $0°$ and $180°$ which satisfy the equation $2 \cos x + 3 \sin x = 2.5$.

20. Prove that $\sin (A + B) \sin (A - B) = \sin^2 A - \sin^2 B$.

21. Expand $\tan (A + B + C)$ by considering $\tan \{A + (B + C)\}$.

22. Use the result of the last question to show that, if
$$A + B + C = 180°, \tan A + \tan B + \tan C = \tan A \tan B \tan C.$$

23. If A, B, C are acute angles such that $\tan A = \frac{1}{2}$, $\tan B = \frac{1}{5}$ and $\tan C = \frac{1}{8}$, show that $A + B + C = 45°$.

24. Find a solution of the equation
$$\cos \theta \cos 2\theta - \sin \theta \sin 2\theta = \frac{1}{2}.$$

25. If A, B, C are the angles of a triangle, show that
$$\cos C = \sin A \sin B - \cos A \cos B.$$

26. Simplify $\sin A \sin (B - C) + \sin B \sin (C - A) + \sin C \sin (A - B)$.

27. Show that
$$(\sin A + \cos A)(\sin B + \cos B) = \sin (A + B) + \cos (A - B).$$

28. In the triangle ABC, show that $a \cos B - b \cos A = 2R \sin (A - B)$.

29. Show that $\dfrac{\sin (\alpha + \beta) + \sin (\alpha - \beta)}{\sin (\alpha + \beta) - \sin (\alpha - \beta)} = \dfrac{\tan \alpha}{\tan \beta}$.

30. Show that $\dfrac{\sin (x + 45°) + \sin (x - 45°)}{\sin (x + 45°) - \sin (x - 45°)} = \tan x$.

31. Assuming that
$$c = b \cos A + a \cos B \quad \text{and that} \quad 0 = b \sin A - a \sin B,$$
by squaring and adding show that $c^2 = a^2 + b^2 - 2ab \cos C$.

32. Show that in any triangle ABC,
$$a \cos (B + x) + b \cos (A - x) = c \cos x \quad \text{(see question 31)}.$$

33. ABC is an acute-angled triangle in which $A = 45°$. The foot of the perpendicular from A to BC is N. Show that
$$BC . AN + BN . NC = AN^2.$$

34. The triangle ABC is such that $AB = 2$ cm, $BC = 3$ cm and $B = 90°$. The point X lies on BC and $BX = 1$ cm. Find $\tan CAX$.

35. ABC is a triangle in which $AB = BC = 7$ cm and $B = 90°$. The point X lies on BC and $BX = 1$ cm. Find $\sin CAX$.

36. If $\tan x = a$ and $\tan y = a + 1$, express $\cot (y - x)$ in terms of a.

37. Given that $\sin (x + y) = \frac{3}{4}$ and $\sin (x - y) = \frac{1}{2}$, prove that $\tan x = 5 \tan y$.

38. Given that $\tan (x + y) = \frac{1}{2}$ and that $\tan x \tan y = \frac{1}{4}$, prove that $\tan x$ and $\tan y$ are the roots of the equation $8t^2 - 3t + 2 = 0$.

39. If $\tan \alpha$ and $\tan \beta$ are the roots of the equation $4t^2 - 3t - 5 = 0$, show that $\tan (\alpha + \beta) = \frac{1}{3}$.

40. If $\tan \alpha$ and $\tan \beta$ are the roots of the equation $4t^2 - 3t - 5 = 0$, find the value of $\tan (\alpha - \beta)$.

32 Multiple Angles

Double angle formulae

The addition formulae may be used to deduce expressions for the ratios of the angle $2x$ in terms of the ratios of the angle x.

Using $\qquad \sin(\theta + \varphi) = \sin\theta \cos\varphi + \sin\varphi \cos\theta$

and putting $\theta = \varphi = x$,

we have $\qquad \sin 2x = \sin x \cos x + \sin x \cos x.$

$\qquad\qquad \therefore \sin 2x = 2 \sin x \cos x.$

Using $\qquad \cos(\theta + \varphi) = \cos\theta \cos\varphi - \sin\theta \sin\varphi$

and putting $\theta = \varphi = x$,

we have $\qquad\qquad \cos 2x = \cos^2 x - \sin^2 x.$

Since $\cos^2 x + \sin^2 x = 1$,

$\qquad \cos^2 x = 1 - \sin^2 x \quad$ and $\quad \sin^2 x = 1 - \cos^2 x.$

These formulae enable us to express $\cos 2x$ in terms of $\cos x$ alone or $\sin x$ alone.

$\qquad \cos 2x = \cos^2 x - (1 - \cos^2 x) = 2\cos^2 x - 1,$

and $\qquad \cos 2x = (1 - \sin^2 x) - \sin^2 x = 1 - 2\sin^2 x.$

Using $\qquad\qquad \tan(\theta + \varphi) = \dfrac{\tan\theta + \tan\varphi}{1 - \tan\theta \tan\varphi}$

and putting $\theta = \varphi = x$,

$$\tan 2x = \frac{2\tan x}{1 - \tan^2 x}.$$

These formulae should be memorized and are listed below for convenience.

$\qquad \sin 2x = 2\sin x \cos x.$

$\qquad \cos 2x = \cos^2 x - \sin^2 x = 2\cos^2 x - 1 = 1 - 2\sin^2 x.$

$\qquad \tan 2x = \dfrac{2\tan x}{1 - \tan^2 x}.$

Example 1. *Find the values of* $\cos 2A$ *and* $\tan 2A$ *given that* $\cos A = \frac{4}{5}$.

$\qquad \cos 2A = 2\cos^2 A - 1 = 2(\frac{4}{5})^2 - 1 = \frac{32}{25} - 1 = \frac{7}{25}.$

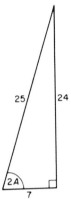

Fig. 32.1

Draw a right-angled triangle as shown in Fig. 32.1 with the hypotenuse 25 and one of the other sides 7.

The length of the third side is $\sqrt{25^2 - 7^2}$, i.e. 24.

So from the diagram when $\cos 2A = \frac{7}{25}$, $\tan 2A = \frac{24}{7}$.

Since we are not told in which quadrant the angle $2A$ lies, $\tan 2A$ may be positive or negative.

$$\therefore \cos 2A = \tfrac{7}{25}, \quad \tan 2A = \pm\tfrac{24}{7}.$$

Example 2. *Calculate the value of tan 22° 30′ in surd form.*

We know that $\tan 45° = 1$; let $\tan 22° 30′ = t$.

Since
$$\tan 45° = \frac{2 \tan 22° 30′}{1 - \tan^2 22° 30′} = \frac{2t}{1 - t^2},$$

$$\frac{2t}{1 - t^2} = 1.$$

$$\therefore t^2 + 2t - 1 = 0.$$

$$\therefore t = \frac{-2 \pm \sqrt{8}}{2} = -1 \pm \sqrt{2}.$$

But $\tan 22° 30′$ is positive and therefore $\tan 22° 30′ = \sqrt{2} - 1$.

Exercise 32a

Simplify the following (questions 1 to 10):

1. $2 \sin 15° \cos 15°$.

2. $\cos^2 50° - \sin^2 50°$.

3. $2 \sin 50° \cos 50°$.

4. $2 \cos^2 15° - 1$.

5. $\dfrac{2 \tan 20°}{1 - \tan^2 20°}.$

6. $1 - 2 \sin^2 3x.$

7. $2 \sin \frac{1}{2}x \cos \frac{1}{2}x.$

8. $\dfrac{2 \tan (x/2)}{1 - \tan^2 (x/2)}.$

9. $\cos^2 4x - \sin^2 4x.$

10. $(\cos x + \sin x)(\cos x - \sin x).$

11. Express $(\cos A + \sin A)^2$ in terms of the double angle.

12. Find the value of $\tan 15°$ in surd form.

13. Find the value of $\cos 2x$ given that $\sin x = \frac{3}{4}$.

14. Find the value of $\tan 2x$ when $\tan x = \frac{1}{2}$.

15. Find the possible values of $\tan \frac{1}{2}x$ when $\tan x = \frac{1}{2}$.

16. Simplify $\dfrac{\cos^2 x - \sin^2 x}{\cos^2 x + \sin^2 x}.$

17. Evaluate $\cos^2 75° - \sin^2 75°$.

18. Show that $\dfrac{1 - \cos 2A}{1 + \cos 2A} = \tan^2 A.$

19. Simplify $\cos^2 (45° - x) - \sin^2 (45° - x)$.

20. Simplify $2 \cos (45° + x) \sin (45° + x)$.

Two useful formulae

From
$$\cos 2A = 2\cos^2 A - 1 = 1 - 2 \sin^2 A, \quad \text{we see that}$$
$$1 + \cos 2A = 2 \cos^2 A$$
and $1 - \cos 2A = 2 \sin^2 A$.

Writing $2A = x$,

and
$$1 + \cos x = 2 \cos^2 \tfrac{1}{2}x,$$
$$1 - \cos x = 2 \sin^2 \tfrac{1}{2}x.$$

These formulae are often very helpful in proving identities.

cos 3A and sin 3A

We have seen how the addition formulae may be used to express $\cos 2A$ and $\sin 2A$ in terms of ratios of the angle A. Formulae for $\cos 3A$ and $\sin 3A$ may be built up in a similar way.

$$\begin{aligned}
\cos 3A &= \cos (2A + A) = \cos 2A \cos A - \sin 2A \sin A \\
&= (2 \cos^2 A - 1) \cos A - (2 \sin A \cos A) \sin A \\
&= 2 \cos^3 A - \cos A - 2 \sin^2 A \cos A \\
&= 2 \cos^3 A - \cos A - 2 \cos A(1 - \cos^2 A)
\end{aligned}$$
$$\therefore \cos 3A = 4 \cos^3 A - 3 \cos A.$$

$$\sin 3A = \sin (2A + A) = \sin 2A \cos A + \cos 2A \sin A$$
$$= (2 \sin A \cos A) \cos A + (1 - 2 \sin^2 A) \sin A$$
$$= 2 \sin A \cos^2 A + \sin A - \sin^3 A$$
$$= 2 \sin A(1 - \sin^2 A) + \sin A - 2 \sin^3 A$$
$$\therefore \mathbf{\sin 3A = 3 \sin A - 4 \sin^3 A.}$$

tan $(A + B + C)$

A formula for $\tan (A + B + C)$ can be found from that for $\tan (A + B)$ and is question 21 of Exercise 31c.

$$\tan (A + B + C) = \frac{\tan A + \tan (B + C)}{1 - \tan A \tan (B + C)}$$

$$= \frac{\tan A + \dfrac{\tan B + \tan C}{1 - \tan B \tan C}}{1 - \tan A \cdot \dfrac{\tan B + \tan C}{1 - \tan B \tan C}}$$

$$= \frac{\tan A + \tan B + \tan C - \tan A \tan B \tan C}{1 - \tan B \tan C - \tan C \tan A - \tan A \tan B}$$

(multiplying numerator and denominator by $(1 - \tan B \tan C)$).

This may be written in the form

$$\tan (A + B + C) = \frac{\sum (\tan A) - \tan A \tan B \tan C}{1 - \sum (\tan A \tan B)}.$$

It is also possible to find formulae for $\sin (A + B + C)$ and $\cos (A + B + C)$ but they cannot be expressed so conveniently.

Deduction (i)

If A, B, C are the angles of a triangle, $\tan (A + B + C) = 0$. The numerator of the fraction is zero and so

$$\tan A + \tan B + \tan C = \tan A \tan B \tan C.$$

Deduction (ii)

The expansion for $\tan (\frac{1}{2}A + \frac{1}{2}B + \frac{1}{2}C)$ is

$$\frac{\tan \dfrac{A}{2} + \tan \dfrac{B}{2} + \tan \dfrac{C}{2} - \tan \dfrac{A}{2} \tan \dfrac{B}{2} \tan \dfrac{C}{2}}{1 - \tan \dfrac{B}{2} \tan \dfrac{C}{2} - \tan \dfrac{C}{2} \tan \dfrac{A}{2} - \tan \dfrac{A}{2} \tan \dfrac{B}{2}}.$$

If A, B, C are angles of a triangle,

$$\tan\left(\tfrac{1}{2}A + \tfrac{1}{2}B + \tfrac{1}{2}C\right) = \tan 90° = \infty.$$

The denominator of the fraction is therefore zero and so

$$\tan \tfrac{1}{2}B \tan \tfrac{1}{2}C + \tan \tfrac{1}{2}C \tan \tfrac{1}{2}A + \tan \tfrac{1}{2}A \tan \tfrac{1}{2}B = 1.$$

Multiplying throughout by $\cot \tfrac{1}{2}A \cot \tfrac{1}{2}B \cot \tfrac{1}{2}C$,

$$\cot \tfrac{1}{2}A + \cot \tfrac{1}{2}B + \cot \tfrac{1}{2}C = \cot \tfrac{1}{2}A \cot \tfrac{1}{2}B \cot \tfrac{1}{2}C.$$

Deduction (*iii*)

If B and C are made equal to A, we have the formula for $\tan 3A$, namely

$$\tan 3A = \frac{3 \tan A - \tan^3 A}{1 - 3 \tan^2 A}.$$

cos 2A and sin 2A in terms of tan A

The formulae which express $\cos 2A$ and $\sin 2A$ in terms of $\tan A$ are useful in many ways, particularly in integration and in the solution of trigonometrical equations.

$$\cos 2A = \cos^2 A - \sin^2 A.$$

Since $\cos^2 A + \sin^2 A = 1$, $\cos 2A$ may be expressed as

$$\frac{\cos^2 A - \sin^2 A}{\cos^2 A + \sin^2 A}.$$

Dividing both numerator and denominator by $\cos^2 A$,

$$\cos 2A = \frac{1 - \tan^2 A}{1 + \tan^2 A}.$$

Similarly, $\sin 2A = 2 \sin A \cos A = \dfrac{2 \sin A \cos A}{\cos^2 A + \sin^2 A}$

$$= \frac{2 \tan A}{1 + \tan^2 A}.$$

These formulae are generally needed in terms of the half angle rather than the double angle, i.e.

$$\cos x = \frac{1 - t^2}{1 + t^2}, \quad \sin x = \frac{2t}{1 + t^2} \quad \text{where } t = \tan \tfrac{1}{2}x.$$

Example 1. *By considering the equation* $2 \cos 3x = 1$, *prove that* $\cos 20°$ *is a root of the equation* $8c^3 - 6c - 1 = 0$. *What are the other roots?*

$$\cos 3x = 4 \cos^3 x - 3 \cos x.$$
$$\therefore 8 \cos^3 x - 6 \cos x - 1 = 0.$$

Putting $c = \cos x$, the equation becomes
$$8c^3 - 6c - 1 = 0.$$
If $\cos 3x = \frac{1}{2}$, a possible value of $3x$ is $60°$ and a possible value of x is $20°$. When $x = 20°$, $c = \cos 20°$ and therefore $\cos 20°$ is one root of the equation $8c^3 - 6c - 1 = 0$.
What other values of x satisfy $\cos 3x = \frac{1}{2}$?
The values of $3x$ from the equation $\cos 3x = \frac{1}{2}$ are
$$60°, \quad 300°, \quad 420°, \quad 660°, \quad 780°, \quad 1020°, \quad \text{etc.}$$

The values of x are
$$20°, \quad 100°, \quad 140°, \quad 220°, \quad 260°, \quad 340°, \quad \text{etc.}$$

There is an infinite number of values of x. How many different values of $\cos x$ are there?
The values of c are
$$\cos 20°, \cos 100°, \cos 140°, \cos 220°, \cos 260°, \cos 340°, \text{etc.}$$
or
$$\cos 20°, -\cos 80°, -\cos 40°, -\cos 40°, -\cos 80°, \cos 20°, \text{etc.}$$
There are only three different values of c, namely
$$\cos 20°, -\cos 40° \quad \text{and} \quad -\cos 80°.$$
The other roots of the equation $8c^3 - 6c - 1 = 0$ are
$$-\cos 40° \quad \text{and} \quad -\cos 80°.$$

Example 2. *Find the solutions of the equations $3 \cos \theta + 4 \sin \theta = 2$ which lie between $0°$ and $360°$.*

   (See also example under 'subsidiary angle' in the last chapter.)

If $t = \tan \dfrac{\theta}{2}$,
$$\cos \theta = \frac{1 - t^2}{1 + t^2} \quad \text{and} \quad \sin \theta = \frac{2t}{1 + t^2}.$$
$$\therefore 3\left(\frac{1 - t^2}{1 + t^2}\right) + 4\left(\frac{2t}{1 + t^2}\right) = 2.$$
$$\therefore 3 - 3t^2 + 8t = 2 + 2t^2.$$
$$\therefore 5t^2 - 8t - 1 = 0.$$
$$\therefore t = \frac{8 \pm \sqrt{64 + 20}}{10} = \frac{8 \pm 9.165}{10}.$$
$$\therefore \tan \frac{\theta}{2} = 1.7165 \quad \text{or} \quad -0.1165.$$

$\dfrac{\theta}{2}$ must lie between $0°$ and $180°$.
$$\therefore \frac{\theta}{2} = 59° \, 47' \quad \text{or} \quad (180° - 6° \, 39')$$
$$= 59° \, 47' \quad \text{or} \quad 173° \, 21'.$$
$$\therefore \theta = 119° \, 34' \quad \text{or} \quad 346° \, 42'.$$

The slight differences in the answers between this and the subsidiary angle method are due to the fact that tables are not accurate enough to give the results to the nearest minute.

Example 3. *Solve the equation* $\cos 2x + 3 \sin x = 2$ *for values of x between* $0°$ *and* $360°$.

This equation may be put in quadratic form by using
$$\cos 2x = 1 - 2 \sin^2 x.$$
The equation becomes $1 - 2 \sin^2 x + 3 \sin x = 2$
or $\qquad\qquad 2 \sin^2 x - 3 \sin x + 1 = 0.$
$$\therefore (2 \sin x - 1)(\sin x - 1) = 0.$$
$$\therefore \qquad\qquad \sin x = \tfrac{1}{2} \quad \text{or} \quad 1.$$
When $\sin x = \tfrac{1}{2}$, $\qquad x = 30°$ or $150°$.
When $\sin x = 1$, $\qquad x = 90°$.
The solutions are $30°$, $90°$ and $150°$.

Exercise 32b: Miscellaneous

1. Find the value of $\cos 2x$ given that $\sin x = \tfrac{12}{13}$.
2. Find the value of $\sin 2x$ given that $\tan x = \tfrac{3}{4}$.
3. Find the value of $\sin 3x$ if $\sin x = \tfrac{1}{3}$.
4. Show that $\cos 4x = 8 \cos^4 x - 8 \cos^2 x + 1.$
5. Show that $\dfrac{1 - \cos 4x}{1 + \cos 4x} = \tan^2 2x.$
6. Show that $(\cos x + \sin x)^4 = 1 + 2 \sin 2x + \sin^2 2x.$
7. Show that $\dfrac{1}{1 - \tan x} - \dfrac{1}{1 + \tan x} = \tan 2x.$
8. Find the value of $\tan 2x$ given that $\tan x = \tfrac{1}{3}$.
9. Find $\tan 3x$ given that $\tan x = 1$.
10. Show that $\sin 3x + \sin x = 4 \sin x \cos^2 x.$
11. If $x = 18°$, use $\sin 3x = \sin (90° - 2x)$ to show that
$$4 \sin^3 x - 2 \sin^2 x - 3 \sin x + 1 = 0.$$
12. Given that $s = 1$ is one root of the equation
$$4s^3 - 2s^2 - 3s + 1 = 0,$$
find the value of $\sin 18°$ in surd form.
13. Express $\tan 4x$ in terms of t, where $t = \tan x$.
14. Given that $\tan x = \tfrac{1}{2}$, find the value of $\tan 4x$.
15. Show that $\sin 4x = 4 \sin x \cos x \cos 2x.$
16. Show that $\sin 8x = 8 \sin x \cos x \cos 2x \cos 4x.$
17. Eliminate x between $\cos x + \sin x = a$, $\cos 2x = b$.
18. Eliminate θ between $\cos \theta = a$, $\sin 2\theta = 2b$.
19. Eliminate θ between $\cos 3\theta + 3 \cos \theta = 4a$
and $\qquad\qquad\qquad 3 \sin \theta - \sin 3\theta = 4b.$

20. Show that $2 \cos^2 A \cos^2 B - 2 \sin^2 A \sin^2 B = \cos 2A + \cos 2B$.

21. Find the solutions between $0°$ and $360°$ of the equation
$$3 \cos x - 2 \sin x = 1.$$

22. Find the solutions between $0°$ and $360°$ of the equation
$$\cos x + 5 \sin x = 1.$$

23. Find the solutions between $0°$ and $360°$ of the equation
$$3 \cos 2x + 5 \sin x = 4.$$

24. Express $\sqrt{\dfrac{1 + \sin \theta}{1 - \sin \theta}}$ in terms of $\tan \dfrac{\theta}{2}$.

25. If $\tan \alpha$ and $\tan \beta$ are the roots of the equation $t^2 - 4t + 2 = 0$, find the value of $\tan 2(\alpha + \beta)$.

26. If $A + B = 45°$, show that $\tan A + \tan B + \tan A \tan B = 1$.

27. In the rectangle $ABCD$ it is given that $AB = 8$ cm, $BC = 6$ cm. If the diagonals meet at O, calculate the tangent of the angle AOB.

28. The diagonals of a rhombus are of lengths a cm and b cm. Find the tangents of the angles of the rhombus.

29. Show that $\dfrac{\cos x - \cos 3x}{\sin 3x - \sin x} = \tan 2x$.

30. Show that $\dfrac{\sin x + \sin 2x}{1 + \cos x + \cos 2x} = \tan x$.

31. Solve, for values of θ between $0°$ and $360°$, the equation
$$\cos x + \sin x = 0.8.$$

32. Express $\dfrac{5 - 13 \sin x}{\tan \frac{1}{2}x - 5}$ in terms of t, where $t = \tan \frac{1}{2}x$.

33. Express $\dfrac{7 - 5 \cos x}{5 - 7 \cos x}$ in terms of t, where $t = \tan \frac{1}{2}x$.

34. A minor arc of a circle subtends an angle whose sine is $\frac{2}{3}$ at the circumference of the circle. Find the tangent of the angle subtended by the arc at the centre of the circle.

35. A minor arc of a circle subtends an angle whose cosine is $\frac{15}{17}$ at the centre of the circle. Find the tangent of the angle subtended by the arc at the circumference of the circle.

36. By expressing $\sin 80°$ as $\sin (60° + 20°)$, show that
$$\sin 80° = \sin 40° + \sin 20°.$$

37. From the identity of question 36, show that
$$4 \cos 20° \cos 40° = 2 \cos 20° + 1.$$

38. Using the result of question 37, show that, if $c = \cos 20°$,
$$8c^3 - 6c - 1 = 0.$$

39. Find the values of x between $0°$ and $360°$ which satisfy
$$2 \sin \frac{x}{2} = \cos \frac{x}{2}.$$

40. Find the values of x between $0°$ and $360°$ which satisfy
$$\cos 2x = \sin x.$$

41. Show that $\cos 4\theta = 4(\cos^4 \theta + \sin^4 \theta) - 3$.

42. Find the values of x between $0°$ and $360°$ which satisfy
$$\sin 3x = \sin x.$$

43. Show that $\operatorname{cosec} \theta + \cot \theta = \cot \frac{1}{2}\theta$.

44. If $\alpha + \beta = 2\gamma$, find a relationship between $\tan \alpha$, $\tan \beta$ and $\tan \gamma$.

45. Find all values of x between $0°$ and $360°$ which satisfy
$$\sin 2x + \cos x = 0.$$

46. Given that α and β are solutions of the equation $8 \cos \theta - \sin \theta = 3$ in the range $0° < \theta < 360°$, find the equation whose roots are $\tan \frac{1}{2}\alpha$ and $\tan \frac{1}{2}\beta$.

47. Show that $\dfrac{\sin 5\theta}{\sin \theta} - \dfrac{\cos 5\theta}{\cos \theta} = 8 \cos^2 \theta - 4$.

48. If $x = 36°$, use $\cos 3x = \cos (180° - 2x)$ to show that $\cos 36°$ is a root of the equation $4c^3 + 2c^2 - 3c - 1 = 0$.

49. Using question 48 and the information that $(c + 1)$ is a factor of $4c^3 + 2c^2 - 3c - 1$, find $\cos 36°$ in surd form.

50. Find all the solutions between $0°$ and $360°$ of the equation
$$\cot 2x = 4 \tan x.$$

33 Sums and Products

Products as sums or differences

The compound angle formulae may be used to express the product of two sines or cosines or of a sine and cosine as a sum or difference.

We know that

$$\sin (A + B) = \sin A \cos B + \cos A \sin B$$

and $\qquad \sin (A - B) = \sin A \cos B - \cos A \sin B.$

Adding: $\qquad \sin (A + B) + \sin (A - B) = 2 \sin A \cos B.$

Subtracting: $\sin (A + B) - \sin (A - B) = 2 \cos A \sin B.$

If we deal with the formulae for $\cos (A + B)$ and $\cos (A - B)$ in a similar way, we have:

$$\cos (A + B) = \cos A \cos B - \sin A \sin B$$

and $\qquad \cos (A - B) = \cos A \cos B + \sin A \sin B.$

Adding: $\qquad \cos (A + B) + \cos (A - B) = 2 \cos A \cos B.$

Subtracting: $\quad \cos (A - B) - \cos (A + B) = 2 \sin A \sin B.$

The four formulae expressing products as sums or differences are:

$$2 \sin A \cos B = \sin (A + B) + \sin (A - B).$$
$$2 \cos A \sin B = \sin (A + B) - \sin (A - B).$$
$$2 \cos A \cos B = \cos (A + B) + \cos (A - B).$$
$$2 \sin A \sin B = \cos (A - B) - \cos (A + B).$$

Notes. The product $2 \sin A \cos B$ is equal to the sum of two sines if $A > B$. The product $2 \sin A \cos B$ is equal to the difference of two sines if $A < B$. When $2 \sin A \sin B$ is expressed as the difference of two cosines, the smaller angle is written first.

Example. *Express $2 \sin 75° \cos 45°$ as a product and hence find $\sin 75°$ in surd form.*

$$2 \sin 75° \cos 45° = \sin (75° + 45°) + \sin (75° - 45°)$$
$$= \sin 120° + \sin 30°$$

$$= \frac{\sqrt{3}}{2} + \frac{1}{2}.$$

$$\therefore \sqrt{2} \sin 75° = \frac{\sqrt{3} + 1}{2} \quad \text{and} \quad \sin 75° = \frac{\sqrt{3} + 1}{2\sqrt{2}}.$$

Exercise 33a

Express the following as sums or differences:

1. $2 \sin 20° \sin 30°$.
2. $2 \sin 40° \cos 20°$.
3. $2 \sin 40° \cos 60°$.
4. $2 \cos 50° \cos 70°$.
5. $2 \cos 80° \cos 20°$.
6. $2 \sin 80° \sin 20°$.
7. $2 \cos 80° \sin 20°$.
8. $2 \cos 10° \sin 20°$.
9. $2 \cos x \cos 3x$.
10. $2 \sin y \sin 5y$.
11. $2 \sin (x + 45°) \sin (x - 45°)$.
12. $2 \cos (x + 45°) \cos (x - 45°)$.
13. $2 \cos (x + 45°) \sin (x - 45°)$.
14. $2 \sin (x + 60°) \sin (x + 30°)$.
15. $2 \cos (x + 60°) \cos (x + 30°)$.
16. $2 \cos (x + 60°) \sin (x + 30°)$.
17. $2 \cos 3x \sin 5x$.
18. $2 \cos (A + B) \cos (A - B)$.
19. $2 \sin (A + B) \cos (A - B)$.
20. $2 \sin (A + B) \sin (A - B)$.

Sums as products

The four formulae proved in the last paragraph may also be used to factorize the sum or difference of two sines or cosines.

The formulae rearranged are:

$$\sin (A + B) + \sin (A - B) = 2 \sin A \cos B.$$
$$\sin (A + B) - \sin (A - B) = 2 \cos A \sin B.$$
$$\cos (A + B) + \cos (A - B) = 2 \cos A \cos B.$$
$$\cos (A - B) - \cos (A + B) = 2 \sin A \sin B.$$

Now put $A + B = x$ and $A - B = y$ so that $A = \frac{1}{2}(x + y)$, $B = \frac{1}{2}(x - y)$.

$$\sin x + \sin y = 2 \sin \tfrac{1}{2}(x + y) \cos \tfrac{1}{2}(x - y).$$
$$\sin x - \sin y = 2 \cos \tfrac{1}{2}(x + y) \sin \tfrac{1}{2}(x - y).$$
$$\cos x + \cos y = 2 \cos \tfrac{1}{2}(x + y) \cos \tfrac{1}{2}(x - y).$$
$$\cos y - \cos x = 2 \sin \tfrac{1}{2}(x + y) \sin \tfrac{1}{2}(x - y).$$

These four formulae are useful in many ways. One example is in differentiating $\sin x$ and $\cos x$ from first principles.

Note the change in order between x and y in the formula for the difference of two cosines.

Example. *Show that* $\cos 40° + \cos 80° = \cos 20°$.
$$\cos 40° + \cos 80° = 2 \cos \tfrac{1}{2}(40° + 80°) \cos \tfrac{1}{2}(80° - 40°)$$
$$= 2 \cos 60° \cos 20°$$
$$= \cos 20° \quad \text{since } \cos 60° = \tfrac{1}{2}.$$

Exercise 33b

Factorize:

1. $\cos 30° + \cos 70°$.
2. $\sin 20° - \sin 60°$.
3. $\sin 30° + \sin 50°$.
4. $\cos 30° + \cos 70°$.

5. $\cos 20° - \cos 80°$.

6. $\sin 3x - \sin x$.

7. $\sin 3x + \sin x$.

8. $\cos x - \cos 3x$.

9. $\cos x + \cos 3x$.

10. $\cos 4x - \cos 2x$.

11. $\cos (x + 45°) + \cos (x - 45°)$.

12. $\cos (45° - x) - \cos (45° + x)$.

13. $\sin (45° + x) + \sin (45° - x)$.

14. $\sin \dfrac{5x}{2} - \sin \dfrac{3x}{2}$.

15. $\cos (x + 20°) + \cos (x + 40°)$.

16. $\sin (A + B) - \sin (A - B)$.

17. $\sin (x + 2y) - \sin x$.

18. $\cos (x + 4y) - \cos x$.

19. $\cos (x + 3y) + \cos (x - 3y)$.

20. $\cos (x - y) - \cos (x - 2y)$.

The factor formulae are also useful in proving identities and in solving equations. Some worked examples are now given.

Example 1. *Show that* $\dfrac{\sin 3x - \sin x}{\cos x - \cos 3x} = \cot 2x.$

$$\sin 3x - \sin x = \cos 2x \sin x.$$
$$\cos x - \cos 3x = 2 \sin 2x \sin x.$$

$$\therefore \ \frac{\sin 3x - \sin x}{\cos x - \cos 3x} = \frac{2 \cos 2x \sin x}{2 \sin 2x \sin x} = \cot 2x.$$

Example 2. *Find all the angles between $0°$ and $360°$ which satisfy the equation* $\sin x + \sin 3x = \sin 2x.$

$$\sin x + \sin 3x = 2 \sin 2x \cos x.$$
$$\therefore \ 2 \sin 2x \cos x = \sin 2x,$$

i.e. $\qquad\qquad \sin 2x = 0 \quad$ or $\quad \cos x = \tfrac{1}{2}.$

If $\sin 2x = 0$,

$$2x = 0°, \ 180° \ 360°, \ 540° \text{ or } 720°.$$
$$\therefore x = 0°, \ 90°, \ 180°, \ 270° \text{ or } 360°.$$

If $\cos x = \tfrac{1}{2}$, $\qquad\qquad x = 60° \text{ or } 300°.$

The possible values of x are $0°, 60°, 90°, 180°, 270°, 300°,$ and $360°$.

Example 3. *Solve for values of x between $0°$ and $360°$ the equation*
$$\cos (x + 30°) - \sin x = \tfrac{1}{2}.$$

There is no formula for expressing the difference of a sine and cosine as a product. We therefore express $\sin x$ as a cosine by using $\sin x = \cos (90° - x)$. The equation becomes

$$\cos (x + 30°) - \cos (90° - x) = \tfrac{1}{2}$$

or $\qquad\qquad 2 \sin 60° \sin (30° - x) = \tfrac{1}{2}.$

$$\therefore \ \sin (30° - x) = \frac{1}{4 \sin 60°} = \frac{1}{2\sqrt{3}} = \frac{\sqrt{3}}{6} = \frac{1.7321}{6} = 0.2887.$$

$$\therefore \ \sin (x - 30°) = -0.2887.$$

The angle $(x - 30°)$ must lie between $-30°$ and $330°$.

$$\therefore \ x - 30° = -16° \, 47' \quad \text{or} \quad 196° \, 47'.$$
$$x = 13° \, 13' \quad \text{or} \quad 226° \, 47'.$$

Example 4. *Factorize* $\sin 2A + \sin 2B + \sin 2C$ *when*
$$A + B + C = 180°.$$
$$\sin 2A + \sin 2B = 2 \sin (A + B) \cos (A - B).$$
When $A + B + C = 180°$,
$$A + B = 180° - C.$$
$$\therefore \sin (A + B) = \sin (180° - C) = \sin C.$$
$$\therefore \sin 2A + \sin 2B = 2 \sin C \cos (A - B).$$
So $\quad \sin 2A + \sin 2B + \sin 2C = 2 \sin C \cos (A - B) + \sin 2C$
$$= 2 \sin C \cos (A - B) + 2 \sin C \cos C$$
$$= 2 \sin C (\cos \overline{A - B} + \cos C)$$
Since $A + B + C = 180°$,
$$\cos C = \cos (180° - A - B)$$
$$= -\cos (A + B).$$
$$\therefore \sin 2A + \sin 2B + \sin 2C = 2 \sin C(\cos \overline{A - B} - \cos \overline{A + B})$$
$$= 2 \sin C(2 \sin A \sin B)$$
$$= 4 \sin A \sin B \sin C.$$

Exercise 33c: Miscellaneous

1. If A, B and C are the angles of a triangle, simplify $\tan 2(A + B)$.
2. If A, B and C are the angles of a triangle, simplify $\cos \frac{1}{2}(A + B)$.
3. Show that $\cos 10° = \cos 50° + \cos 70°$.
4. Show that $\sin 10° = \sin 70° - \sin 50°$.
5. Show that $\sin 2A - \sin 4A + \sin 6A = \sin 4A(2 \cos 2A - 1)$.
6. If $x = \dfrac{\pi}{5}$, find the value of $\dfrac{\cos x \sin 4x}{\sin 3x}$.
7. Show that $\sin^2 x - \sin^2 y = \sin (x + y) \sin (x - y)$.
8. Show that $\cos^2 y - \cos^2 x = \sin (x + y) \sin (x - y)$.
9. Show that $\cos 76° = 2 \sin 7° \sin 83°$.
10. Show that $\cos 62° = 2 \sin 14° \sin 76°$.
11. Solve for values of x from $0°$ to $180°$ inclusive the equation $\cos x = \cos 3x$.
12. Solve the equation $\cos (x - 10°) = \cos (20° - x)$ when x is acute.
13. Show that $1 + 2 \cos 4x + \cos 8x = 4 \cos 4x \cos^2 2x$.
14. Show that $\cos x + \cos \left(x + \dfrac{2\pi}{3}\right) + \cos \left(x + \dfrac{4\pi}{3}\right) = 0$.
15. Show that $\sin x + \sin \left(x + \dfrac{2\pi}{3}\right) + \sin \left(x + \dfrac{4\pi}{3}\right) = 0$.
16. Show that $1 + \cos 2A + \sin 2A = 2 \cos A (\cos A + \sin A)$.
17. Show that $2 \sin 18° \cos 12° + 2 \sin 72° \sin 12° = 1$.
18. Solve the equation $\cos x = \sin 2x$ when x is acute.

19. If $\cos (A + B) = \frac{1}{5}$ and $\cos (A - B) = \frac{3}{5}$, find the value of $\cos A \cos B$.

20. If $\cos (A + B) = \frac{1}{5}$ and $\cos (A - B) = \frac{3}{5}$, find the value of $\tan A \tan B$.

21. If $\cos (A + B) = \frac{1}{5}$ and $\cos (A - B) = \frac{3}{5}$, find the value of $\tan A - \tan B$.

22. Solve the equation $\cos 2x = \cos (30° - x)$ for values of x between $0°$ and $180°$.

23. Integrate $\cos^2 x$ by expressing it in terms of $\cos 2x$.

$$\left(\text{Assume} \int \cos nx \, dx = \frac{1}{n} \sin nx + c. \right)$$

24. Integrate $\sin^2 x$ by expressing it in terms of $\cos 2x$.

25. Find the integral of $2 \sin x \sin 3x$ by expressing it as a difference.

26. Find the integral of $2 \cos x \cos 3x$ by expressing it as a sum.

27. Find the integral of $2 \sin 3x \cos x$ by expressing it as a sum.

28. Show that $\tan (x + y) + \tan (x - y) = \dfrac{2 \sin 2x}{\cos 2x + \cos 2y}$.

29. Show that $\dfrac{\sin B - \sin C}{\sin B + \sin C} = \cot \frac{1}{2}(B + C) \tan \frac{1}{2}(B - C)$.

30. In the triangle ABC, show that $\dfrac{a}{b + c} = \dfrac{\cos \dfrac{B + C}{2}}{\cos \dfrac{B - C}{2}}$.

31. Factorize $\cos A + \cos B + \cos C - 1$ given that A, B and C are the angles of a triangle.

32. Factorize $\sin 2A + \sin 2B + \sin 2(A + B)$.

33. If $A + B + C = 180°$, show that
$$\sin 4A + \sin 4B + \sin 4C = -4 \sin 2A \sin 2B \sin 2C.$$

34. If $A + B + C = 180°$, show that
$$\cos^2 A + \cos^2 B + \cos^2 C = 1 - 2 \cos A \cos B \cos C.$$

35. If $A + B + C = 180°$, show that
$$\cos 2A + \cos 2B + \cos 2C + 1 = -4 \cos A \cos B \cos C.$$

36. Solve for values of x between $0°$ and $180°$ inclusive, the equation
$$\sin x + \sin 5x = \sin 3x.$$

37. Show that $\dfrac{\cos x + \cos y}{\cos x - \cos y} = \cot \frac{1}{2}(x + y) \cot \frac{1}{2}(y - x)$.

38. Show that $\sin (\alpha - \beta) + 2 \sin \alpha + \sin (\alpha + \beta) = 4 \sin \alpha \cos^2 \dfrac{\beta}{2}$.

39. If A, B and C are the angles of a triangle, show that
$$\cos 2A + \cos 2B - \cos 2C - 1 = -4 \sin A \sin B \cos C.$$

40. If A, B and C are the angles of a triangle, show that
$$\sin A - \sin B + \sin C = 4 \sin \tfrac{1}{2}A \cos \tfrac{1}{2}B \sin \tfrac{1}{2}C.$$

41. In the triangle ABC, show that
$$a - b + c = 8R \sin \tfrac{1}{2}A \cos \tfrac{1}{2}B \sin \tfrac{1}{2}C.$$

42. In the triangle ABC, show that if $s = \tfrac{1}{2}(a + b + c)$,
$$s - b = 4R \sin \tfrac{1}{2}A \cos \tfrac{1}{2}B \sin \tfrac{1}{2}C.$$

43. Solve for values of x between $0°$ and $180°$ inclusive the equation
$$\cos x + \cos 2x = \cos \tfrac{1}{2}x.$$

44. Express as a sum or difference of sines and cosines the expression
$$4 \cos x \cos 3x \cos 5x.$$

45. Integrate $\cos x \cos 3x \cos 5x$.

46. Express $4 \sin x \sin 3x \cos 5x$ as a sum or difference of sines and cosines.

47. Integrate $\sin x \sin 3x \cos 5x$.

48. Solve for values of x between $0°$ and $180°$ the equation
$$\cos x + \cos 2x = \sin x + \sin 2x.$$

49. Show that $\dfrac{1 + 2 \cos x + \cos 2x}{\sin x + \sin 2x} = \dfrac{\cos \dfrac{3x}{2} + \cos \dfrac{x}{2}}{\sin \dfrac{3x}{2}}.$

50. If $A + B + C = 90°$, show that
$$\cos A \cos B \cos C = \cos A \sin B \sin C + \cos B \sin A \sin C$$
$$+ \cos C \sin A \sin B.$$

34 Half Angle Formulae: Area of Triangle

Any triangle may be solved by the use of the sine and cosine formulae provided that enough measurements are given to determine the triangle. The cosine formula is, however, unwieldy and not adapted to logarithmic calculation. There are two cases of solution for which the cosine formula has until now been necessary and in this chapter more suitable formulae for these two cases will be considered. The two cases are: (i) three sides given; (ii) two sides and the included angle given. To solve a triangle given three sides, the half angle formulae are generally the most suitable and proofs of these formulae follow.

The half angle formulae

From the cosine formula,

$$\cos A = \frac{b^2 + c^2 - a^2}{2bc}.$$

Since

$$\cos A = 2 \cos^2 \frac{A}{2} - 1,$$

$$2 \cos^2 \frac{A}{2} - 1 = \frac{b^2 + c^2 - a^2}{2bc}.$$

and

$$2 \cos^2 \frac{A}{2} = \frac{b^2 + c^2 - a^2}{2bc} + 1$$

$$= \frac{b^2 + c^2 + 2bc - a^2}{2bc}$$

$$= \frac{(b + c)^2 - a^2}{2bc}$$

$$= \frac{(b + c + a)(b + c - a)}{2bc}.$$

If $2s = a + b + c$, s is the semi-perimeter of the triangle and

$$2s - 2a = b + c - a.$$

$$\therefore 2 \cos^2 \frac{A}{2} = \frac{2s(2s - 2a)}{2bc}$$

and
$$\cos \frac{A}{2} = \sqrt{\frac{s(s-a)}{bc}}.$$

Starting from the formula for cos A and putting

$$\cos A = 1 - 2 \sin^2 \frac{A}{2},$$

$$1 - 2 \sin^2 \frac{A}{2} = \frac{b^2 + c^2 - a^2}{2bc}.$$

$$\therefore 2 \sin^2 \frac{A}{2} = 1 - \frac{b^2 + c^2 - a^2}{2bc}$$

$$= \frac{a^2 - (b^2 - 2bc + c^2)}{2bc}$$

$$= \frac{a^2 - (b - c)^2}{2bc}$$

$$= \frac{(a - b + c)(a + b - c)}{2bc}.$$

But $a - b + c = 2s - 2b$ and $a + b - c = 2s - 2c.$

$$\therefore 2 \sin^2 \frac{A}{2} = \frac{(2s - 2b)(2s - 2c)}{2bc}$$

and

$$\sin \frac{A}{2} = \sqrt{\frac{(s-b)(s-c)}{bc}}.$$

The third formula is found by division.

$$\tan \frac{A}{2} = \frac{\sin \frac{A}{2}}{\cos \frac{A}{2}} = \sqrt{\frac{(s-b)(s-c)}{s(s-a)}}.$$

Corresponding formulae of course apply for the ratios of the other half angles, e.g.

$$\cos \frac{C}{2} = \sqrt{\frac{s(s-c)}{ab}}.$$

The formulae are grouped below for reference.

$$\cos\frac{A}{2} = \sqrt{\frac{s(s-a)}{bc}}.$$

$$\sin\frac{A}{2} = \sqrt{\frac{(s-b)(s-c)}{bc}}.$$

$$\tan\frac{A}{2} = \sqrt{\frac{(s-b)(s-c)}{s(s-a)}}.$$

In solving a triangle given three sides, any one of these formulae may be used.

Example. *Find the angle C of the triangle ABC, given that $a = 14.2$, $b = 12.1$ and $c = 20.7$.*

$2s = a + b + c$
$\quad = 14.2 + 12.1 + 20.7 = 47.$
$\therefore s = 23.5 \quad$ and $\quad s - c = 2.8.$

$$\cos\frac{C}{2} = \sqrt{\frac{s(s-c)}{ab}} = \sqrt{\frac{23.5 \times 2.8}{14.2 \times 12.1}}$$

$\therefore \dfrac{C}{2} = 51°\,46' \quad$ and $\quad C = 103°\,32'.$

No.	Log
23.5	1.3711
2.8	0.4472
	1.8183
14.2	1.1523
12.1	1.0828
	$\overline{1}.5832 \div 2$
$\cos\dfrac{C}{2}$	$\overline{1}.7916$

The other angles may now be found using the sine formula.

Exercise 34a

All examples refer to the triangle *ABC*.

1. Find *A* given that $b = c = 14.2$, $a = 12.4$.
2. Find *B* given that $a = 10$, $b = 14$, $c = 17$.
3. Find *C* given that $a = 3.4$, $b = 4.2$, $c = 5.4$.
4. Find *A* given that $a = 12.24$, $b = 4.82$, $c = 9.24$.
5. Find *C* given that $a = 10.42$, $b = 8.78$, $c = 9.36$.

Two sides and the included angle

The formula used for solving a triangle given two sides and the included angle is

$$\tan\frac{B-C}{2} = \frac{b-c}{b+c}\cot\frac{A}{2}.$$

Consider the fraction $\dfrac{b - c}{b + c}$.

Express this in terms of angles using
$$b = 2R \sin B, \quad c = 2R \sin C.$$

$$\frac{b - c}{b + c} = \frac{2R(\sin B - \sin C)}{2R(\sin B + \sin C)} = \frac{2 \cos \dfrac{B + C}{2} \sin \dfrac{B - C}{2}}{2 \sin \dfrac{B + C}{2} \cos \dfrac{B - C}{2}}$$

(using the product formulae)

$$= \frac{\tan \dfrac{B - C}{2}}{\tan \dfrac{B + C}{2}}.$$

$$\therefore \tan \frac{B - C}{2} = \frac{b - c}{b + c} \tan \frac{B + C}{2}.$$

Since $\dfrac{A}{2} + \dfrac{B}{2} + \dfrac{C}{2} = 90°$,

$$\tan \frac{B + C}{2} = \cot \frac{A}{2}.$$

$$\therefore \tan \frac{B - C}{2} = \frac{b - c}{b + c} \cot \frac{A}{2}.$$

Example. *Given that b = 12.2, c = 10.4 and A = 40°, find B and C.*

$$\tan \frac{B - C}{2} = \frac{12.2 - 10.4}{12.2 + 10.4} \cot 20°$$

$$= \frac{1.8}{22.6} \cot 20°.$$

$$\therefore \frac{B - C}{2} = 12° 21' \quad \text{and} \quad B - C = 24° 42'.$$

No.	Log
1.8	0.2553
cot 20°	0.4389
	0.6942
22.6	1.3541
$\tan \dfrac{B - C}{2}$	$\overline{1}.3401$

But $B + C = 180° - A = 140°$.
By addition $2B = 164° 42'$ and $B = 82° 21'$.
By subtraction, $2C = 115° 18'$ and $C = 57° 39'$.

Exercise 34b

All examples refer to the triangle ABC.

1. Find B and C given that $b = 4.8$, $c = 3.6$ and $A = 47°$.

2. Find A and B given that $a = 14.2$, $b = 12.3$ and $C = 112°$.

3. Find A and C given that $a = 7.82$, $c = 9.12$ and $B = 94° \, 48'$.

4. Find A and B given that $a = 12.24$, $b = 8.62$ and $C = 60° \, 12'$.

5. Find A and C given that $a = 4.82$, $c = 6.42$ and $B = 100°$.

The area of a triangle

The area of a triangle is equal to half the product of base and altitude.
  In Fig. 34.1,

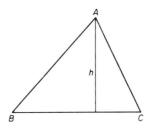

Fig. 34.1

$$\Delta \text{ (the area)} = \tfrac{1}{2}ah.$$

But
$$\frac{h}{c} = \sin B.$$

$$\therefore h = c \sin B \quad \text{and} \quad \Delta = \tfrac{1}{2}ac \sin B.$$

By the sine formula,
$$ac \sin B = bc \sin A = ab \sin C.$$
$$\therefore \Delta = \tfrac{1}{2}bc \sin A = \tfrac{1}{2}ac \sin B = \tfrac{1}{2}ab \sin C.$$

So the area of a triangle is equal to half the product of any two sides and the sine of the included angle.

Example. *The area of a triangle is 10 cm². Given that $a = 14$ cm, $b = 10$ cm, find the angle C.*

$$\Delta = \tfrac{1}{2}ab \sin C.$$
$$\therefore 10 = \tfrac{1}{2}(14)(10) \sin C.$$
$$\therefore \sin C = \tfrac{1}{7} = 0.1429.$$
$$\therefore C = 8° \, 13' \quad \text{or} \quad (180° - 8° \, 13'),$$

i.e.
$$C = 8° \, 13' \quad \text{or} \quad 171° \, 47'.$$

Heron's formula

The area of a triangle may also be expressed in terms of the sides as follows.

$$\Delta = \tfrac{1}{2}ab \sin C = \tfrac{1}{2}ab(2 \sin \tfrac{1}{2}C \cos \tfrac{1}{2}C)$$

$$= ab \sqrt{\frac{(s-a)(s-b)}{ab} \cdot \frac{s(s-c)}{ab}}$$

$$= \sqrt{s(s-a)(s-b)(s-c)}.$$

This is called Heron's formula for the area of a triangle.

Example. *Find the area of a triangle in which a = 14 cm, b = 12 cm and c = 10 cm.*

$$s = \tfrac{1}{2}(a+b+c) = 18 \text{ cm}$$
$$\therefore s - a = 4 \text{ cm}; \quad s - b = 6 \text{ cm}; \quad s - c = 8 \text{ cm}$$
$$\therefore \Delta = \sqrt{18.4.6.8} = \sqrt{3^2.8^2.6} = 24\sqrt{6}$$
$$= 58.8 \text{ cm}^2 \text{ (to 3 s.f.).}$$

The in-radius

The radius of the inscribed circle of a triangle, i.e. the circle which touches the three sides, may also be expressed in terms of the area.

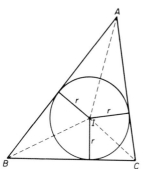

Fig. 34.2

In Fig. 34.2, I is the centre of the in-circle. Let r be its radius.
The area of the triangle $BIC = \tfrac{1}{2}ar$.
The area of the triangle $CIA = \tfrac{1}{2}br$.
The area of the triangle $AIB = \tfrac{1}{2}cr$.
The sum of these three areas is Δ.

$$\therefore \Delta = \tfrac{1}{2}r(a+b+c) = \tfrac{1}{2}r(2s) = rs.$$

$$\therefore r = \frac{\Delta}{s}.$$

Example. *Find the in-radius of the triangle whose sides are 14 cm, 12 cm and 10 cm.*

It has already been shown in a previous example that

$$\Delta = 24\sqrt{6} \text{ cm}^2 \quad \text{and that} \quad s = 18 \text{ cm}.$$

$$\therefore r = \frac{\Delta}{s} = \frac{24\sqrt{6}}{18} = \frac{4\sqrt{6}}{3} = 3.27 \text{ cm (to 3 s.f.)}.$$

Exercise 34c

1. Find the area of the triangle in which $b = 4.8$ cm, $c = 6.4$ cm and $A = 50°$.
2. Find the area of the triangle in which $a = 4.8$ cm, $b = 6.2$ cm and $c = 7.4$ cm.
3. Find the in-radius of the triangle in which $a = 4.8$ cm, $b = 6.2$ cm and $c = 7.4$ cm.
4. Find the area of the triangle in which $B = 42°$, $C = 54°$ and $a = 8$ cm.
5. Given that ABC is an acute-angled triangle, find the angle C given that the area of the triangle is 12 cm^2 and that $a = 8$ cm, $b = 6$ cm.
6. Find the area of the triangle in which $a = 12.8$ cm, $b = 11.4$ cm and $c = 10.2$ cm.
7. Find a formula for the area of a triangle in terms of a, B and C.
8. Find the in-radius of the triangle in which $B = 42°$, $C = 54°$ and $a = 8$ cm.
9. Show that $(s - a) \tan \frac{1}{2}A = r$.
10. Show that $(s - a) \tan \frac{1}{2}A = (s - b) \tan \frac{1}{2}B = (s - c) \tan \frac{1}{2}C$.

Three-dimensional problems

The formulae proved in this chapter have applications in the solution of three-dimensional problems. Before some illustrative examples are worked, definitions for the angle between a line and a plane and the angle between two planes are given.

The angle between a line and a plane

The angle between a line PO, which meets a plane $ABCD$ at O, and the plane $ABCD$ is the angle between PO and its projection on the plane. If PN is perpendicular to the plane, the angle between the line and the plane is the angle PON, as shown in Fig. 34.3.

The angle between two planes

The angle between two planes which meet in the line AB is the angle between two lines, one in each plane, perpendicular to the line of intersection, AB, of the planes (see Fig. 34.4).

The angle between the two planes shown is ROQ.

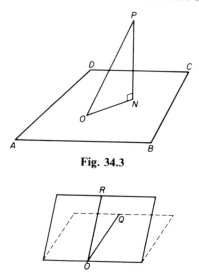

Fig. 34.3

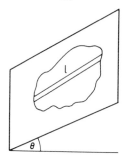

Fig. 34.4

The projection of an area

One further property of the area of a triangle is often useful in finding the angle between two planes. If a triangle has area A and the plane of the triangle makes an angle θ with the horizontal, the area of the projection of the triangle on the horizontal plane is $A \cos \theta$. This in fact is true for any area, whether it be a triangle or an irregular figure.

Consider the area of any closed figure and take an element along a line of greatest slope. Suppose the length of the element as shown in Fig. 34.5 is l and that its breadth if δx. The area of the figure is $\sum l \delta x$, where $\sum$ stands for the sum of all such elements. When the element is projected, its breadth remains unaltered but the length becomes $l \cos \theta$. The new area is therefore $\sum l \cos \theta . \delta x$ or $A \cos \theta$.

Fig. 34.5

Example 1. *AP, BQ, CR, DS are parallel edges of a cube of side 4 cm. Find*

(i) *the angle RAD;*

(ii) *the angle between RA and the plane ABCD;*

(iii) *the angle between the planes DRA and ABCD.*

(i)
$$RD^2 = DC^2 + CR^2$$
$$= 4^2 + 4^2 = 32.$$
$$\therefore RD = \sqrt{32} = 4\sqrt{2}.$$

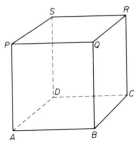

Fig. 34.6

AD is perpendicular to the plane *RCDS*.

$$\therefore \text{ the angle } ADR = 90°.$$

From the triangle *ARD*,

$$\tan R\hat{A}D = \frac{4\sqrt{2}}{4} = 1.414.$$

$$\therefore RAD = 54° \ 44'.$$

(ii) The foot of the perpendicular from *R* to the plane *ABCD* is *C*. Therefore the angle required is *RAC*.

In the triangle *RAC*, the angle $RCA = 90°$, $RC = 4$ cm and $AC = 4\sqrt{2}$ cm.

$$\therefore \tan RAC = \frac{4}{4\sqrt{2}} = \frac{1}{\sqrt{2}} = 0.7071$$

and
$$RAC = 35° \ 16'.$$

(iii) The line of intersection of the planes is *AD*. The angle required is therefore *RDC*, which equals 45°.

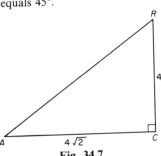

Fig. 34.7

Example 2. *A square ABCD of side 6 cm is held with the corner A in a horizontal plane so that B is 3 cm above the plane and D is 4 cm above the plane. Find the inclination of the plane ABCD to the horizontal.*

Suppose that B', C', D' are the projections of the points B, C, D on the horizontal plane.

From the triangle ADD' in Fig. 34.8,

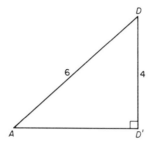

Fig. 34.8

$$AD'^2 = 6^2 - 4^2 = 20.$$
$$\therefore AD' = \sqrt{20} = 2\sqrt{5} \text{ cm.}$$

From the triangle ABB' in Fig. 34.9,

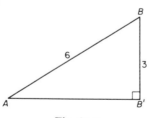

Fig. 34.9

$$AB'^2 = 6^2 - 3^2 = 27.$$
$$\therefore AB' = \sqrt{27} = 3\sqrt{3} \text{ cm.}$$

We now set out to find the length of the third side of the triangle $AB'D'$ in order to calculate its area.

$$BD^2 = 6^2 + 6^2 \quad \text{(diagonal of square)}$$
$$= 72.$$
$$\therefore BD = \sqrt{72} = 6\sqrt{2} \text{ cm.}$$

From Fig. 34.10,

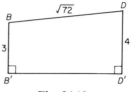

Fig. 34.10

$$B'D'^2 = BD^2 - (4 - 3)^2$$
$$= 72 - 1$$
$$= 71.$$

So the lengths of the sides of the triangle $AB'D'$ are $\sqrt{20}$, $\sqrt{27}$ and $\sqrt{71}$, as shown in Fig. 34.11.

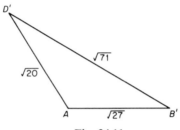

Fig. 34.11

The area of the triangle $AB'D'$ may be found by Heron's formula but it is perhaps easier to use the cosine formula in this case.

$$\cos B'AD' = \frac{27 + 20 - 71}{2\sqrt{27}.\sqrt{20}} = -\frac{24}{2.3\sqrt{3.2}\sqrt{5}} = -\frac{2}{\sqrt{15}}.$$
$$\sin^2 B'AD' = 1 - \cos^2 B'AD' = 1 - \tfrac{4}{15} = \tfrac{11}{15}.$$
The area of $B'AD' = \tfrac{1}{2}\sqrt{27}.\sqrt{20} \sin B'AD' = \tfrac{1}{2}\sqrt{27}.\sqrt{20}.\sqrt{\tfrac{11}{15}}$
$$= 3\sqrt{11} \text{ cm}^2.$$

But the area of ABD is 18 cm^2
So if θ is the angle between the two planes,

$$\cos \theta = \frac{3\sqrt{11}}{\cdot 18} = \frac{\sqrt{11}}{6} = \frac{3.317}{6} = 0.5528,$$

and $\theta = 56° 26'$.

Example 3. *Three points A, B and F lie in a horizontal plane. It is given that A is west of F and that B is south-west of F. The angles of elevation of the top T of a vertical mast FT from A and B are $42°$ and $38°$ respectively. Find the bearing of A from B.*

Suppose the height of the mast is h. Then from Fig. 34.12,
$$AF = h \cot 42° \quad \text{and} \quad BF = h \cot 38°.$$

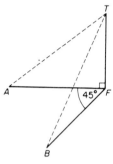

Fig. 34.12

Consider the triangle AFB.

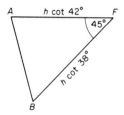

Fig. 34.13

From the formula $\dfrac{a - b}{a + b} \cot \frac{1}{2}C = \tan \frac{1}{2}(A - B)$, we have

$$\left(\frac{h \cot 38° - h \cot 42°}{h \cot 38° + h \cot 42°}\right) \cot 22° 30' = \tan \frac{A - B}{2}.$$

$$\therefore \left(\frac{1.2799 - 1.1106}{1.2799 + 1.1106}\right) \cot 22° 30' = \tan \frac{A - B}{2}.$$

$$\therefore \tan \frac{A - B}{2} = \frac{0.1693}{2.3905} \cot 22° 30'.$$

$$\therefore \frac{A - B}{2} = 9° 42' \quad \text{and} \quad A - B = 19° 24'.$$

Since $A + B = 135°$
$$2A = 154° 24' \quad \text{and} \quad A = 77° 12'.$$

The bearing of A from B is $270° + 77° 12'$ or $347° 12'$.

No.	Log
0.1693	$\bar{1}.2287$
cot 22° 30'	0.3828
	$\bar{1}.6115$
2.3905	0.3785
$\tan \dfrac{A - B}{2}$	$\bar{1}.2330$

Exercise 34d: Miscellaneous

1. The sides of a triangle are 32 cm, 48 cm and 56 cm. Calculate its area.

2. The triangle ABC is such that $AB = 4$ cm and the angle $B = 50°$. Given that the area of the triangle is 2.2 cm², calculate the length of BC.

3. The area of an isosceles triangle is 6 cm² and each of the equal sides is 4 cm long. Calculate the possible angles between the equal sides.

4. $ABCD$ is a tetrahedron in which $BC = CD = DB = 6$ cm and $AB = AC = AD = 8$ cm. Calculate the length of the perpendicular from A to the plane BCD.

5. An equilateral triangle ABC has sides 5 cm long. A point D is 6 cm from each of A, B and C. Calculate the angle DC makes with the plane ABC.

6. Find the smallest angle of the triangle whose sides are 4 cm, 5 cm and 6 cm.

7. Calculate the largest angle of the triangle whose sides are 6.2 cm, 8.4 cm and 9.2 cm.

8. In the triangle ABC of area 48 cm², it is given that $a = 12$ cm and $s = 18$ cm. Calculate the value of sin A.

9. Given that $a = 25.5$, $b = 12.5$ and $c = 26$, find sin A.

10. $ABCD$ is a square of side 6 cm lying in a horizontal plane. The points P, Q, R, S are 4 cm above, A, B, C, D respectively. Calculate
    (i) the length of AS;
    (ii) the angle AS makes with the horizontal;
    (iii) the angle the plane ABS makes with the horizontal.

11. A field in the shape of a convex quadrilateral $ABCD$ is divided into two equal parts by a fence BD. Given that $BC = 90$ m, $CD = 84$ m, $BD = 120$ m and $DA = 100$ m, calculate the angle BDA.

12. A paddock is in the shape of a trapezium $ABCD$ in which $AB = 120$ m, $AD = 60$ m, the angle $A = 45°$ and the angle $B = 60°$. Given that DC is parallel to AB, calculate the area of the paddock.

13. In the triangle ABC, show that $\dfrac{a}{b+c} = \dfrac{\sin \dfrac{A}{2}}{\cos \dfrac{B-C}{2}}$.

14. The triangle ABC is held with AB horizontal and C 4 cm above the level of AB. Given that $AB = 10$ cm, $BC = 8$ cm and $CA = 6$ cm, calculate the angle the plane ABC makes with the horizontal.

15. A rectangle $ABCD$ in which $AB = 4$, $BC = 3$, has AB horizontal and the plane makes an angle of $30°$ with the horizontal. Find the angle AC makes with the horizontal.

16. A square $ABCD$ is rotated through an angle of $45°$ about the side AB. Find the angle between the two positions of the diagonal AC.

17. Show that the area of the triangle ABC is equal to
$$\tfrac{1}{4}(b^2 \sin 2C + c^2 \sin 2B).$$

18. Show that $\sin \frac{1}{2}A \sin \frac{1}{2}B \sin \frac{1}{2}C = \dfrac{\Delta^2}{abcs}$.

19. Show that $\cos \frac{1}{2}A \cos \frac{1}{2}B \cos \frac{1}{2}C = \dfrac{\Delta s}{abc}$.

20. Show that $\sin A \sin B \sin C = \dfrac{8\Delta^3}{a^2b^2c^2}$.

21. Show that $\tan \frac{1}{2}A \tan \frac{1}{2}B \tan \frac{1}{2}C = \dfrac{\Delta}{s^2}$.

22. The sides of a triangle are given by $a = 3$, $b = 5$ and $c = 6$. Show that $\tan \frac{1}{2}B = 2 \tan \frac{1}{2}A$.

23. If $a = 14$, $b = 12$ and $c = 10$, show that $\sin \frac{1}{2}B = 2 \sin \frac{1}{2}A \sin \frac{1}{2}C$.

24. Show that $R = \dfrac{abc}{4\Delta}$.

25. Given that $a + b = 2c$, show that $2 \cos \frac{1}{2}C = \dfrac{\sqrt{3c}}{\sqrt{ab}}$.

26. Show that $r = 4R \sin \frac{1}{2}A \sin \frac{1}{2}B \sin \frac{1}{2}C$.

27. A square $ABCD$ of side 8 cm is held with A in a horizontal plane, B 4 cm above the horizontal plane and C 6 cm above it. Find the angle the plane $ABCD$ makes with the horizontal.

28. The points P, Q, F lie in a horizontal plane with P south of F and Q west of F. The angles of elevation of the top T of a vertical mast FT from P and Q are $30°$ and $60°$ respectively. Find the bearing of Q from P.

29. The points A, B and F lie in a horizontal plane. The point A is south of F and the bearing of B from F is $220°$. The angles of elevation of the top T of a vertical mast FT from A and B are $47°$ and $42°$ respectively. Calculate the bearing of B from A.

30. A hillside is a plane which slopes at an angle α to the horizontal. A track on the hillside makes an angle β with a line of greatest slope. Find the inclination of the track to the horizontal.

Answers

Exercise 1a (p. 4)

1. $\frac{1}{8}$.   **2.** 9.   **3.** 1.   **4.** 3.   **5.** $\frac{1}{3}$.   **6.** 8.

7. $\frac{1}{8}$.   **8.** 5.   **9.** $\frac{1}{5}$.   **10.** 2.   **11.** $\frac{1}{2}$.   **12.** 7.

13. $\frac{1}{7}$.   **14.** 3.   **15.** $\frac{1}{3}$.   **16.** 9.   **17.** $\frac{1}{9}$.   **18.** 27.

19. $\frac{1}{27}$.   **20.** $\frac{1}{100}$.   **21.** $\frac{1}{1000}$.   **22.** 1.   **23.** $\frac{1}{10}$.   **24.** $\frac{4}{3}$.

25. $\frac{25}{16}$.   **26.** $\frac{2}{3}$.   **27.** $\frac{3}{2}$.   **28.** $\frac{9}{4}$.   **29.** $\frac{81}{16}$.   **30.** $\frac{144}{121}$.

31. $\frac{11}{12}$.   **32.** $\frac{12}{11}$.   **33.** 8.   **34.** $\frac{1}{8}$.   **35.** 4.   **36.** $\frac{1}{4}$.

37. 16.   **38.** $\frac{1}{16}$.   **39.** 2.   **40.** $\frac{1}{2}$.   **41.** 32.   **42.** $\frac{1}{32}$.

43. 1.   **44.** $\frac{1}{10}$.   **45.** 5.   **46.** $\frac{1}{5}$.   **47.** 25.   **48.** $\frac{1}{25}$.

49. $\frac{1}{121}$.   **50.** $\frac{1}{27}$.

Exercise 1b (p. 6)

1. 4.   **2.** 8.   **3.** 8.   **4.** 6.   **5.** $\frac{2}{3}$.   **6.** 2.

7. $\frac{9}{2}$.   **8.** 9.   **9.** 3.   **10.** 5.   **11.** $\frac{3}{2}$.   **12.** 3.

13. 3.   **14.** 2.   **15.** -2.   **16.** $-\frac{1}{2}$.   **17.** $\frac{1}{3}$.   **18.** -1.

19. $\frac{2}{3}$.   **20.** $-\frac{2}{3}$.   **21.** 0 or 2.   **22.** 1 or 2.   **23.** 0 or 1.   **24.** 0 or 2.

25. 1 or 2.   **26.** $\dfrac{1}{y}$.   **27.** $y^{-\frac{1}{2}}$.   **28.** $z^3 y^{-1}$.   **29.** $z^{\frac{1}{2}} y^{-\frac{1}{2}}$.   **30.** $z^{\frac{1}{2}} y^{-\frac{3}{2}}$.

31. $y^3 z^{-3}$.   **32.** $y^{-2} z^{-2}$.   **33.** $z^6 y^{-4}$.   **34.** zy^{-1}.   **35.** $z^3 y^{-\frac{3}{2}}$.   **36.** ab.

37. 10^x.   **38.** $\dfrac{x+1}{x}$.   **39.** $\dfrac{1}{x}$.   **40.** 2^n.   **41.** $3\frac{1}{2}$.   **42.** 7.

43. 11.   **44.** $10x^2$.   **45.** $2^n + 8^n$.   **46.** 5.   **47.** $\dfrac{1}{y}$.   **48.** $y^{\frac{1}{2}} + 1$.

49. 5^{2n}.   **50.** x^6.

Exercise 1c (p. 8)

1. $2\sqrt{2}$.   **2.** $4\sqrt{2}$.   **3.** $2\sqrt[3]{2}$.   **4.** $2\sqrt[3]{4}$.   **5.** $2\sqrt[5]{2}$.   **6.** $10\sqrt{5}$.

7. $10\sqrt{10}$.   **8.** $5\sqrt[3]{2}$.   **9.** $4\sqrt[3]{2}$.   **10.** $3\sqrt[3]{2}$.   **11.** $\dfrac{\sqrt{2}}{2}$.   **12.** $\dfrac{3\sqrt{2}}{4}$.

13. $\dfrac{\sqrt{2}}{4}$.   **14.** $\dfrac{\sqrt{2}}{5}$.   **15.** $\sqrt{2}$.   **16.** $\dfrac{\sqrt{2}}{4}$.

17. $\sqrt{2} + 1$.   **18.** $\sqrt{2} - 1$.   **19.** $2 + \sqrt{3}$.

20. $2 - \sqrt{3}$.   **21.** $2 - \sqrt{2}$.   **22.** $3 + \sqrt{3}$.

23. 1.   **24.** 1.   **25.** $7 + 3\sqrt{3}$.   **26.** $3\sqrt{3} - 5$.

27. $5 - 3\sqrt{2}$.   **28.** $5 + 3\sqrt{3}$.   **29.** 1.   **30.** $a - 1$.

31. $\dfrac{\sqrt{3} + 1}{2}$.   **32.** $\dfrac{2\sqrt{3} + 1}{11}$.   **33.** $\dfrac{\sqrt{7} + \sqrt{5}}{2}$.

34. $\dfrac{\sqrt{7}-\sqrt{5}}{2}$.

35. $\sqrt{3}-\sqrt{2}$.

36. $\dfrac{2\sqrt{3}-\sqrt{2}}{10}$.

37. $\dfrac{\sqrt{7}+2}{3}$.

38. $\dfrac{\sqrt{6}-1}{5}$.

39. $\dfrac{2\sqrt{5}+\sqrt{3}}{17}$.

40. $\dfrac{\sqrt{5}+\sqrt{2}}{3}$.

Exercise 1d (p. 9)

1. 4. **2.** $6yz$. **3.** 2^{2x+3}. **4.** 3. **5.** $b^{-\frac{1}{3}}$. **6.** $\frac{512}{125}$.
7. 6. **8.** x^{-7} **9.** 0 or 1. **10.** $-\frac{1}{5}$. **11.** 1000. **12.** 100.
13. 2^{-5}. **14.** $x+y$. **15.** 4. **16.** $4\sqrt{2}$ cm.

17. $\dfrac{25\sqrt{3}}{4}$ cm². **18.** $\sqrt{6}$ cm. **19.** 0 or -1.

20. $5-2\sqrt{6}$. **21.** $a+b-2\sqrt{ab}$. **22.** $\sqrt{5}-\sqrt{3}$.
23. $3-\sqrt{5}$. **24.** 0.366. **25.** 5.83. **26.** 1.5.
27. $\sqrt{14}$ cm. **28.** 4. **29.** 9. **30.** $2\sqrt{2}$ cm.

Exercise 2a (p. 12)

1. 2.2. **2.** 2.2. **3.** 0.32. **4.** 4.0. **5.** 5.0. **6.** 15.9.
7. 25.1. **8.** 159. **9.** 50.1. **10.** 251. **11.** 0.30. **12.** 0.40.
13. 0.48. **14.** 0.70. **15.** 0.86. **16.** 1.30. **17.** $\bar{1}$.40. **18.** $\bar{2}$.48.
19. 2.70. **20.** 5.86.

Exercise 2b (p. 13)

1. 1. **2.** 1. **3.** 3. **4.** 3. **5.** 2. **6.** 5.
7. 4. **8.** $\frac{1}{2}$. **9.** $1\frac{1}{2}$ **10.** 2. **11.** 1. **12.** 4.
13. 1. **14.** 2. **15.** $5\log x$. **16.** 2. **17.** 1. **18.** 3.
19. 2. **20.** $\frac{1}{3}$.

Exercise 2c (p. 15)

1. 1.26. **2.** 1.46. **3.** 1.64. **4.** 1.86. **5.** 1.50. **6.** 0.792.
7. 1.23. **8.** 0.672. **9.** 0.845. **10.** 1.19.

Exercise 2d (p. 15)

1. $1000x^{-2}$. **2.** 93.0. **3.** 0. **4.** 2 or 2.32. **5.** 0.464.
6. 1.39. **7.** $\frac{5}{6}$. **8.** $1\frac{1}{2}$. **9.** $1\frac{1}{3}$. **10.** -3. **11.** $-1\frac{1}{2}$.
12. 4. **14.** $\dfrac{100}{y}$. **15.** 1.29. **16.** 4. **17.** 14.2. **18.** 0.563.

19. (i) 2; (ii) $\frac{1}{2}$; (iii) 4. **20.** (i) 1.0791; (ii) 0.6990.

21. -1.47. **22.** $2\log 2 + 3\log 3$. **23.** $\sqrt{\dfrac{6}{x}}$.
24. 1.71. **25.** 8.63.

Exercise 3a (p. 18)

1. $x = 1, y = 2; x = 2, y = 1$.
2. $x = 3, y = 2; x = -2, y = -3$.
3. $x = 1, y = 1; x = -\frac{1}{4}, y = 3\frac{1}{2}$.
4. $x = 2, y = 1; x = -1, y = 2$.
5. $x = 3, y = 4; x = 4, y = 3$.
6. $x = 2, y = 3; x = -2, y = -3$.
7. $x = 1, y = \frac{1}{2}$.
8. $x = 1, y = 1; x = 2, y = -1$.
9. $x = 4, y = -1; x = 1, y = -4$.
10. $x = 4, y = 1; x = -4, y = -1$.

Exercise 3b (p. 20)

1. $x = 0, y = 3, z = 0$.
2. $x = 1, y = -1, z = 1$.
3. $x = 2, y = 1, z = 0$.
4. $x = 1, y = 2, z = 3$.
5. $x = 1, y = 2, z = -1$.
6. Dependent.
7. Inconsistent.   8. Dependent.
9. Inconsistent.   10. Inconsistent.

Exercise 3c (p. 20)

1. $x = 1, y = 3; x = -3\frac{1}{3}, y = 5\frac{1}{6}$.
2. $x = 3, y = 2, z = 1$.
3. $x = 3, y = -\frac{1}{3}$.
4. No.
5. $x = 4, y = 6; x = -4, y = -6$.
6. $x = b, y = a$.
7. $x = 2, y = 2; x = 6, y = \frac{2}{3}$.
8. $x = 2, y = -1; x = \frac{1}{2}, y = -4$.
9. $x = \frac{1}{2}, y = \frac{1}{3}; x = \frac{1}{3}, y = \frac{1}{2}$.
10. $x = 2, y = 4, z = 6$.
11. 36 cm^3 or $35\frac{23}{27} \text{ cm}^3$.
12. 40 000, 48 000, 24 000.
13. 432.
14. 72 cm^3 or 69.9 cm^3.   15. 20 km.

Exercise 4a (p. 23)

1. -7.
2. 1.
3. 30.
4. -15.
5. $a^3 - 3a$.
6. -4.
7. -5.
8. -7.
9. 161.
10. -2.
11. -1.
12. 6.
13. 5, 8.
14. $-4, 1$.
15. $(x + y)(x + 2y)(x + 3y)$.
16. 3, 5.
17. $k = 1; (x - 2)(x - 3)$.
18. $1, -4$.
19. $k = -5$; 3 and -2.
20. $3, -2$.

Exercise 4b (p. 25)

1. $\pm 2, 3$.
2. $1, \dfrac{1 \pm \sqrt{5}}{2}$.
3. -2.
4. 1 or -2.

5. $2, \dfrac{1 \pm \sqrt{13}}{2}$.
6. 6.
8. Odd.
9. 31, 12.

10. $3, 7, -2$.   12. 2.
13. -1.
15. Odd.
17. $\pm 2, -1$.
18. -1.
19. $-(a - b)(b - c)(c - a)$.
20. $3(a - b)(b - c)(c - a)$.
21. A multiple of 2 but not a multiple of 4.
22. $(x - a - b - c)$.
23. $(a - b)(b - c)(c - a)$.
24. $\dfrac{f(a) - f(-a)}{2a}x + \dfrac{f(a) + f(-a)}{2}$.   25. $-(a - b)(b - c)(c - a)$.

Exercise 5a (p. 28)

1. $y = 10x$.   **2.** 2.
3. (i) the square root of A; (ii) $\propto \sqrt[3]{V}$.
4. No.   **5.** Directly as square.   **6.** $R = kV^2$.
7. $8\frac{1}{2}$.   **8.** 32.   **9.** 9:1.   **10.** 8:1.   **11.** 4000 cm³.
12. 6.   **13.** 3.375 kg.   **14.** 32 m.   **15.** $\sqrt{2}$:1.   **16.** $V \propto S^{\frac{3}{2}}$.
17. 1:2500.   **18.** 4:1.   **20.** Directly as each.

Exercise 5b (p. 30)

1. 12.   **2.** 9.   **3.** 64.   **4.** 0.
5. (i) inversely as the square root of V; (ii) $\dfrac{1}{A^2}$.
6. Directly.   **7.** Directly as square.   **8.** 250 cm³.
9. 16 m/s.   **10.** Inversely as square.   **11.** 11.
12. $v \propto \dfrac{1}{u}$.   **13.** $y \propto \dfrac{1}{x^2}$.   **14.** $yx^n = k$.   **15.** $22\frac{1}{2}$ days.
16. $3xz^2 = 16y$.   **19.** Ratio 1:8.   **20.** 12.

Exercise 5c (p. 33)

1. 32.   **3.** 32 km/h.   **4.** £130.   **5.** 24.
6. Trebled.   **7.** 1800 joules.   **8.** 6.25 kg.   **9.** 5.6 kg.
10. $V \propto \dfrac{x^2}{t}$.   **11.** 580 newtons.   **12.** 8 newtons.   **13.** 900:1.
14. 9.1% reduction.   **15.** 27%.   **16.** $2\sqrt{2}$:1.
17. Quadrupled.   **18.** 600 cm³.   **19.** Halved.
20. Proportional to square root.   **21.** 13.5 m.
22. $z = ax + by$; 5.   **23.** $\sqrt{\dfrac{9lT}{2}}$.   **24.** $V = 16xy^2$.
25. 2.4 m.

Exercise 6a (p. 38)

1. 99.   **2.** -70.   **3.** $2n - 1$.   **4.** 400.   **5.** No.
6. 9.   **7.** 14, 23.   **8.** 6.   **9.** 465.   **10.** -410.
11. $3.9 + \dfrac{n}{10}$.   **12.** 2450.   **13.** $2n + 1$.   **14.** $n - 1$.   **15.** 23.
16. $n(n + 4)$.   **17.** 6, 8, 10.   **18.** $2n - 7$.   **19.** -2.   **20.** 35.

Exercise 6b (p. 40)

1. 2×5^{19} **2.** 3×2^{16}.   **3.** $\dfrac{1}{3^{13}}$.   **4.** $(-7)^9$.   **5.** 16.
6. 18.   **7.** 30.   **8.** 20, 80.   **9.** $3^{12} - 1$.
10. $\frac{1}{4}(5^{20} - 1)$.   **11.** $3(2^{2n} - 1)$.   **12.** $12(\frac{1}{2})^{n-1}$.   **13.** $5a(2a)^{n-1}$.
14. $2^n - 1$.   **15.** 6.   **16.** 2.   **17.** 4, 9.
18. 126.   **19.** 15, 75, 375.   **20.** $3(2^n - 1)$.

Exercise 6c (p. 42)

1. n^2. **3.** 23 **4.** $\dfrac{n(3n-1)}{2}$. **5.** £5000.

6. 6, 8, 10. **7.** 10, 40, 200. **8.** $2y - x$. **9.** $72\frac{1}{2}$.

10. 867. **12.** 15(log 2 + 7 log 3). **13.** $\frac{3}{2}$.

15. $\dfrac{25 - xn}{n}$. **16.** 77 m. **17.** $\frac{3}{4}\{1 - (\frac{1}{3})^n\}$; 4.

18. 11 583. **19.** 9 cm, $13\frac{1}{2}$ cm. **20.** 90. **21.** $\frac{5}{2}(3x + y)$.

22. 9200 m. **23.** $n(2n + 7)$. **24.** 10, -3. **25.** 33.1.

Exercise 7a (p. 46)

1. $2\frac{1}{4}$. **2.** Not between $\dfrac{3 \pm \sqrt{17}}{2}$.

3. Between 1 and 2. **4.** x cannot lie between 1 and 2.

7. ± 12. **10.** 0. **11.** $-\frac{1}{2}$. **12.** $c \geqslant 9$.

13. Between ± 6. **14.** -3. **15.** 2. **16.** $\pm\sqrt{24}$.

17. 1, 2; $k = 2$. **18.** $+1, +3$; -4 or $-1, -3$; $+4$.

Exercise 7b (p. 48)

1. $2b^2 = 9ac$. **2.** $3b^2 = 16ac$. **3.** $-\frac{14}{15}$.

4. $-\frac{19}{25}$; imaginary. **5.** $x^2 - 9x + 9 = 0$. **6.** $7x^2 + 2x - 4 = 0$.

7. $x^2 - x - 7 = 0$. **8.** $c = 0$.

9. $ax^2 + x(b - 2a) + (a + c - b) = 0$.

10. $a^2y^2 + 3aby + (2b^2 + ac) = 0$. **12.** $19\frac{3}{8}$. **15.** 433.

16. All except between $-\frac{4}{9}$ and 4. **17.** $a^2x^2 + x(2ac - b^2) + c^2 = 0$.

20. 322. **22.** $x^2 - 2x - 8 = 0$. **23.** $x^4 - 13x^2 + 36 = 0$.

Exercise 8a (p. 51)

1. (i) 6. (ii) 720. (iii) 39 916 800. **2.** (i) 1716. (ii) 286. (iii) 70.

3. (i) $\dfrac{8!}{5!}$. (ii) $\dfrac{12!}{8!}$. (iii) $\dfrac{20!}{16!4!}$. (iv) $\dfrac{n!}{3!(n-3)!}$.

4. (i) 24. (ii) 12. (iii) 6. **5.** 20 **6.** (i) 120. (ii) 720. (iii) 720.

7. (i) 60. (ii) 5040. (iii) 10 080. **8.** 720. **9.** 1000.

10. 30.

Exercise 8b (p. 53)

1. (i) 20. (ii) 35. (iii) 35. **2.** 10. **3.** (i) 6. (ii) 4. (iii) 3.

4. 56. **5.** 1365. **6.** 210. **7.** 10. **8.** 10.

9. $\frac{1}{2}n(n - 3)$. **10.** 15; 45.

Exercise 8c (p. 55)

1. 1 814 400.   **2.** 420.   **3.** 120; 24.   **4.** $\dfrac{13!}{2!}$.

5. 224.   **6.** 1960; 1540.   **7.** 84   **8.** 20 160.
9. 126.   **10.** 50.   **11.** 240.

12. $\dfrac{16!}{12!\,4!}$; $\dfrac{8!}{4!\,4!}$.   **13.** 420.   **14.** 120.

15. 28.   **16.** 36.   **17.** (i) 15 600. (ii) 26^3.
18. 6 084 000.   **19.** 32.   **20.** 60; 24.   **21.** 30.
22. 30.   **23.** $2^n(n+1)-1$.   **24.** 3^{13}.

Exercise 9a (p. 59)

2. $1 + 5x + 10x^2 + 10x^3 + 5x^4 + x^5$.

3. $16 - 32x + 24x^2 - 8x^3 + x^4$.   **4.** $x^3 + 3x + \dfrac{3}{x} + \dfrac{1}{x^3}$.

5. 280.   **6.** 405.   **7.** 160.   **8.** 13.   **9.** 15.   **10.** 2.
11. $_nC_r x^r a^{n-r}$.   **12.** $_nC_r(-2)^{n-r}$.   **13.** 56.   **14.** 82.
15. 104 060 401.   **16.** 970 299.   **17.** 1.0721.

18. 0.96060.   **19.** $1 - nx + \dfrac{n(n+1)}{2}x^2$.   **20.** $_{2n}C_n$.

Exercise 9b (p. 61)

1. 0.94, 0.94.   **2.** 1.03.
3. $1 - \frac{1}{3}x - \frac{1}{9}x^2 - \frac{5}{81}x^3$.   **4.** $1 - 3x + 6x^2 - 10x^3$.
5. 1.6%.   **6.** $1 - 2x + 2x^2 - 2x^3$.   **7.** $1 + 4x + 7x^2 + 10x^3$.
8. $x^4 + 4x^3y + 6x^2y^2 + 4xy^3 + y^4$; 16.322.

9. $\dfrac{63 \times 3^5}{8}$.   **10.** -160.   **11.** 1.913.   **12.** 36.

13. $1 + x + \frac{1}{2}x^2$; 3.32.   **14.** $1 + 2x + 3x^2$; $(n+1)$.
15. $1 - x - \frac{1}{2}x^2$; 3.162.   **16.** $a = -2, b = 1$.

17. $\dfrac{n(n-3)}{2}$.   **18.** $2n + 1$.

19. $1 - 4x + 18x^2 - 40x^3 + 91x^4$.   **20.** $1 - \frac{5}{3}x - \frac{25}{9}x^2 - \frac{625}{81}x^3$; 1.82.
21. $n = 5, a = 3$.   **22.** $_{3n}C_n$.   **24.** $1 + x - \frac{1}{2}x^2 + \frac{1}{2}x^3$; 10.0995.
25. $-x^2 - 4x^3 - 11x^4$.

Exercise 10a (p. 66)

1. (i) $\frac{1}{27}$. (ii) $\frac{26}{27}$. (iii) $\frac{2}{27}$. (iv) $\frac{13}{54}$. (v) $\frac{8}{27}$. (vi) $\frac{19}{27}$. (vii) $\frac{1}{54}$. (viii) $\frac{2}{9}$. (ix) $\frac{2}{9}$. (x) $\frac{7}{9}$.
2. (i) $\frac{1}{2}$. (ii) $\frac{8}{25}$. (iii) $\frac{17}{25}$. (iv) $\frac{7}{50}$. (v) $\frac{2}{50}$. (vi) $\frac{1}{3}$. (vii) $\frac{1}{3}$. (viii) $\frac{17}{30}$. (ix) $\frac{2}{5}$. (x) 0.
3. (i) $\frac{2}{5}$. (ii) $\frac{7}{19}$. (iii) $\frac{2}{3}$. (iv) $\frac{3}{7}$. (v) 1.
4. (i) $\frac{4}{9}$. (ii) $\frac{1}{2}$. (iii) $\frac{6}{13}$. (iv) $\frac{7}{13}$. (v) 0.
5. (i) $\frac{3}{5}$. (ii) $\frac{2}{3}$. (iii) $\frac{10}{17}$. (iv) 5.

Exercise 10b (p. 71)

1. (i) $\frac{4}{9}$. (ii) $\frac{1}{9}$. (iii) $\frac{4}{9}$.

2. (i) $\frac{1}{36}$. (ii) $\frac{1}{9}$. (iii) $\frac{1}{4}$.

3. $\frac{1}{36}, \frac{1}{9}, \frac{5}{18}, \frac{1}{3}, \frac{1}{4}$.

4. (i) $\frac{1}{27}$. (ii) $\frac{8}{27}$. (iii) $\frac{19}{27}$.

5. (i) $\frac{1}{221}$. (ii) $\frac{188}{221}$. (iii) $\frac{32}{221}$.

6. (i) $\frac{1}{17}$. (ii) $\frac{19}{34}$. (iii) $\frac{13}{34}$.

7. (i) $\frac{13}{204}$. (ii) $\frac{13}{204}$. (iii) $\frac{13}{204}$. (iv) $\frac{1}{4}$.

8. (i) $\frac{5}{51}$. (ii) $\frac{5}{51}$. (iii) $\frac{1}{3}$.

9. (i) $\frac{11}{16}$. (ii) $\frac{83}{192}$.

10. (i) $\frac{11}{18}$. (ii) $\frac{107}{216}$.

Exercise 10c (p. 75)

1. (i) 10^{-7}. (ii) 0.11. (iii) 2.6×10^{-2}.

2. (i) $(0.8)^6$. (ii) $(0.2)^6$. (iii) 0.082.

3. (i) 0.21. (ii) 0.087.

4. (i) 0.0256. (ii) 0.017. (iii) 0.138.

5. (i) 1.7×10^{-8}. (ii) 8.3×10^{-7}. (iii) 0.32. (iv) 0.16. (v) one.

6. (i) 0.18. (ii) 0.2916.   **7.** (i) 0.096. (ii) 0.0256. (iii) 0.0064.

8. (i) 1.5×10^{-3}. (ii) 2×10^{-5}.

Exercise 10d (p. 76)

1. (i) 0.16. (ii) 0.0064. (iii) 0.0016.

2. (i) 0.09. (ii) 0.081. (iii) 0.48. (iv) 51.

3. (i) 0.09. (ii) 0.066. (iii) 0.35.

4. (i) 10^{-2}. (ii) 10^{-2}. (iii) 1, taking $1 - 10^{-2} = 1$; 0.94 is a better approximation.

5. (i) 0.147. (ii) 0.07203. (iii) 0.04.

Exercise 11 (p. 89)

1. (i) Reflection in x-axis. (ii) Enlargement factor 3. (iii) Reduction factor $\frac{1}{2}$. (iv) Reflection in y-axis, and stretching factor 2 parallel to x-axis. (v) Reflection in origin and enlargement factor 2. (vi) Reflection in origin and stretching factor 2 parallel to x-axis. (vii) Reflection in $y = x$ and enlargement factor 2. (viii), (xi) Shear parallel to x-axis. (ix), (x) Shear parallel to x-axis and stretch parallel to y-axis (factors $\frac{1}{2}$, 2). (xii) Shear parallel to y-axis.

2. (i) $\begin{pmatrix} 3 & 0 \\ 0 & 3 \end{pmatrix}$. (ii) $\begin{pmatrix} 3 & 0 \\ 0 & 1 \end{pmatrix}$. (iii) $\begin{pmatrix} 1 & 0 \\ 0 & -1 \end{pmatrix}$. (iv) $\begin{pmatrix} 2 & 0 \\ 0 & -2 \end{pmatrix}$.

(v) $\begin{pmatrix} 2 & 1 \\ 1 & -1 \end{pmatrix}$. (vi) $\begin{pmatrix} 2 & -1 \\ 2 & -2 \end{pmatrix}$.

3. (i) $\begin{pmatrix} 1 & 0 \\ 0 & -1 \end{pmatrix}$. (ii) $\begin{pmatrix} 0 & 1 \\ 1 & 0 \end{pmatrix}$. (iii) $\begin{pmatrix} 5 & 0 \\ 0 & 5 \end{pmatrix}$. (iv) $\begin{pmatrix} -1 & 0 \\ 0 & -1 \end{pmatrix}$.

4. (i) $\begin{pmatrix} 0 & 2 & 2 & 0 \\ 0 & 0 & 2 & 2 \end{pmatrix}, \begin{pmatrix} \frac{1}{2} & 0 \\ 0 & \frac{1}{2} \end{pmatrix}$ (ii) $\begin{pmatrix} 0 & \frac{1}{4} & \frac{1}{4} & 0 \\ 0 & 0 & \frac{1}{4} & \frac{1}{4} \end{pmatrix}, \begin{pmatrix} 4 & 0 \\ 0 & 4 \end{pmatrix}.$

(iii) $\begin{pmatrix} 0 & 0 & -1 & -1 \\ 0 & 1 & 1 & 0 \end{pmatrix}, \begin{pmatrix} 0 & -1 \\ 1 & 0 \end{pmatrix}.$

(iv) $\begin{pmatrix} 0 & 2 & 2 & 0 \\ 0 & 0 & -2 & -2 \end{pmatrix}, \begin{pmatrix} -\frac{1}{2} & 0 \\ 0 & \frac{1}{2} \end{pmatrix}.$

(v) $\begin{pmatrix} 0 & 2 & 2 & 0 \\ 0 & 0 & 3 & 3 \end{pmatrix}, \begin{pmatrix} \frac{1}{3} & 0 \\ 0 & \frac{1}{3} \end{pmatrix}.$

(vi) $\begin{pmatrix} 0 & -1 & -2 & -1 \\ 0 & -1 & 0 & 1 \end{pmatrix}, \begin{pmatrix} -\frac{1}{2} & -\frac{1}{2} \\ -\frac{1}{2} & \frac{1}{2} \end{pmatrix}.$

6. (i) 1. (ii) 1. (iii) 10. (iv) 2. (v) 0. (vi) 1.

7. $\begin{pmatrix} 0 & -1 \\ 1 & 0 \end{pmatrix}$

    (i) Rotation 90° anticlockwise sense.
    (ii) Rotation 45° anti-clockwise.
    (iii) Rotation 45° clockwise.

8. $\begin{pmatrix} -\dfrac{1}{2} & -\dfrac{\sqrt{3}}{2} \\ \dfrac{\sqrt{3}}{2} & -\dfrac{1}{2} \end{pmatrix}, \begin{pmatrix} -1 & 0 \\ 0 & -1 \end{pmatrix}.$

    Rotation thro' (i) 180°. (ii) 60°. (iii) 120°. (iv) 360°. (v) 300°anticlockwise.
    (vi) 60° clockwise.

9. $\sin 2\theta = 2 \sin \theta \cos \theta.$

Exercise 12a (p. 98)

1. $\begin{pmatrix} 1 \\ 0 \end{pmatrix}, \begin{pmatrix} 3 \\ 1 \end{pmatrix}, \begin{pmatrix} 4 \\ -1 \end{pmatrix}, \begin{pmatrix} -2 \\ 3 \end{pmatrix}.$ (i) $\begin{pmatrix} 2 \\ 1 \end{pmatrix}, \begin{pmatrix} 3 \\ -1 \end{pmatrix}, \begin{pmatrix} -3 \\ 3 \end{pmatrix}.$

(ii) $\begin{pmatrix} -2 \\ -1 \end{pmatrix}, \begin{pmatrix} 1 \\ -2 \end{pmatrix}, \begin{pmatrix} -5 \\ 2 \end{pmatrix}.$

2. $\mathbf{i}$, $2\mathbf{i} + 2\mathbf{j}$, $3\mathbf{i} + 6\mathbf{j}$, $\mathbf{i} + 2\mathbf{j}$. (i) $\mathbf{i} + 2\mathbf{j}$, $2\mathbf{i} + 6\mathbf{j}$, $2\mathbf{j}$. (ii) $-2\mathbf{i} - 6\mathbf{j}$, $-\mathbf{i} - 4\mathbf{j}$, $-2\mathbf{i} - 4\mathbf{j}$. (iii) $\overrightarrow{AB} = -\overrightarrow{BA}$. (iv) $2\overrightarrow{AB} = \overrightarrow{DC}$.

3. $\begin{pmatrix} \frac{3}{2} \\ 1 \end{pmatrix}, \begin{pmatrix} 6 \\ 4 \end{pmatrix}.$ **4.** (i) 5. (ii) $\sqrt{13}$. (iii) $\sqrt{17}$. (iv) $\sqrt{2}$.

9. (i) $\mathbf{i} + 2\mathbf{j}$. (ii) $-2\mathbf{i} + 3\mathbf{j}$; $3 : 1.$

Exercise 12b (p. 101)

1. (i) 1.85. (ii) 0.77. (iii) 2.8. (iv) 1.47.
2. (i) 0.68. (ii) 2.84. (iii) 1.37. (iv) 1.39.

3. 2.19. (i) 2.19. (ii) 4.38. (iii) 8.77.
4. 142°.   **5.** 5.   **6.** 1.5, 4.6.   **7.** 10:7.
8. (i) 41 minutes. (ii) 41 minutes, 282°. (iii) 233°, 168 km h^{-1}.
9. 018°, 3.4 km h^{-1}.
10. (i) 226 km away, bearing 058°. (ii) 237°.

Exercise 13 (p. 107)

1. (i) $2\mathbf{i} + 2\mathbf{j}$. (ii) $\mathbf{i} + 2\mathbf{j}$. (iii) $\mathbf{i} - \mathbf{j}$.
2. (i) $2\mathbf{i} + 4\mathbf{j}$. (ii) $2\mathbf{j}$. (iii) $2\mathbf{i}$.
3. (i) $10\mathbf{i} + 7\mathbf{j}$. (ii) $9\mathbf{i} + 8\mathbf{j}$. (iii) $5\mathbf{i} + 12\mathbf{j}$.
4. (i) $\mathbf{r} = \mathbf{i} + \mathbf{j} + \lambda(\mathbf{i} + 3\mathbf{j})$. (ii) $\mathbf{r} = \mathbf{i} - \mathbf{j} + \lambda(\mathbf{i} - 3\mathbf{j})$.
   (iii) $\mathbf{r} = \mathbf{i} + \lambda(-\mathbf{i} + \mathbf{j})$.
5. (i) Yes. (ii) No. (iii) No.

7. (i) $2\mathbf{i} + 3\mathbf{j}$. (ii) $4\mathbf{i} + \mathbf{j}$.   **10.** $\begin{pmatrix} -1 \\ 3\sqrt{3} \end{pmatrix}, \begin{pmatrix} -1 \\ -3\sqrt{3} \end{pmatrix}$.
11. 10.   **12.** $3\frac{3}{4}$.
15. $\mathbf{c} = \frac{1}{2}(\mathbf{a} + \mathbf{b})$.   **16.** $\mathbf{b} = \frac{1}{3}(2\mathbf{a} + \mathbf{d})$, $\mathbf{c} = \frac{1}{3}(\mathbf{a} + 2\mathbf{d})$.
17. $\mathbf{b} - \mathbf{a} = \lambda(\mathbf{d} - \mathbf{c})$.   **18.** $\mathbf{c} - \mathbf{a} = \lambda[\frac{1}{2}(\mathbf{c} + \mathbf{d}) - \mathbf{b}]$.
19. $\mathbf{d} - \mathbf{c} = \lambda[\frac{1}{2}(\mathbf{b} + \mathbf{d}) - \mathbf{a}]$.   **20.** $\mathbf{b} + \mathbf{c} - \mathbf{a}$.
21. $\mathbf{c} + 2(\mathbf{d} - \mathbf{a})$.   **22.** $m : (m + n)$.
24. AQ.   **25.** $\frac{1}{3}(2\mathbf{a} + \mathbf{c})$.
26. (i) $\frac{1}{3}(\mathbf{a} + \mathbf{b})$. (ii) $\frac{1}{3}(2\mathbf{b} - \mathbf{a})$. (iii) $\frac{2}{5}(\mathbf{a} + \mathbf{b})$. (iv) $\frac{2}{5}(2\mathbf{b} - \mathbf{a})$.
27. (i) $\frac{1}{4}(2\mathbf{a} + \mathbf{b})$. (ii) $\frac{1}{4}(3\mathbf{b} - 2\mathbf{a})$.
28. $\frac{2}{7}(\mathbf{a} + \mathbf{b})$.

Exercise 14 (p. 113)

1. (i) $\begin{pmatrix} 2 \\ 2 \\ -1 \end{pmatrix}$. (ii) $\begin{pmatrix} -2 \\ -2 \\ 1 \end{pmatrix}$. (iii) $\begin{pmatrix} -5 \\ -3 \\ 4 \end{pmatrix}$. (iv) $\begin{pmatrix} -4 \\ -4 \\ 2 \end{pmatrix}$; 3, 3, $\sqrt{50}$, 6.

2. (i) 3; $\frac{1}{3}, \frac{2}{3}, \frac{2}{3}$. (ii) 3; $\frac{2}{3}, -\frac{1}{3}, -\frac{2}{3}$. (iii) $\sqrt{3}$; $-\frac{1}{\sqrt{3}}, \frac{1}{\sqrt{3}}, \frac{1}{\sqrt{3}}$.
   (iv) 5; 0, $\frac{3}{5}, -\frac{4}{5}$. (v) 3; $-\frac{1}{3}, \frac{2}{3}, -\frac{2}{3}$.

3. (i) $\begin{pmatrix} 4 \\ 3 \\ 5 \end{pmatrix}$. (ii) $\begin{pmatrix} 7 \\ 4 \\ 9 \end{pmatrix}$. (iii) $\begin{pmatrix} 12 \\ 4 \\ 7 \end{pmatrix}$. (iv) $\begin{pmatrix} 4 \\ 8 \\ 13 \end{pmatrix}$.

4. (i) $\begin{pmatrix} 3 \\ -1 \\ 3 \end{pmatrix}$. (ii) $\begin{pmatrix} 4 \\ 1 \\ 5 \end{pmatrix}$.   **5.** $\begin{pmatrix} 1 \\ 0 \\ 1 \end{pmatrix}$.   **6.** $\begin{pmatrix} 0 \\ -1 \\ 1 \end{pmatrix}$.

8. 3, 2,   **9.** 4, -1.   **10.** 2, -1, -1.

14. $\mathbf{r} = \begin{pmatrix} 3 \\ 0 \\ -2 \end{pmatrix} + \lambda \begin{pmatrix} 2 \\ 1 \\ -1 \end{pmatrix}$. (i) No. (ii) $\begin{pmatrix} -1 \\ -2 \\ 0 \end{pmatrix}$.

15. $\begin{pmatrix} 3 \\ 1 \\ 0 \end{pmatrix}$. **16.** $\sqrt{24}, \sqrt{13}, \sqrt{21}; \dfrac{4\sqrt{14}}{21}$.

Exercise 15a (p. 121)

1. (i) 5. (ii) 1. (iii) -2. (iv) 13. (v) 6. (vi) 6.

2. $\dfrac{2}{\sqrt{41}}$. **3.** (i) $\dfrac{1}{\sqrt{2}}$. (ii) $\dfrac{5}{\sqrt{26}}$. (iii) $\dfrac{3}{\sqrt{10}}$.

4. **a** and **c**, **b** and **d**.
5. (i) $\mathbf{r}.(\mathbf{i} + \mathbf{j}) = 0$, $x + y = 0$. (ii) $\mathbf{r}.(2\mathbf{i} + 5\mathbf{j}) = 29$; $2x + 5y = 29$.
(iii) $\mathbf{r}.(5\mathbf{i} - 2\mathbf{j}) = 0$; $5x - 2y = 0$.

6. $\dfrac{1}{5\sqrt{2}}$. **7.** $\dfrac{5}{\sqrt{221}}$. **8.** $(\mathbf{r} - \mathbf{a}).(\mathbf{r} - \mathbf{b}) = 0$.

9. $\tfrac{4}{5}\mathbf{i} + \tfrac{3}{5}\mathbf{j}$. **10.** $\pm\dfrac{1}{2\sqrt{5}}\left((1 \pm 2\sqrt{3})\mathbf{i} - (2 \mp \sqrt{3})\mathbf{j}\right)$.

Exercise 15b (p. 124)

1. (i) -27. (ii) 10. (iii) -38. (iv) 9. (v) 225. (vi) 25. (vii) $-\tfrac{3}{5}$. (viii) $-\tfrac{38}{75}$.
  (ix) $\tfrac{2}{3}$.
2. (i) 13. (ii) -5. (iii) 1. (iv) 14. (v) 18. (vi) 6.

  (vii) $\dfrac{13}{6\sqrt{7}}$. (viii) $\dfrac{-5}{6\sqrt{3}}$. (ix) $\dfrac{1}{2\sqrt{21}}$.

3. $\sqrt{\dfrac{6}{17}}$.

4. -1 **6.** $\pm\tfrac{1}{3}(\mathbf{i} + 2\mathbf{j} - 2\mathbf{k})$.

7. $\pm\dfrac{1}{\sqrt{3}}(\mathbf{i} + \mathbf{j} - \mathbf{k})$. **8.** $\pm\dfrac{3}{\sqrt{2}}(\mathbf{j} - \mathbf{k})$.

10. (i) $\mathbf{r}.(3\mathbf{i} + 2\mathbf{j} + \mathbf{k}) = 0$. (ii) $\mathbf{r}.(\mathbf{i} + \mathbf{j} + \mathbf{k}) = 9$. (iii) $\mathbf{r}.\mathbf{i} = 1$.

11. $x + 2y - 4z = -5$. **12.** $\dfrac{4}{\sqrt{78}}, \dfrac{5}{\sqrt{273}}, \dfrac{8}{3\sqrt{14}}$.

13. $45°$. **14.** (i) $\dfrac{1}{\sqrt{2}}$. (ii) $\dfrac{7}{\sqrt{247}}$. (iii) $\dfrac{26}{\sqrt{1022}}$.

15. $(\mathbf{r} - \mathbf{a}).(\mathbf{r} - \mathbf{b}) = 0$.
16. $x^2 + y^2 + z^2 - 3x + y - z + 2 = 0$.
18. $(\mathbf{r} - \mathbf{a}).\mathbf{u} = 0$.

Exercise 16a (p. 135)

1. 4.   **2.** 6.   **3.** 10.   **4.** 20.
5. 24.   **6.** 20.
7. 3.   **8.** 3.   **9.** $6x' + 4$.   **10.** $2ax' + b$.
11. 6.   **12.** 8.   **13.** 19.   **14.** 26.   **15.** 10.   **16.** 13.
17. 4 m s^{-2}.   **18.** 4 m s^{-2}.   **19.** 11 m s^{-2}.   **20.** $6t + 6$, 6.

Exercise 16b (p. 137)

1. 1.   **2.** 6.   **3.** 17.   **4.** $y = 3x - 10$.
5. $y = 4x + 3$.   **6.** $y = 2x - 2$.   **7.** $y = x - 3$.
8. $y = -2x - 4$.   **9.** $y = 2x - 4$.   **10.** $(\frac{3}{5}, 0)$.
11. $(0, -3)$.   **12.** $(\frac{1}{2}, -\frac{1}{4})$.   **13.** $(1, -1)$.
14. 8 m s^{-2}.   **15.** 6.   **16.** $(1\frac{1}{2}, 1)$.   **17.** $(1\frac{1}{2}, 2)$.
19. 7 m s^{-1}.   **20.** 11 m s^{-2}.   **21.** $y = 2x$.
22. $y = 2tx - t^2$; $y = -2x - 1$ and $y = 6x - 9$.
23. -2; -2.   **24.** $-2, 3, 0$.   **25.** $\frac{1}{2}, 3, 0$.

Exercise 17a (p. 140)

1. $2x + 1$.   **2.** $8x - 6$.   **3.** $4x - 1$.   **4.** $4x - 2$.
5. $1 + x$.   **6.** $2x$.   **7.** $2x + 1$.   **8.** $8x + 12$.
9. $18x - 24$.   **10.** $-1 - 2x$.   **11.** $24x + 14$.   **12.** $2x + 3$.
13. $10x + 14$.   **14.** $4x$.   **15.** $10x - 7$.

Exercise 17b (p. 146)

1. $9x^2 - 4x - 4$.   **2.** $20x^3 - 12x$.

3. $15x^4 - 12x^2 - 7$.   **4.** $-\dfrac{4}{x^3}$.   **5.** $-\dfrac{4}{x^2} - \dfrac{2}{x^3}$.

6. $4x^3 - 4x$.   **7.** $\dfrac{1}{2\sqrt{x}} - \dfrac{1}{2\sqrt{x^3}}$.   **8.** $\dfrac{1}{2\sqrt{x}} - \dfrac{1}{\sqrt{x^3}}$.

9. $1 + \dfrac{1}{2\sqrt{x}} - \dfrac{1}{2\sqrt{x^3}}$.   **10.** $6x^2 - 2x + 2$.   **11.** $9t^2 - 7$.

12. $-2 - 15y^2$.   **13.** $-\dfrac{1}{z^2} + \dfrac{2}{z^3}$.   **14.** $1 - \dfrac{1}{2\sqrt{x^3}}$.

15. $3u^2 + 12u + 11$.   **16.** $3z^2 - 3$.   **17.** $-\dfrac{1}{v^2}$.   **18.** $5t^4 - 1$.

19. $-\dfrac{2}{y^3} + \dfrac{3}{y^4}$.   **20.** $1 - \dfrac{6}{x^2}$.   **21.** $5 - 2t + 3t^2$.

22. $-\dfrac{2}{t^3} - \dfrac{1}{t^2}$.   **23.** $2t + 1$.   **24.** $1 - \dfrac{1}{t^2}$.

25. $6t^2 + 4t - 1$.   **26.** $-\dfrac{6}{y^7}$.   **27.** $15y^{14} - 13y^{12}$.

28. $\dfrac{7}{2} y^{\frac{5}{2}}$.

29. $-\dfrac{5}{y^6} + \dfrac{4}{y^5}$.

30. $3y^2 + 6y + 3$.

31. -3.

32. 16.

33. $(1, -2); (-1, 2)$.

34. $(1, 1); (-1, -1)$.

35. $y = 7x - 16$.

36. $y = \pm 16$.

37. -4.

38. $x + y = 2$.

39. $y + 2x = 3$.

40. -1.

41. (i) $\dfrac{1}{2\sqrt{x}}$; (ii) $\dfrac{1}{2y}$.

42. -2.

43. 14.

44. $12t^2 - 12t$.

45. $(0, 0); (\frac{2}{3}, -\frac{4}{27})$.

46. $6x - 6.$,

47. $6t - 6$.

48. $48x^2 - 6$.

49. $\dfrac{8}{x^3}$.

50. $4\frac{1}{2}$.

Exercise 18a (p. 155)

1. -4; min.

2. 10; max.

3. -11; min.

4. 0, min.; 4, max.

5. 6, min.; -6, max.

6. 3; min.

7. 9, max.; 8, min.

8. 54, max.; -54, min.

9. 128, max.; -128, min.

10. 0, max.; -1, min.

11. 324.

12. 400 cm^2.

13. $27\pi \text{ cm}^2$.

14. $54\pi \text{ cm}^2$.

15. 800 m^2.

16. 108 cm^2.

17. $y - 2 = m(x - 1)$; 4 square units.

18. $\frac{32}{3}\pi \text{ m}^3$.

19. $\frac{2000}{27} \text{ cm}^3$.

20. $ty + x = at^2$; $t = -\frac{1}{2}$.

Exercise 18b (p. 157)

1. 0.4 cm s^{-1}.

2. $0.4\pi \text{ cm s}^{-1}$.

3. $4\pi \text{ cm}^2 \text{ s}^{-1}$.

4. $16\pi \text{ cm}^2 \text{ s}^{-1}$.

5. $0.8 \text{ cm}^2 \text{ s}^{-1}$.

6. $30\pi \text{ cm}^3 \text{ s}^{-1}$.

7. $1.6 \text{ cm}^2 \text{ s}^{-1}$.

8. 12 m s^{-2}.

10. 4 m s^{-2}.

Exercise 18c (p. 157)

1. $2, -1, 2$.

2. 13, max.; -14, min.

3. 4 m s^{-2}.

4. 1.

5. (i) 11 m s^{-1}; (ii) 29 m s^{-1}.

6. $0, (-\frac{1}{2}, -4)$.

8. $400\,000 \text{ kg}$.

9. 0, max.; -27, min.

11. $-4, 8$.

12. $-4, 6$.

13. ± 1.

14. $(2, 6); (3, 6)$.

15. 0 or $\frac{2}{3}$.

17. $3.2\pi \text{ cm}^2 \text{ s}^{-1}$.

18. $\dfrac{4}{\pi} \text{ cm s}^{-1}$.

20. 288 cm^2.

Exercise 19a (p. 162)

(For the sake of brevity, constants of integration are omitted.)

1. $\frac{1}{6}x^6$.

2. $\frac{1}{4}x^4 - x^2$.

3. $\frac{1}{4}x^4 + 6x$.

4. $-\dfrac{1}{2x^2}$.

5. $\frac{1}{3}ax^3 + \frac{1}{2}bx^2 + cx$.

6. $\frac{1}{3}x^3 + \frac{3}{2}x^2 + 2x$.

7. $2\sqrt{x}$.

8. $-\dfrac{1}{3x^3} - \dfrac{1}{2x^2}$.

9. $\frac{1}{2}x^2 - \dfrac{1}{x}$.

10. $2\sqrt{x} + \frac{2}{3}\sqrt{x^3}$.

11. $\frac{1}{7}y^7$.

12. $\frac{1}{4}y^4 - y^3 + y$.

13. $\frac{1}{5}y^5 - \frac{2}{3}y^3 + 6y$.

14. $-\dfrac{1}{3y^3}$.

15. $\frac{1}{3}y^3 - \frac{1}{2}y^2$.

16. $y - \dfrac{1}{y}$.

17. $\frac{2}{5}\sqrt{y^5} + 2\sqrt{y}$.

18. $-\dfrac{1}{y} - \dfrac{1}{2y^2}$.

19. $\dfrac{5y^2}{2} - y$.

20 $\frac{1}{2}y^2 + y$.

21. $y = 2x - 7$.

22. $y = 2x + \frac{1}{2}x^2 - \frac{3}{2}$.

23. $y = 4x + 7$.

24. $y = \frac{3}{2}x^2 - 2x + \frac{3}{2}$.

25. $y = \frac{2}{3}x^3 - \frac{1}{2}x^2$.

Exercise 19b (p. 163)

1. $t^2 + 5t$.

2. $t^3 + t$.

3. $2t^3 + 4$.

4. $\frac{1}{12}t^4 + \frac{1}{6}t^3 + 5t$.

5. $\frac{1}{20}t^5 + \frac{1}{12}t^4 + 8t$.

6. $66\frac{2}{3}$ m.

7. 20 m s^{-1}.

8. 36 m.

9. 4 m.

10. 12 m s^{-1}.

11. 50 m s^{-1}.

12. 40 m s^{-1}.

13. 180 m.

14. 36 m.

15. 4 m.

16. 20 m.

17. 4 s.

18. 3.2 s.

19. 3 s.

20. 44 m.

Exercise 19c (p. 170)

1. $\frac{1}{2}$.

2. $8\frac{5}{6}$.

3. $5\frac{1}{3}$.

4. $2\frac{2}{3}$.

5. $\frac{5}{6}$.

6. $18\frac{3}{4}$.

7. $\frac{4}{3}$.

8. $8\frac{2}{3}$.

9. $12\frac{2}{5}$.

10. 4.

11. $8\frac{1}{2}$.

12. $1\frac{1}{3}$.

13. $10\frac{2}{3}$.

14. $4\frac{2}{3}$.

15. $\frac{1}{2}$.

16. $10\frac{2}{3}$.

17. $\frac{1}{4}$.

18. $21\frac{1}{3}$.

19. $\dfrac{8\sqrt{2}}{3}$.

20. $\frac{1}{12}$.

Exercise 19d (p. 176)

1. 0.

2. $\dfrac{5\pi}{2}$.

3. $\dfrac{348\pi}{5}$.

6. $\dfrac{\pi}{30}$.

7. $\dfrac{64\pi}{5}$.

8. $\dfrac{256\pi}{5}$.

9. $\frac{3}{4}$.

10. $\dfrac{10\pi}{21}$.

11. $\dfrac{158\pi}{3}$.

12. $\dfrac{\pi h}{3}(R^2 + Rr + r^2)$.

13. $21\frac{1}{3}$.

14. $\frac{3}{4}$.

15. $\dfrac{130\pi}{21}$.

16. $y = \frac{1}{2}x^2 + x + 2\frac{1}{2}$.

17. 7.

18. $6y = 2x^3 - 3x^2 + 7$.

19. 2, 3.

20. 22 m s^{-1}.

21. 8π.

22. $\frac{4}{3}\pi ab^2$.

23. $\frac{16}{3}x^3 - 12x^2 + 9x$.

24. $\dfrac{\pi}{2}$.

25. $\dfrac{7\pi}{3}$.

26. $(\frac{3}{4}, \frac{3}{10})$.

27. $(\frac{124}{75}, \frac{254}{105})$.

28. $(0, -\frac{2}{5})$.

29. $(1, \frac{2}{5})$.

30. $(\frac{3}{8}, \frac{3}{5})$.

31. $(2, \frac{4}{3})$.

32. $(\frac{67}{44}, -\frac{203}{110})$.

33. $(\frac{5}{6}, 0)$.

34. $(\frac{1}{2}, \frac{2}{5})$.

35. $(\frac{8}{5}, 2)$.

Exercise 20a (p. 184)

1. $\sin x + x \cos x$. **2.** $\cos^2 x - \sin^2 x$. **3.** $\dfrac{2}{(x + 1)^2}$.

4. $\dfrac{x^2 + 4x + 1}{(x + 2)^2}$.

5. $\dfrac{\sin x - (x + 1) \cos x}{\sin^2 x}$. **6.** $\dfrac{-(x + 1) \sin x - \cos x}{(x + 1)^2}$.

7. $3x^2 + 4x + 1$. **8.** $\dfrac{x^2 + 2x}{(x + 1)^2}$. **9.** $-\dfrac{1}{(1 + x)^2}$.

10. $-\dfrac{2x}{(x^2 + 1)^2}$. **11.** $3x^2 \cos x - x^3 \sin x$.

12. $\sec^2 x$. **13.** $-\csc^2 x$. **14.** $-\csc x \cot x$.

15. $\sec x \tan x$. **16.** $-\dfrac{2x + 1}{(x^2 + x + 1)^2}$. **17.** $\dfrac{1 - x^2}{(x^2 + x + 1)^2}$.

18. $\dfrac{x - 1}{2\sqrt{x^3}}$. **19.** $-\dfrac{1 + x}{2(x - 1)^2 \sqrt{x}}$. **20.** $-\dfrac{1}{2\sqrt{x^3}}$.

21. 1. **22.** 1. **23.** $1 - \dfrac{1}{\sqrt{2}}$.

24. $1 - \dfrac{1}{\sqrt{2}}$. **25.** $\dfrac{\sqrt{3} - 1}{2}$.

Exercise 20b (p. 188)

1. $3(x + 1)^2$. **2.** $9(3x - 1)^2$. **3.** $-8(1 - 2x)^3$.

4. $5(2x + 1)(1 + x + x^2)^4$. **5.** $\dfrac{1}{\sqrt{2x + 1}}$. **6.** $-\dfrac{1}{\sqrt{(2x + 1)^3}}$.

7. $-\dfrac{2}{\sqrt{1 - 4x}}$. **8.** $\dfrac{2}{\sqrt{(1 - 4x)^3}}$. **9.** $4(3x - 2)(3x^2 - 4x + 1)$.

10. $7 \cos 7x$. **11.** $-3 \sin 3x$. **12.** $3 \sin^2 x \cos x$.

13. $-12 \cos^2 4x \sin 4x$. **14.** $\dfrac{1}{2\sqrt{x}} \cos\sqrt{x}$. **15.** $-2x \sin (x^2 + 1)$.

16. $-(2x + 1) \sin (x^2 + x + 1)$. **17.** $-\dfrac{x}{\sqrt{x^2 + 1}} \sin \sqrt{x^2 + 1}$.

18. $-3x^2 \sin x^3$. **19.** $2 \sin (x + 1) \cos (x + 1)$.

20. $6 \sin^2 (2x + 1) \cos (2x + 1)$.

Exercise 20c (p. 189)

1. $-\dfrac{x}{y}$.   **2.** $\dfrac{8}{3y^2}$.   **3.** -1.   **4.** $-\sqrt{\dfrac{y}{x}}$.   **5.** $-\dfrac{x+y}{x}$.

6. $-\dfrac{x^2}{y^2}$.   **7.** $-\dfrac{y}{x+2y}$.   **8.** $-\dfrac{y\cos x}{\sin x + 2y}$.   **9.** $\sec y$.

10. $-\operatorname{cosec} y$.   **11.** $-\tfrac{1}{4}$.   **12.** ∞.   **13.** $-\tfrac{1}{3}$.   **14.** 4.

15. $-\tfrac{5}{4}$.

Exercise 20d (p. 193)

1. $\sin 2x + 2x \cos 2x$.   **2.** $\dfrac{\sin 2x - 2x \cos 2x}{\sin^2 2x}$.

3. $\tfrac{1}{2}$.   **4.** 1.   **5.** $\tfrac{1}{2}$.   **6.** $4x^3 \cos 2x - 2x^4 \sin 2x$.

7. $\tfrac{1}{5}$ or $\tfrac{11}{20}$.   **8.** $\tfrac{2}{45}$.   **9.** $\tfrac{3}{2}$.   **10.** $\dfrac{\pi}{2}$.   **11.** $\tfrac{7}{54}$.

12. $-\dfrac{2\sin x + 3}{(2 + 3\sin x)^2}$.   **13.** $y = \sqrt{2}$.   **14.** $\sqrt{10} - 1$.

15. $\dfrac{1}{1 - \sin x}$.   **16.** $x \cos x$.   **17.** $-\tfrac{6}{5}$; $-\tfrac{112}{125}$.   **18.** $\dfrac{3\cos 3x}{2\sqrt{\sin 3x}}$.

19. $3x^2 \sin 2x + 2x^3 \cos 2x$.

20. $\dfrac{\sqrt{3}}{4}$.   **21.** $\dfrac{-\cos x \cos 3x - 3 \sin x \sin 3x}{\sin^2 x}$.

22. 2.   **23.** $\dfrac{2x^2 + a^2}{\sqrt{x^2 + a^2}}$.   **24.** $\dfrac{\pi}{2}$.

25. $\dfrac{2\cos x \sin 2x - \cos 2x \sin x}{\cos^2 2x}$.   **26.** 1.

27. $3\tfrac{1}{3}$.   **28.** 10.   **29.** $13\tfrac{1}{3}$.

30. $\dfrac{\sin x}{2} - \dfrac{\sin 3x}{6}$.   **31.** $\dfrac{\sin 2x}{4} + \dfrac{\sin 2x}{8}$.   **32.** $\dfrac{\sqrt{3}}{16}$.

33. 2.   **34.** $p\cos^2(px + q) - p\sin^2(px + q)$.

35. $-\dfrac{1}{a}\cos(ax + b)$.   **36.** $\dfrac{2(1 - x^2)}{(x^2 - x + 1)^2}$.   **37.** $81\tfrac{11}{15}$.

38. π.   **39.** $\dfrac{2x - 1}{2y + 1}$.   **40.** $y + 2x = 3$.   **41.** $\dfrac{x}{2} + \dfrac{\sin 2x}{4}$.

42. $\dfrac{\pi}{4}$.   **43.** $\dfrac{\sin 3x}{12} + \dfrac{3\sin x}{4}$.   **44.** $\tan x - 1$.   **45.** $\dfrac{\pi^2}{4}$.

Exercise 21 (p. 202)

1. (i) 13.55; (ii) 13.50; (iii) 13.5.   **2.** 53.
3. 65.7 m.   **4.** 3.909.   **5.** 0.5235; 3.141.   **6.** 43.8 m s^{-1}.
7. 1.17 m^3.   **8.** 0.8815.   **9.** $\frac{4}{3}\pi r^3$.   **10.** 312 cm^2.

Exercise 22a (p. 211)

1. $\sqrt{10}$.   **2.** $\sqrt{89}$.   **3.** $\sqrt{41}$.   **4.** $\sqrt{29}$.   **5.** $\sqrt{8}$.
6. $(\frac{5}{2}, 3)$.   **7.** $(4, 0)$.   **8.** $(-\frac{1}{2}, 1)$.   **9.** $(7, -\frac{1}{2})$.   **10.** $(-2, -4)$.
13. 10.   **14.** 10.   **15.** $(5, 2)$.   **20.** $(1, 1)$.

Exercise 22b (p. 216)

1. $y = x + 3$.   **2.** $y = x + 1$.   **3.** $y + x = 5$.   **4.** $y + x = 5$.
5. $(1\frac{1}{3}, 3\frac{2}{3})$.   **6.** $(10, -5)$.   **7.** $x + y = 5$.   **8.** $(-3, 8)$.
9. $6\sqrt{2}, 2\sqrt{5}, 2\sqrt{5}$.   **10.** $-\frac{4}{5}$.   **11.** $3x + 5y = 13$.
12. $3x + 2y = 1$.   **13.** $y = 2x$.   **14.** $\frac{1}{2}$.
15. $y - x = 1$.   **16.** $x + y = 5$.   **17.** $y = 2x + 2$.
18. $y = 2x - 1$.   **19.** $x = 2y$.

Exercise 22c (p. 219)

1. $(4, 10)$.   **2.** $y + 2x = 8$.   **3.** $19x = 17y$.

4. $\left(1 - \dfrac{2}{m}, 2 - m\right)$.   **6.** $2\frac{1}{2}$.

7. $y + 2x = 3$.   **8.** $3x + 2y = 4$.   **9.** $x + y = 5$.
10. $x + y = 5$.   **12.** 5.   **13.** $a = 0.33, b = 0.5$.
14. $a = 0.1, n = 2.0$.   **15.** $a = 0.2, b = 1.0$.   **16.** $y = 3x - 4$.
17. $n = \frac{1}{2}, a = 2$.   **18.** 2.632 should be 2.24.
19. $(3, 8)$.   **20.** $(\frac{3}{5}, -\frac{4}{5})$.   **23.** $x = 2, y = 1$.
24. $\tan^{-1} \frac{1}{3}$.   **25.** -1.   **26.** $-\frac{1}{11}$.
27. $x + 5y = 32$.   **29.** $x + 4y = 10$.   **30.** $3x - 2y = 7$.

Exercise 23a (p. 227)

1. $(\frac{27}{7}, \frac{2}{7})$.   **2.** $(13\frac{2}{3}, 19\frac{1}{3})$.   **3.** $y + 3x = 13$.
4. $4x + 3y = 12$.   **5.** $x + y = 2$.   **6.** $1\frac{2}{5}$.
7. $2\frac{1}{5}$.   **8.** $2\frac{1}{2}$.   **9.** $\dfrac{x}{\sqrt{10}} + \dfrac{3y}{\sqrt{10}} = \dfrac{2}{\sqrt{10}}$.

10. $(2, 7)$.   **11.** $my + x = a$.   **12.** $xt + \dfrac{y}{t} = 2c$.

13. 6.   **14.** $(\frac{18}{13}, \frac{25}{13})$; 1.   **15.** $45°$; $\sqrt{2}$.
16. $x \cos \theta + y \sin \theta = a$.   **17.** $(4, 6)$; $(6, 7)$.
19. 1 and -1.   **20.** $y = 3x - 9$.

Exercise 23b (p. 233)

1. (a) $\dfrac{3}{\sqrt{13}}$; (b) $\dfrac{1}{\sqrt{5}}$; (c) $\frac{11}{3}$; (d) $\dfrac{6}{\sqrt{5}}$; (e) $\dfrac{35}{\sqrt{34}}$.

2. (a) opp.; (b) same; (c) opp.; (d) same; (e) opp.

3. Above.　　**4.** Below.　　**6.** $90°$.　　**7.** $45°$.　　**8.** 7.

Exercise 23c (p. 235)

1. $y = 7x$.　　**2.** $x + y = 3$.　　**3.** $x + 4y + 3 = 0$.

4. $x = y$.　　**5.** 3.84.　　**6.** $(10, 16)$.

7. $3x + 2y = 6$; $\dfrac{6}{\sqrt{13}}$; $(\frac{18}{13}, \frac{12}{13})$.　　**8.** $(2,3)$.　　**9.** $(2, -2)$.

10. $\dfrac{a}{\sqrt{2}}$.　　**11.** $y = 0$.　　**12.** $\tan^{-1} \frac{1}{5}$.

13. $\tan^{-1} \frac{9}{37}$.　　**14.** $x + yt^2 = 2ct$.　　**15.** $(\frac{11}{5}, \frac{13}{5})$.

16. $y = 3$ and $3x + 4y = 15$.　　**17.** $\dfrac{5a^2}{2}$; $\sqrt{5}a$.

18. $x + ymt = c(m + t)$.　　**19.** $(a - p)x + (b - q)y = 0$.

20. $y(m + t) - 2x = 2amt$.　　**21.** $(6, 0)$.　　**22.** No.　　**24.** ±24.

Exercise 24 (p. 239)

1. $x + 3y = 12$.　　**2.** $x + y = a + b$.

3. $x^2 + y^2 = 16$.　　**4.** $x^2 + y^2 - 2x - 6y - 6 = 0$.

5. $x^2 + y^2 - 3x - 3y + 4 = 0$.

6. $x^2 + y^2 - (a + b)x - (a + b)y + 2ab = 0$.

7. $3x^2 + 3y^2 + 6x + 4y - 13 = 0$.　　**8.** $x^2 + y^2 = 1$.

9. $3x^2 + 4y^2 = 2$.　　**10.** $xy = 4$.　　**11.** $y^2 = 2x - 1$.

12. $y^2 + 1 = 2x$.　　**13.** $x + y = 0$.

14. $2x^2 + 2y^2 - 6x - 6y + 7 = 0$.　　**15.** $3x^2 + 4y^2 = 12$.

16. $x^2 + y^2 = 16$.　　**17.** $x^2 + y^2 = 16$.　　**18.** $2x + 3y = 2$.

19. $xy - 3y + x = 1$.　　**20.** $2x - 3y = \pm4$.　　**21.** $x^2 + y^2 = 32$.

22. $2y = 4x - 1$.　　**23.** $4y^2 - 4y - 8x + 5 = 0$.

24. $y^2 = 2x$.　　**25.** $y^2 - 2x - 2y + 2 = 0$.

26. $2y + 3x = 2xy$.　　**27.** $\alpha y + \beta x = 2xy$.　　**28.** $3y = 2x(1 - x)$.

29. $xy = 4$.　　**30.** $4xy = c^2$.　　**31.** $y^2 = 8x$.

32. $(x - x_1)(x - x_2) + (y - y_1)(y - y_2) = 0$.

33. $(x - \alpha)^2 + (y - \beta)^2 = r^2$.　　**34.** $x^2 + y^2 = 9$.

35. $x^2 + 4y^2 = 16$.　　**36.** $(x + y)(y - x - 1) = 0$.

Exercise 25a (p. 243)

1. (a) $x^2 + y^2 - 2x = 8$; (b) $x^2 + y^2 - 4x + 2y = 0$;
   (c) $x^2 + y^2 - 6x - 4y = 3$; (d) $x^2 + y^2 + 2x + 2y = 0$;
   (e) $x^2 + y^2 - 2x - 8y + 12 = 0$.
2. (a) $(-1, -2)$; 1. (b) $(2, 1)$; 3. (c) $(\frac{3}{2}, 0)$; $\frac{1}{2}\sqrt{57}$. (d) $(0, 2)$; 2.
   (e) $(\frac{7}{8}, 1)$; $\frac{1}{8}\sqrt{145}$.   3. $x^2 + y^2 - 4x + 2y = 5$.
4. $x^2 + y^2 - 13x - 12y + 36 = 0$.   5. $x^2 + y^2 - 5x - 3y + 6 = 0$.
6. $x^2 + y^2 - 9x - 16y + 42 = 0$.   7. $x = 3$.

Exercise 25b (p. 246)

1. (a) $x + y = 4$; (b) $x = 6$; (c) $2x + 3y = 5$; (d) $x + y = 3$.
2. $4y = 3x \pm 10$.   3. $4y = \pm 3x$.   4. $2\sqrt{17}$.
5. (a) $(2, 3)$; (b) $(1, 2)$; (c) $(1, 1)$; (d) $(1, -1)$; (e) $(3, 0)$.

Exercise 25c (p. 249)

1. (a) 2; (b) 4; (c) 3; (d) $1\frac{1}{2}$.
2. (a) $4x + 9y = 13$; (b) $3x + y = 4$; (c) $x - y = 1$;
   (d) $6x - y + 1 = 0$; (e) $2x = y + 1$.
3. $6x = 2y + 3$.   4. $3x^2 + 3y^2 = 4$.   5. $(0, 1)$; $(-\frac{4}{5}, \frac{3}{5})$.

Exercise 25d (p. 249)

1. $x^2 + y^2 - 2x - 4y + 4 = 0$.   2. $(-3, -4)$.
3. $x^2 + y^2 - 4x - 6y + 4 = 0$.   5. $f^2 = c$.
6. $x^2 + y^2 - 3x + y = 0$.   8. $x^2 + y^2 - 6x = 0$.
9. $x^2 + y^2 - 2x = 0$.   10. $x^2 + y^2 + 6x - 8y = 0$.
11. $x^2 + y^2 - 4x - 4y + 4 = 0$; $x^2 + y^2 - 20x - 20y + 100 = 0$.
12. $3x^2 + 3y^2 - 8x - 16y + 20 = 0$.   14. $5x + y = 6$.
15. $y = 3x - 10$.   16. $(\frac{5}{4}, 0)$.   17. $x^2 + y^2 = 8$.
18. $x^2 + y^2 + 2y - 4x = 0$.   20. $9x^2 + 9y^2 = 10x + 25y$.
21. $45°$.   22. $x = 2$.   23. $x^2 + y^2 - 6x - 4y + 9 = 0$.
24. 7 or -1.   25. $4x + 5y = 11$.
26. $x^2 + y^2 - 8x - 10y + 16 = 0$.
28. $x + 4y = 1$, $x - 4y = 2$.
29. $3x^2 + 3y^2 - 5x - 14y + 13 = 0$.
30. $\left(\dfrac{k^2 + 1}{k^2 - 1}, 0\right)$; $\dfrac{2k}{k^2 - 1}$.

Exercise 26 (p. 258)

1. (a) $x^2 + y^2 = 4$; (b) $\dfrac{x^2}{4} + \dfrac{y^2}{9} = 1$; (c) $\dfrac{x^2}{9} - \dfrac{y^2}{4} = 1$; (d) $y^2 = 16ax$;
   (e) $x(y - 1) = 1$.
2. (a) $(4\cos\theta, \sin\theta)$; (b) $(4\sec\theta, 3\tan\theta)$;
   (c) $(4t, 2t^2)$; (d) (m^2, m^3); (e) $\dfrac{1}{1 + m^3}, \dfrac{m}{1 + m^3}$.

3. (a) $x + yt^2 = 2ct$; (b) $x \cos \theta + y \sin \theta = 2$; (c) $my = x + am^2$;
   (d) $3x \cos \theta + 2y \sin \theta = 6$; (e) $2x \sec \theta - 3y \tan \theta = 6$.

4. (a) $t^3 x - ty = c(t^4 - 1)$; (b) $x \sin \theta - y \cos \theta = 0$;
   (c) $y + mx = 2am + am^3$; (d) $3y \cos \theta - 2x \sin \theta = 5 \cos \theta \sin \theta$;
   (e) $3x \tan \theta + 2y \sec \theta = 13 \sec \theta \tan \theta$.

5. $y(m + t) - 2x = 2amt$; $my - x = am^2$.

6. $ymt + x = c(m + t)$; $ym^2 + x = 2cm$.   **7.** $(a, 2a)$; tangent.

8. $-\tan \dfrac{\theta}{2}$.   **10.** $y \cos m + x \sin m = a \cos m \sin m$.

11. $3my = 2x + am^3$.   **12.** $2y + 3tx = 2at^2 + 3at^4$.

14. $x^2 + xy = y$.   **16.** $xt(t + 2) - y = t^2$.

17. $t(t + 1)(t + 2)y + (t + 1)x = t(t^3 + 2t^2 + 1)$.

18. $y(1 - 2t) + tx(3t - 2) = -t^2(1 - t)^2$.

21. $2y = x + xy$.   **22.** $2(t + 1)^2 x - (t + 2)^2 y = t^2$.

23. $(y - x)^2 = 4x$.   **24.** $y(t - 1) - tx = (t - 1)^2$.

Exercise 27 (p. 267)

1. (a) $85° 57'$; (b) $139° 12'$; (c) $38° 58'$; (d) $4° 1'$; (e) $76° 47'$.

2. (a) 0.561; (b) 1.850; (c) 0.0186; (d) 3.876; (e) 3.142.

3. (a) $90°$; (b) $60°$; (c) $270°$; (d) $240°$; (e) $112\frac{1}{2}°$.

4. (a) $\dfrac{7\pi}{6}$; (b) $\dfrac{3\pi}{2}$; (c) $\dfrac{\pi}{5}$; (d) $\dfrac{5\pi}{12}$; (e) $\dfrac{3\pi}{4}$.

5. (i) 0.506; (ii) 0.377; (iii) 0.200.

6. 8.60 cm, 8.80 cm.   **7.** 1.86 km.

8. (i) 15.7 cm; (ii) 10.5 cm; (iii) 7.86 cm.

9. (i) 1; (ii) 1; (iii) $-\dfrac{1}{\sqrt{2}}$; (iv) -1; (v) $\dfrac{\sqrt{3}}{2}$; (vi) $\frac{1}{2}$; (vii) $-\frac{1}{2}$;

   (viii) $\dfrac{1}{\sqrt{2}}$.

10. (i) $\cos x$; (ii) $-\sin x$; (iii) $-\cos x$; (iv) $\sin x$; (v) $-\tan x$.

11. 15.3 cm.   **12.** 4π.   **13.** $\dfrac{2\pi}{3}$.   **14.** $\dfrac{5\pi}{12}$.

15. (i) $\dfrac{2\pi}{3}$; (ii) $\dfrac{\pi}{3}$.   **16.** $180°$.   **17.** 16.8 m/sec.   **18.** 11.1.

19. 23.5 cm², 21.4 cm², 2.0 cm².   **20.** 22.4 cm².

21. 12.5 cm.   **24.** 2.   **25.** $\frac{2}{3}$.   **26.** $\frac{1}{2}$.   **27.** 2π.   **28.** $\frac{2}{3}$.

29. $\dfrac{\pi}{180}$.   **30.** $\dfrac{\pi}{6}, \dfrac{\pi}{3}, \dfrac{\pi}{2}$.   **31.** 45.5 cm.   **32.** 40 cm.

33. $x - \frac{1}{2} \sin 2x = \dfrac{\pi}{3}$.   **34.** 6.10 cm².   **35.** 4.89 cm.

36. 1.0006.

Exercise 28 (p. 275)

1. 4th   **2.** 2nd.   **3.** 4th.   **4.** 4th.   **5.** 3rd.   **6.** 3rd.

7. 4th.   **8.** 1st.   **9.** 2nd.   **10.** 4th.   **11.** $-\dfrac{\sqrt{3}}{2}$.   **12.** $\sqrt{3}$.

13. 1.   **14.** -2.   **15.** $\dfrac{1}{\sqrt{3}}$.   **16.** $-\sin \alpha$.   **17.** $\tan \alpha$.   **18.** $\tan \alpha$.

19. $-\cos \alpha$. **20.** $\operatorname{cosec} \alpha$. **21.** $\tan \alpha$.   **22.** $-\tan \alpha$. **23.** $\cos 2\alpha$. **24.** $\sin 3\alpha$.
25. $-\sec \alpha$. **26.** 5.828.   **27.** -0.268. **28.** 0.866.   **29.** -0.5.   **30.** -0.707.
31. $-\frac{24}{25}, -\frac{7}{24}$.   **32.** $-\sqrt{2}, +1$.   **33.** $\sqrt{2}, -1$.   **34.** $-\sqrt{3}, \sqrt{\frac{3}{2}}$.
35. $-\sqrt{2}, -1$.   **36.** 1.   **37.** 1.   **38.** 1.
39. -1.   **40.** -1.   **41.** $395°, 575°$.   **42.** $510°$.
43. $570°, 690°$.   **44.** $-\frac{11}{15}$.   **45.** (i) $\frac{209}{156}$; (ii) $-\frac{209}{156}$.
46. $270°$ and $315°$.   **47.** $\pm\frac{1}{2}\left(c + \dfrac{1}{c}\right)$.   **48.** $\frac{1}{2}(z \pm \sqrt{z^2 - 4})$.

49. $\frac{1}{2}\left(u + \dfrac{1}{u}\right)$.

Exercise 29 (p. 282)

11. $34.3°$.   **12.** $68.5°$.   **13.** $119.5°$.   **14.** $52.3°$.   **15.** $62.3°$.   **16.** $30°$.
17. 13.1 and 56.   **18.** $90°$ and $180°$.   **19.** 38.2.   **20.** $40.5°$.

Exercise 30b (p. 286)

1. $a^2(1 + b^2) = 1$.   **2.** $(1 + b)^2 = a^2(1 + b^2)$.
3. $(c - bd)^2 + (ad - c)^2 = (a - b)^2$.   **4.** $b^2 - 2ab + 1 = 0$.
5. $(a - 1)^2 + (b - 1)^2 = 1$.   **6.** $ab = 1$.
7. $ab^2 - 2b + a = 0$.   **8.** $4(a - b)^2 + (a^2 - b^2)^2 = 16$.

Exercise 30c (p. 287)

1. $0°, 60°$.   **2.** $0°, 60°, 180°$.   **3.** $45°, 63° 26'$.
4. $45°, 60°, 120°$.   **5.** $30°, 90°, 150°$.   **6.** $30°, 45°, 150°$.
7. $0°, 180°$.   **8.** $120°, 180°$.

Exercise 30d (p. 292)

1. 1.29 cm.   **2.** $-\frac{3}{5}, \frac{4}{5}$.   **3.** $60°, 120°$.   **5.** $30°, 150°$.
7. $45°, 60°, 120°, 135°$.   **10.** 4.06 cm.   **11.** $15°, 75°$.
16. $a^2b^2 = (1 - a^2)(a - b)^2$.   **17.** $\dfrac{\sqrt{1 + p^2}}{p}$.

18. $\cot x$.   **19.** $\sin x$.   **21.** $(b - a)^2 + (q - p)^2 = 1$.
22. $30°, 150°$.   **23.** $27.3\ m$.
26. $123° 34', 5.34$ cm.
29. $30°, 270°$.   **32.** $\frac{24}{25}$.

Exercise 31a (p. 295)

1. 1. **2.** 0. **3.** 1. **4.** 0. **5.** $\dfrac{\sqrt{3}+1}{2\sqrt{2}}$.

6. $\dfrac{1-\sqrt{3}}{2\sqrt{2}}$. **7.** $-\dfrac{\sqrt{3}+1}{2\sqrt{2}}$. **8.** $\dfrac{\sqrt{3}-1}{2\sqrt{2}}$.

Exercise 31b (p. 298)

1. $\sin 30°$. **2.** $\cos 60°$. **3.** $\cos(\theta - 2\varphi)$. **4.** $\sin 3A$.
5. $\cos 30°$. **6.** $\sin(A + 2B)$. **7.** $\cos 5\varphi$. **8.** $\cos B$.
9. $\sin A$. **10.** $\sin(\theta - 45°)$. **11.** 13. **12.** $\sqrt{a^2 + b^2}$.
13. -25. **14.** $-\sqrt{a^2 + b^2}$. **15.** $45°$. **16.** $60°$.
17. $\frac{4}{5}, \frac{3}{4}$. **18.** 7. **19.** $-\frac{31}{17}$. **20.** $-\frac{117}{44}$.

Exercise 31c (p. 299)

1. $2 \sin A \cos B$. **2.** $2 \cos A \cos 2B$. **3.** $2 \sin \theta \cos \theta$.

4. $\cos^2 \theta - \sin^2 \theta$. **5.** $\dfrac{2 \tan \theta}{1 - \tan^2 \theta}$. **7.** $\frac{63}{65}$.

8. 1. **9.** 0. **10.** $\sqrt{29}; \frac{2}{5}$. **11.** $\sqrt{73}, \frac{3}{8}$.

12. $\frac{1}{2}$. **13.** $\sqrt{3}$. **14.** $\sqrt{a^2 + b^2}$. **17.** $17\sqrt{2}$ cm.

18. $15° \, 34', 111° \, 18'$. **19.** $10° \, 12', 102° \, 26'$.

21. $\dfrac{\tan A + \tan B + \tan C - \tan A \tan B \tan C}{1 - \tan B \tan C - \tan A \tan C - \tan A \tan B}$. **24.** $20°$.

26. 0. **34.** $\frac{4}{7}$. **35.** $\frac{3}{5}$. **36.** $1 + a + a^2$.

40. $\pm\sqrt{89}$.

Exercise 32a (p. 302)

1. $\sin 30°$. **2.** $\cos 100°$. **3.** $\sin 100°$. **4.** $\cos 30°$. **5.** $\tan 40°$.
6. $\cos 6x$. **7.** $\sin x$. **8.** $\tan x$. **9.** $\cos 8x$. **10.** $\cos 2x$.
11. $1 + \sin 2A$. **12.** $2 - \sqrt{3}$. **13.** $-\frac{1}{8}$. **14.** $\frac{4}{3}$.

15. $-2 \pm \sqrt{5}$. **16.** $\cos 2x$. **17.** $-\dfrac{\sqrt{3}}{2}$. **19.** $\sin 2x$.

20. $\cos 2x$.

Exercise 32b (p. 307)

1. $-\frac{119}{169}$. **2.** $\frac{24}{25}$. **3.** $\frac{23}{27}$. **8.** $\frac{3}{4}$. **9.** -1.

12. $\dfrac{\sqrt{5}-1}{4}$. **13.** $\dfrac{4t - 4t^3}{1 - 6t^2 + t^4}$. **14.** $-\frac{24}{7}$. **17.** $a^2 + \dfrac{b^2}{a^2} = 2$.

18. $a^2 + \dfrac{b^2}{a^2} = 1$. **19.** $a^{\frac{2}{3}} + b^{\frac{2}{3}} = 1$. **21.** $40° \, 12', 252° \, 24'$.

22. $0°$, $157° \, 22'$, $360°$.   **23.** $19° \, 28'$, $30°$, $150°$, $160° \, 32'$.

24. $\dfrac{1 + \tan \theta/2}{1 - \tan \theta/2}$.   **25.** $\frac{8}{15}$.   **27.** $-\frac{24}{7}$.   **28.** $\pm \dfrac{2ab}{a^2 - b^2}$.

31. $100° \, 32'$, $349° \, 28'$.   **32.** $\dfrac{5t - 1}{1 + t^2}$.   **33.** $\dfrac{6t^2 + 1}{6t^2 - 1}$.

34. $4\sqrt{5}$.   **35.** $\frac{1}{4}$.   **39.** $53° \, 8'$.

40. $30°$, $150°$, $270°$.   **42.** $0°$, $45°$, $135°$, $180°$, $225°$, $315°$, $360°$.

44. $\tan \alpha + \tan \beta - 2 \tan \gamma =$
$$\tan \gamma(\tan \alpha \tan \gamma + \tan \beta \tan \gamma - 2 \tan \alpha \tan \beta).$$

45. $90°$, $210°$, $270°$, $330°$.   **46.** $11t^2 + 2t - 5 = 0$.

49. $\dfrac{\sqrt{5} + 1}{4}$.   **50.** $18° \, 26'$, $161° \, 34'$, $198° \, 26'$, $341° \, 34'$.

Exercise 33a (p. 311)

1. $\cos 10° - \cos 50°$.   **2.** $\sin 60° + \sin 20°$.   **3.** $\sin 100° - \sin 20°$.

4. $\cos 120° + \cos 20°$.   **5.** $\cos 100° + \cos 60°$.   **6.** $\cos 60° - \cos 100°$.

7. $\sin 100° - \sin 60°$.   **8.** $\sin 30° + \sin 10°$.   **9.** $\cos 4x + \cos 2x$.

10. $\cos 4y - \cos 6y$.   **11.** $-\cos 2x$.   **12.** $\cos 2x$.

13. $\sin 2x - 1$.   **14.** $\cos 30° - \cos (2x + 90°)$.

15. $\cos (2x + 90°) + \cos 30°$.   **16.** $\sin (2x + 90°) - \sin 30°$.

17. $\sin 8x + \sin 2x$.   **18.** $\cos 2A + \cos 2B$.

19. $\sin 2A + \sin 2B$.   **20.** $\cos 2B - \cos 2A$.

Exercise 33b (p. 311)

1. $2 \cos 50° \cos 20°$.   **2.** $-2 \cos 40° \sin 20°$.   **3.** $2 \sin 40° \cos 10°$.

4. $2 \cos 50° \cos 20°$.   **5.** $2 \sin 50° \sin 30°$.   **6.** $2 \cos 2x \sin x$.

7. $2 \sin 2x \cos x$.   **8.** $2 \sin 2x \sin x$.   **9.** $2 \cos 2x \cos x$.

10. $-2 \sin 3x \sin x$.   **11.** $2 \cos x \cos 45°$.   **12.** $2 \sin x \sin 45°$.

13. $2 \sin 45° \cos x$.   **14.** $2 \cos 2x \sin \dfrac{x}{2}$.

15. $2 \cos (x + 30°) \cos 10°$.   **16.** $2 \cos A \sin B$.

17. $2 \cos (x + y) \sin y$.   **18.** $-2 \sin (x + 2y) \sin 2y$.

19. $2 \cos x \cos 3y$.   **20.** $-2 \sin \left(x - \dfrac{3y}{2}\right) \sin \dfrac{y}{2}$.

Exercise 33c (p. 313)

1. $-\tan 2C$.   **2.** $\sin \dfrac{C}{2}$.   **6.** $\frac{1}{2}$.   **11.** $0°$, $90°$, $180°$.

12. $15°$.   **18.** $30°$.   **19.** $\frac{2}{5}$.   **20.** $\frac{1}{2}$.   **21.** 2.

22. $10°$, $130°$.   **23.** $\frac{1}{2}x + \frac{1}{4} \sin 2x + c$.

24. $\frac{1}{2}x - \frac{1}{4}\sin 2x + c.$

25. $\dfrac{\sin 2x}{2} - \dfrac{\sin 4x}{4} + c.$

26. $\dfrac{\sin 2x}{2} + \dfrac{\sin 4x}{4} + c.$

27. $c - \dfrac{\cos 2x}{2} - \dfrac{\cos 4x}{4}.$

31. $4\sin\dfrac{A}{2}\sin\dfrac{B}{2}\sin\dfrac{C}{2}.$

32. $4\cos A\cos B\sin(A + B).$

36. $0°, 30°, 60°, 120°, 150°, 180°.$ **43.** $40°, 180°.$

44. $\cos x + \cos 3x + \cos 7x + \cos 9x.$

45. $\frac{1}{4}\sin x + \frac{1}{12}\sin 3x + \frac{1}{28}\sin 7x + \frac{1}{36}\sin 9x + c.$

46. $\cos x - \cos 3x - \cos 7x + \cos 9x.$

47. $\frac{1}{4}\sin x - \frac{1}{12}\sin 3x - \frac{1}{28}\sin 7x + \frac{1}{36}\sin 9x + c.$

48. $30°, 150°.$

Exercise 34a (p. 318)

1. $52° 10'.$ **2.** $55° 24'.$ **3.** $89° 54'.$ **4.** $117° 34'.$ **5.** $57° 36'.$

Exercise 34b (p. 320)

1. $B = 84° 41', C = 48° 19'.$ **2.** $A = 36° 43', B = 31° 17'.$

3. $A = 38° 34', C = 46° 38'.$ **4.** $A = 76° 34', B = 43° 14'.$

5. $A = 33° 11', C = 46° 49'.$

Exercise 34c (p. 322)

1. $11.8\text{ cm}^2.$ **2.** $14.8\text{ cm}^2.$ **3.** $1.61\text{ cm}.$ **4.** $17.4\text{ cm}^2.$ **5.** $30°.$

6. $55.4\text{ cm}^2.$ **7.** $\dfrac{a^2\sin B\sin C}{2\sin(B + C)}.$ **8.** $1.75\text{ cm}.$

Exercise 34d (p. 329)

1. $767\text{ cm}^2.$ **2.** $1.44\text{ cm}.$ **3.** $48° 36', 131° 24'.$

4. $7.21\text{ cm}.$ **5.** $61° 15'.$ **6.** $41° 24'.$

7. $76° 28'.$ **8.** $\frac{72}{97}.$ **9.** $\frac{24}{25}.$

10. (i) 7.21; (ii) $33° 42'$; (iii) $33° 42'.$ **11.** $39°.$

12. $3670.$ **14.** $56° 26'.$ **15.** $17° 28'.$

16. $30°.$ **27.** $34°.$ **28.** $341° 34'.$

29. $276° 32'.$ **30.** $\sin^{-1}(\sin\alpha\cos\beta).$

Index

Abscissa 207
Acceleration 155
Angle
  between a line and a plane 322
  between two lines 117, 232
  between two planes 123
  compound 294
  double 302
  general 270
  half angle formulae 316
  multiple 301
  negative 270
  subsidiary 296
Arc, length of 264
Area
  of triangle 212, 320, 321
  sign of 169
  under acceleration–time curve 167
  under velocity–time curve 166, 201
Arrangements 50
Associative Law 116
Astroid 256

Binomial distribution 72, 74
Binomial Theorem 57
  for negative and fractional indices 60

Centre of gravity 222
Change, rate of 156
Circle
  centre 242
  equation, cartesian 242
    vector 121
  intersection of two 248
  parametric form 252
  radius 243
Circular measure 263
Combinations 52
  two important formulae 53
Commutative Law 82, 116
Constant, arbitrary 160
Coordinates 207
Cosine formula
  by vectors 120
Curve tracing 255

Dependent, independent events 64, 68
Derivatives 139

Differential coefficient 144
Discriminant 44
Distance between two points 207
Direction cosines 110
Distributive Law 116

Elimination 285
Ellipse 254
Equations
  dependent 19
  inconsistent 19
  linear in three variables 19
  of a straight line 17, 119, 214
  parametric 252
  roots of 44
  simultaneous, one linear and one quadratic 17
  solutions by graphs 45
  vector, of a straight line 119
  with an unknown index 14
  with given roots 47

Factorial 50
Function
  derived 139
  explicit 188
  implicit 188
  of a function 184

Geometric distribution 76
Gradient
  of a curve 130
  of a straight line 129, 213
Graphs
  cosine 279
  distance–time 131
  quadratic 45
  sine 278
  tangent 280
Gravity, centre of
  of a triangle 222

Heron's formula 321
Hyperbola
  rectangular 254

Identities
  using sine and cosine formula 283

Indices
  negative and fractional 3
  positive 3
Inflexion, point of 150
In-radius 321
Integral
  definite 165
  of cos x 181
  of sin x 181
Integration
  approximate methods of 196
  by substitution 191
  of a function of a function 190
Intercept 224

Law, linear 216
Limit
  of sin x/x 179
Line, straight
  intercept form 224
  of given gradient 136, 215
  (p, α) form 224
  (r, θ) form 226
  through the intersection of two straight
    lines 234
  through two points 223
Loci 237
Logarithmic theory 12
Logarithms 11

Mappings 82
Matrices
  addition 81
  definition 81
  multiplication 82
  singular 86
Maximum and minimum 149
Mean
  arithmetic 37
  geometric 39
Midpoint 209
Midpoint theorem 103
Mutually exclusive events 65

Normal 257
Notation
  increment 143

Ordinate 207

Parameter 226, 239 252
Pascal's triangle 54, 58
Permutations 50
Perpendicular
  length of 225, 228
  sign of 230

Probability
  additional law 65
  definition 63
  independent events 64, 68
  multiplication law 67
  mutually exclusive events 65
Problems, three dimensional 112, 125, 322
Products, to differentiate 181
Progressions
  arithmetic 36
  geometric 39
Projection
  of a region 323
  of a straight line 208
Proportion
  direct 27
  inverse 29

Quadratic equations
  equal roots 45
  product of roots 44
  sum of roots 44
  symmetric functions 47
Quotients, to differentiate 182

Radian 263
Ratio
  of any angle 274
  trigonometric 271
Rationalization 7
Remainder 22
Remainder theorem 22

Scalars 92
Scalar product 121
Sector 265
Selection 52
Sequences 36
Set notation 69
Simpson's rule 198
Sine formula 287
Solid of revolution 170
Sum,
  of an A.P. 37
  of a G.P. 40
Surds 6
Symmetry 47, 63

Tangent
  at a point 137
  condition for 245
  formula 296
  length of 247
  of gradient m 246

Transformation 82
Trapezoidal rule 197
Tree diagram 68, 70, 71, 74
Turning values 150

Variation
  joint 32
  the sum of two parts 32

Vector
  addition 92, 99
  angle between two 99, 117
  components 94
  magnitude 95, 109
  position 96
  unit 94, 117
Velocity 132, 155, 163
Volume of solid of revolution, 170